T0132671

Recent Advances in

MARINE BIOTECHNOLOGY

VOLUME 7
SEAFOOD SAFETY AND HUMAN HEALTH

Recent Advances in MARINE BIOTECHNOLOGY

VOLUME 7
SEAFOOD SAFETY AND HUMAN HEALTH

Editors
Milton Fingerman
Department of Ecology, Evolution and Organismal Biology
Tulane University, U.S.A.

Rachakonda Nagabhushanam
Department of Ecology, Evolution and Organismal Biology
Tulane University, U.S.A.

CIP data will be provided on request.

SCIENCE PUBLISHERS, INC.
Post Office Box 699
Enfield, New Hampshire 03748
United States of America

Internet site: *http://www.scipub.net*

sales@scipub.net (marketing department)
editor@scipub.net (editorial department)
info@scipub.net (for all other enquiries)

ISBN Set 1-57808-010-X
ISBN Vol. 7 1-57808-204-8

Published by Science Publishers Inc., NH, U.S.A.
Printed in India.

Preface

This volume is the seventh in the ongoing series we have entitled, "Recent Advances in Marine Biotechnology." For the first volume, "Endocrine and Reproductive Biology" was selected as the subtitle. Volume 2 is subtitled "Environmental Marine Biotechnology," and Volume 3 is subtitled, "Biofilms, Bioadhesion, Corrosion and Biofouling." For Volume 4 we chose the subtitle "Aquaculture." The subtitle of Volume 5 is "Immunobiology and Pathology." Volume 6 deals with the important subject "Bio-organic Compounds: Chemistry and Biomedical Applications." Now, for Volume 7 we have turned our attention to another important topic, "Seafood Safety and Human Health."

Finfish and shellfish constitute an important percentage of the diet of many humans worldwide. As the world's human population continues to grow, there is more and more urgency to increase the yield of seafood from the oceans. However, naturally occurring toxins and pathogenic organisms are capable of contaminating this food supply. Consequently, there is the need for answers to such questions as how to detect the presence of these toxins and what environmental conditions favor the microorganisms that are the sources of these toxins. Obviously, the safety of the seafood that is consumed by humans is a major concern to the public, health providers, government officials, and scientists. Marine biotechnology has a major role at the forefront in assuring that our seafood is safe, and has begun to provide impressive successes in assuring that it will be so.

Volume 7 presents the most recent information on seafood safety and human health, written by a highly talented group of people who not only are able to communicate through their writing, but also are responsible for many of the advances that are described herein. As with the first six volumes of this series, we, the editors, have been most fortunate in attracting a highly talented, internationally respected group of investigators to serve as authors. We intentionally set out to present a truly international scope to this volume. Consequently, appropriate authors from several countries were sought, and to everyone's benefit, they accepted our invitation.

We, the editors, take pleasure in thanking the authors for their cooperation and excellent contributions, and for keeping to the publication schedule. The efforts of these individuals made our task much less onerous that it might have been. Also, we wish to especially thank our wives, Maria

Esperanza Fingerman and Rachakonda Sarojini, for their constant, undiminishing encouragement and support during the production of this series. We trust, you, the readers, will agree with us that the efforts of these authors will serve collectively to provide a major thrust toward a better understanding of seafood safety and its relationship to human health.

Milton Fingerman
Rachakonda Nagabhushanam

Contents

Preface	v
The Contributors	ix
Brevetoxins: Pharmacology, Toxicokinetics and Detection *Mark A. Poli*	1
Azaspiracid, a New Marine Toxin Isolated from Mussels: Chemistry and Histopathology *Emiko Ito and Masayuki Satake*	33
Enterotoxin of *Vibrio cholerae* *Yoshio Iijima and Takeshi Honda*	41
Exophysiology and Biosynthesis of Marine Biotoxins (TTX and PSP) *Deng-Fwu Hwang and Ya-Hui Lu*	53
Effects of 4-Aminopyridine in the Treatment of Saxitoxin Poisoning *Fat-Chun Tony Chang, Bernard Benton and David Spriggs*	71
Development of Commercial Immunoassays for Seafood Poisonings *Paulo Vale*	89
Highly Sensitive Bioassays for Paralytic Shellfish Toxins *Franca Guerrini, Clementina Bianchi, Lorenzo Beani,* *Laurita Boni and Rossella Pistocchi*	107
Bioassays for Paralytic Shellfish Toxins in Crustacea *Jason Doyle and Lyndon Llewellyn*	121
Antimicrobial Peptides for Fish Disease Control *Aleksander Patrzykat and Robert E.W. Hancock*	141
Fish Protein Hydrolysates and their Potential Use in the Food Industry *Hordur G. Kristinsson and Barbara A. Rasco*	157

Shelf Life Extension and Value Addition of Fishery Products: A Critical Evaluation 183
Vazhiyil Venugopal

Recent Advances in Surimi Technology 241
Herbert O. Hultin

DNA-based Diagnostics in Sea Farming 253
Carlos R. Osorio and Alicia E. Toranzo

Index 311

The Contributors

Lorenzo Beani
Dipartimento di Medicina Clinica
Sperimentale: Sezione di Farmacologia,
Università di Ferrara,
Via Fossato di Mortara
17-19, 44100 Ferrara, Italy

Bernard Benton
U.S. Army Edgewood Chemical
Biological Center,
Aberdeen Proving Ground,
MD 21010-5400, U.S.A.

Clementina Bianchi
Dipartimento di Medicina Clinica
Sperimentale: Sezione di Farmacologia,
Università di Ferrara,
Via Fossato di Mortara
17-19, 44100 Ferrara, Italy

Laurita Boni
Centro Interdipartimentale di
Ricerca per le Scienze Ambientali,
Università di Bologna,
Via Tombesi dall'Ova 55,
48100 Ravenna, Italy

Fat-Chun Tony Chang
Pharmacology Division
U.S. Army Medical Research
Institute of Chemical Defense,
Aberdeen Proving Ground,
MD 21010-5400, U.S.A.

Jason Doyle
Australian Institute of Marine Science,
PMB 3, Townsville MC,
Queensland, Australia 4810

Franca Guerrini
Centro Interdipartimentale di
Ricerca per le Scienze Ambientali,
Università di Bologna,
Via Tombesi dall'Ova 55,
48100 Ravenna, Italy

Robert E.W. Hancock
Department of Microbiology and Immunology,
University of British Columbia,
Vancouver, B.C., Canada V6T 1Z3

Takeshi Honda
Department of Bacterial Infections,
Research Institute for
Microbial Diseases,
Osaka University,
3-1 Yamadaoka,
Suita 565-0871, Japan

Herbert O. Hultin
Department of Food Science,
University of Massachusetts/
Amherst, Gloucester Marine Station,
Gloucester, Massachusetts 01930, U.S.A.

Deng-Fwu Hwang
Department of Food Science,
National Taiwan Ocean University,
Keelung, Taiwan, R.O.C.

Yoshio Iijima
Laboratory of Bacteriology,
Kobe Institute of Health,
4-6 Minatojima-nakamachi,
Chuo-Ku,
Kobe 650-0046, Japan

Emiko Ito
Research Center for Pathogenic Fungi and
Microbial Toxicoses,
Chiba University,
1-8-1 Inohana, Chuo-ku,
Chiba 260-8673, Japan

Hordur G. Kristinsson
Department of Food Science
and Human Nutrition,
University of Florida,
P.O. Box 110370,
Gainesville, Florida 32611, U.S.A.

Lyndon Llewellyn
Australian Institute of Marine Science,
PMB 3, Townsville MC,
Queensland, Australia 4810

Ya-Hui Lu
Department of Food Science,
National Taiwan Ocean University,
Keelung, Taiwan, R.O.C.

Carlos R. Osorio
Departmento de Microbiología y Parasitología,
Facultad de Biologia,
Universidad de Santiago de Compostela,
15782 Santiago de Compostela, Spain

Aleksander Patrzykat
Department of Microbiology and Immunology,
University of British Columbia,
Vancouver, B.C., Canada V6T 1Z3

Rossella Pistocchi
Centro Interdipartimentale di
Ricerca per le Scienze Ambientali,
Università di Bologna,
Via Tombesi dall'Ova 55,
48100 Ravenna, Italy

Mark A. Poli
United States Army Medical Research
Institute of Infectious Diseases,
Fort Detrick,
Maryland 21702-5011, U.S.A.

Barbara A. Rasco
Department of Food Science
and Human Nutrition,
Washington State University,
P.O. Box 646376,
Pullman, Washington 99164, U.S.A.

Masayuki Satake
Faculty of Agriculture,
Tohoku University,
Tsutsumidori-Amamiya,
Aobaku, Sendai 981-8555, Japan

David Spriggs
Pharmacology Division
U.S. Army Medical Research
Institute of Chemical Defense,
Aberdeen Proving Ground,
MD 21010-5400, U.S.A.

Alicia E. Toranzo
Departmento de Microbiología y Parasitología,
Facultad de Biologia,
Universidad de Santiago de Compostela,
15782 Santiago de Compostela, Spain

Paulo Vale
Ecotoxicology Unit,
Instituto de Investigação
das Pescas e do Mar,
Av. Brasília, 1449-006
Lisboa, Portugal

Vazhiyil Venugopal
Food Technology Division,
Bhabha Atomic Research Centre,
Mumbai 400 085, India.

Brevetoxins: Pharmacology, Toxicokinetics and Detection

Mark A. Poli

United States Army Medical Research Institute of Infectious Diseases, Fort Detrick, Maryland 21702-5011, U.S.A.

Introduction

Red tides—blooms of toxic dinoflagellates involving mass mortalities of inshore fishes—were noted in the Gulf of Mexico as early as 1844 (Lasker and Smith 1954). These events, characterized by discolored water and dead and dying fish, can cause significant revenue losses to the coastal seafood and tourism industries. In addition, red tide blooms have been associated with morbidity and mortality in invertebrates and birds (Steidinger *et al.* 1973), bottlenose dolphins (Gunter *et al.* 1948, Geraci 1989) and Florida manatees (Layne 1965, O'Shea *et al.* 1991, Bossart 1998, Landsberg and Steidinger 1998). Human intoxication (neurotoxic shellfish poisoning, NSP) typically occurs through ingestion of filter-feeding mollusks such as clams or oysters (Steidinger *et al.* 1973, Baden 1983, Steidinger *et al.* 1998), although intoxication from molluscivorous whelks has been reported (Poli *et al.* 2000). In addition, toxins can become aerosolized during near-shore blooms and cause respiratory irritation in beach-goers or potentially severe bronchoconstriction in asthmatics (Asai *et al.* 1982, Pierce 1986, Pierce *et al.* 1990).

After a massive red tide along the Florida coast in 1947, Davis identified the causative organism as the unarmored dinoflagellate *Gymnodinium breve*. A decade later T.J. Starr (1958) published his notes suggesting that a lethal toxin elucidated by *G. breve* was the cause of these fish kills. Since then, brevetoxins have been isolated from cultures of raphidophycean flagellates such as *Heterosigma akashiwo* (Khan *et al.* 1997), *Chattonella marina* (Ahmed *et al.* 1995, Khan *et al.* 1995, Hallegraeff *et al.* 1998), *C. antiqua* (Khan *et al.* 1996a), and *Fibrocapsa japonica* (Khan *et al.* 1996b). Toxic blooms of these organisms also cause significant damage to commercial fish-farming operations in many parts of the world. *Gymnodinium breve*, however, remains the organism most associated with brevetoxin production and associated fish kills. The short note by Starr in 1958 pinpointing this association initiated nearly 50 years of intense scientific scrutiny of a family of potent neurotoxins that are not only a serious problem for the Gulf Coast seafood and tourist industries, but also have become sensitive and specific probes for structure and function of the voltage-sensitive sodium channel.

Early work on the brevetoxins was reviewed by others. Baden (1983) published an excellent comprehensive review of marine food-borne dinoflagellate toxins. Steidinger and Baden (1984) reviewed the subject of toxic marine dinoflagellates. Strichartz *et al.* (1987) and Wu and Narahashi (1988) reviewed the electrophysiological effects and mechanism of action of marine toxins. Finally, Baden (1989) specifically reviewed the research on brevetoxins up through the mid-to-late 1980s. This review will focus on advances from the late 1980s to the present. A brief introduction to the toxins will be followed by a discussion of pharmacology, toxicokinetics and detection methods for these fascinating marine toxins.

The Brevetoxins

Over time, the organism's name has evolved from *Gymnodinium breve* to *Ptychodiscus brevis* to *Gymnodinium brevis* and, most recently, to *Karenia brevis* in honor of the eminent phycologist Karen Steidinger (Daugbjerg *et al.* 2000). Toxin nomenclature has undergone a similar metamorphosis (reviewed in Baden 1983). This review uses the nomenclature set out in Poli *et al.* (1986). A family of nine related compounds, they are denoted PbTx-1 through—10 (***P****tychodiscus* ***b****revis* **t**oxins 1-10). Devoid of nitrogen, they are complex multi-ring methylated polyethers containing a lactone functionality in the first ring. There are two sub-groups, based upon ring structure. All known brevetoxins are derivatives of one of these structural backbones (Figure 1). The PbTx-1-type brevetoxins consist of the namesake PbTx-1, an aldehyde-reduced form (PbTx-7), and the α-methylene-reduced form (PbTx-10). The PbTx-2-type toxins include PbTx-2, the aldehyde-reduced form (PbTx-3), the acetylated form (PbTx-5), the 27,28 epoxide of PbTx-2 (PbTx-6), a chloromethyl ketone derivative (PbTx-8) and the α-methylene-reduced form of PbTx-3 (PbTx-9). All are naturally-occurring toxins, with the exception of PbTx-8 (Golik *et al.* 1982) which is likely an artifact of the chloroform used in the extraction process. The term PbTx-4 is not used. It was a place holder for Shimizu's GB-4 which was originally thought to be a new brevetoxin. However, it has since been shown to be the methyl hemiacetal of PbTx-2 formed by reaction with methanol during the isolation procedure (Shimizu, Y. personal communication). PbTx-10 is postulated to occur naturally due to structural correlation to PbTx-9 and

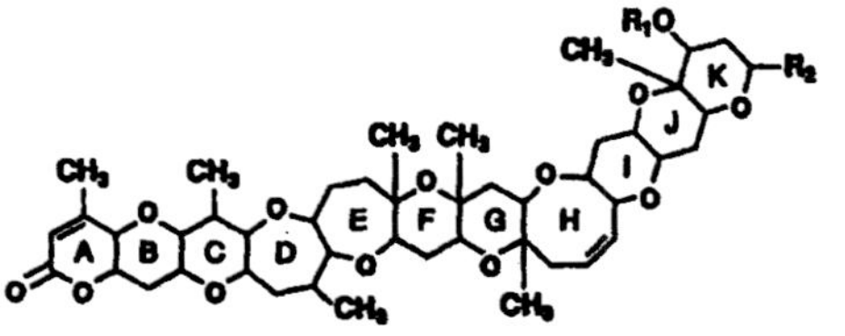

PbTx - 2- type

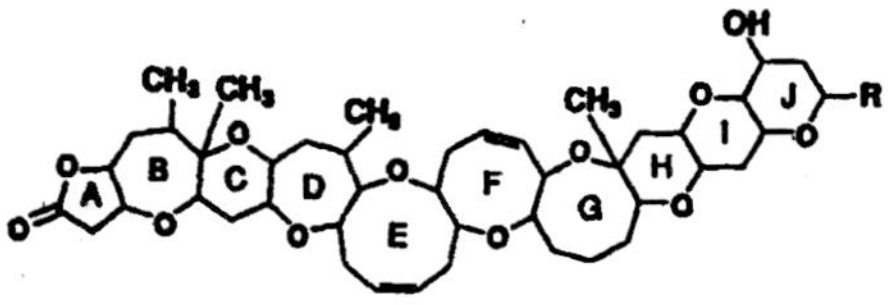

PbTx-2: R_1=H, R_2=CH_2C (=CH_2) CHO
PbTx-3: R_1=H, R_2=CH_2C (=CH_2) CH_2OH
PbTx-5: R_1=Ac, R_2=CH_2C (=CH_2) CHO
PbTx-6: R_1=H, R_2=CH_2C (=CH_2) CHO (27, 28 epoxide)
PbTx-8: R_1=H, R_2=CH_2COCH_2Cl
PbTx-9: R_1=H, R_2=CH_2CH (CH_3) CH_2OH

PbTx-1-type

PbTx-1: R=CH_2C (CH_2) CHO
PbTx-7: R=CH_2C CH_2) CH_2OH
PbTx-10: R=CH_2CH (CH_3) CH_2OH

Figure 1. Structures of the PbTx-1-type and PbTx-2-type brevetoxins.

the presence of PbTx-9 in stationary cell cultures. It has been synthesized *in vitro* from PbTx-1 by reduction with $NaBH_4$ (Baden 1989). Substitution of tritiated $NaBH_4$ for the unlabeled reagent in this reaction results in the synthesis of tritiated forms of PbTx-3, PbTx-9, PbTx-7 and PbTx-10 (Poli *et al.* 1986, Baden 1989). These radiolabeled probes have been invaluable tools for delineating the molecular pharmacology of brevetoxins (see below).

Pharmacology

Early studies (reviewed in Poli 1985, Baden 1989) clearly demonstrated that the pharmacological effects of brevetoxins are mediated by modulation of ion flux through the voltage-gated sodium channels. That these effects are exerted at nanomolar concentrations suggested binding to a specific receptor site rather than nonspecific membrane perturbations. Preliminary work (Catterall and Risk 1981, Catterall and Gainer 1985) demonstrated that PbTx-2 did not displace specifically bound neurotoxins from three of the four known receptor sites on neuroblastoma cells. Based upon this work, they suggested that brevetoxins bind to a new receptor site on the voltage-gated sodium channel.

The elucidation of brevetoxin structures by Lin *et al.* (1981) revealed the close structural relationship between PbTx-2 and PbTx-3 and suggested the aldehyde function of PbTx-2 could be reduced to the alcohol function of PbTx-3 *in vitro*. Baden *et al.* (1984) successfully reduced PbTx-2 to PbTx-3 with NaB^3H_4 to yield the first radiolabeled brevetoxin probe. This probe had no loss in potency, and it allowed for the first time direct investigation of brevetoxin binding to excitable tissues.

The Brevetoxin Receptor Site in the Context of Sodium Channel Structure and other Neurotoxin Receptor Sites

Much has been learned in the past few years about the structure and function of the voltage-gated sodium channel. In the interest of brevity and because the channel itself is not the subject of this chapter, only a brief summary relevant to the current topic will be presented here. More information and references can be found in several excellent recent reviews on this subject (Marban *et al.* 1998, Denac *et al.* 2000, Catterall 2000, Cestele and Catterall 2000).

The sodium channel consists of various subunits, depending upon the tissue of origin. The channel from mammalian brain is a complex of an ∝ subunit (260 kDa), a β1 subunit (36 kDa) and a β2 subunit (33 kDa). The channel from skeletal muscle contains only the α and β1 subunits; that from electric eel electroplax only the α subunit. Only the α subunit is required for ion conductance.

The α subunit is widely recognized to be a protein of modular architecture and similar in structure to the principal subunits of the Ca^{+2} and K^{+} channels. It contains four internally homologous domains denoted I-IV. Within each domain are six α-helical transmembrane segments (S1-S6) connected by internal and external polypeptide chains (Figure 2). The four domains associate within the membrane with infoldings of the S5-S6 extracellular connecting loops forming the pore. Within the pore, a set of four amino acids (aspartic acid, glycine, lysine and alanine), occupying equivalent positions in each of the four domains, come together to form a narrow bottleneck and are likely intimately involved in channel conductance and ion selectivity (Denac *et al.* 2000). The S4 segments carry multiple positive charges by virtue of a number of conserved arginine or lysine residues

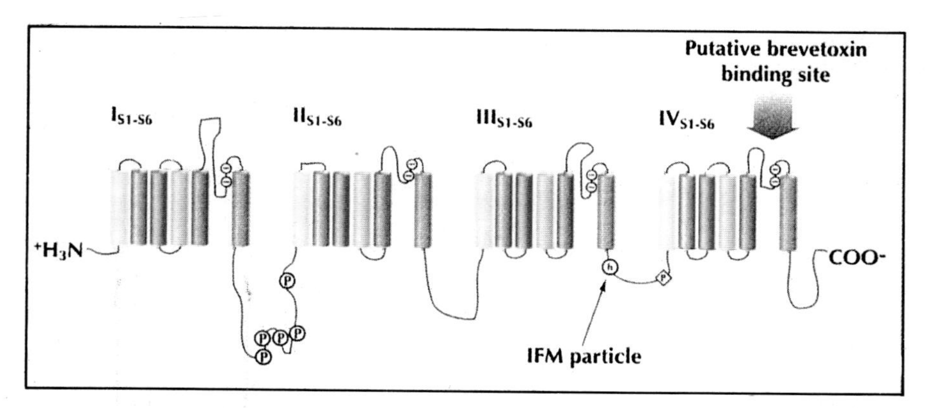

Figure 2. A cartoon representing the secondary structural features of the voltage-gated sodium channel α-subunit. (Reprinted from Gawley *et al.* (1995), with permission from Elsevier Science.)

occupying every third position along the chain interspersed by two hydrophobic residues. These charges physically traverse the membrane in a corkscrew manner in response to the electric field, and are believed to form the voltage sensor. The negative internal transmembrane electric field pulls these charges into the membrane, locking them into a "cocked" position. Current models propose that these positive charges be stabilized within the membrane by the formation of ion pairs with negative charges in adjacent transmembrane segments. Depolarization releases the S4 segments to "unscrew" outward, initiating a conformational change that opens the channel (Catterall 2000 and references therein). A small cytoplasmic loop between domains III and IV, containing the hydrophobic amino acids isoleucine, phenylalanine and methionine (IFM), comprises the fast inactivation gate. These amino acids act as a "tethered ball," which physically blocks the pore by binding to receptor residues in the cytoplasmic opening in response to conformational changes linked to channel activation.

The β subunits are not directly involved with ion conductance, but rather modulate channel gating and ion conductance by stabilizing relevant α subunit conformations. In addition, they possess immunoglobulin-like folds in their secondary structure analogous to cell-adhesion molecules. It is postulated that they may interact with extracellular proteins through these folds and thereby influence the location and density of sodium channels in neuronal tissues (Bonhaus *et al.* 1996, Catterall 2000).

The sodium channel is the target for several classes of naturally occurring neurotoxins. These compounds modulate channel function in a variety of ways by binding to specific receptor sites on the α-subunit and interacting with specific portions of the channel protein. These toxins have been invaluable probes for channel structure and function and their specific sites have, in some cases, been localized on the channel protein.

Receptor site 1

The guanidinium-containing heterocyclic toxins tetrodotoxin (TTX) and saxitoxin (STX) occupy this site, as well as the μ-conotoxins. TTX is the fugu toxin contained in the flesh and gonads of puffer fish, but is also found in various species of mollusks, frogs, octopi and crabs. STX is one of a family of related neurotoxins produced by marine dinoflagellates; accumulation in filter-feeding mollusks leads to paralytic shellfish poisoning (PSP) in humans. The μ-conotoxins are present in the venom of piscivorous cone snails. These toxins block sodium conductance by entering the extracellular opening of the pore and blocking access by monovalent cations (Cestele and Catterall 2000 and references therein). The binding of site 1 toxins is independent of membrane potential. Each of these toxins competitively inhibits binding of the others. Site-directed mutagenesis studies suggest TTX and STX bind to a common site within the channel near the extracellular pore opening, possibly formed from peptide segments of the four S5-S6 loops (Lipkind and Fozzard 1994). Conotoxins are believed to bind to a substantially overlapping site (Stefan *et al.* 1994, Dudley *et al.* 1995), but interact with different charged residues on the protein surface.

Receptor site 2

This site binds several groups of plant toxins: the grayanotoxins (from rhododendrons and relatives), the veratrum alkaloids (from lilacs and relatives), and aconitine (from *Aconitum napellus*). In addition, it binds batrachotoxin, a toxic principal from the skin secretion of the Colombian arrow frog. These toxins bind preferentially to the activated state of the sodium channel, causing persistent activation at resting membrane potential via a block of fast inactivation and a shift in the voltage dependence of activation to more negative membrane potentials (Catterall 1980 and references therein). These compounds are lipid-soluble, and are capable of accessing the channel from within the membrane. Because these toxins bind preferentially to open channels, other toxins that preferentially open the channels allosterically enhance their binding and actions. Among these are the α-scorpion toxins and sea anemone toxins (Catterall *et al.* 1981), pyrethroid insecticides (Lombet *et al.* 1988) and brevetoxins (Trainer *et al.* 1993)(see below).

Photoaffinity labeling and peptide mapping experiments (Trainer *et al.* 1996) as well as site-directed mutagenesis experiments (Wang and Wang 1998) point to the transmembrane segment S6 of domain I as the likely location of receptor site 2. In addition, another S6 segment in domain IV has also been found to be important for batrachotoxin binding (Denac *et al.* 2000), suggesting that site 2 may span more than one domain on the intracellular region of the pore. This site is believed to be separate but adjacent to (and potentially overlapping) the receptor site for local anesthetics (Linford *et al.* 1998). This puts it in the physical vicinity of the proposed inactivation gate, which dovetails nicely with its observed mechanism of action. This concept is further supported by binding studies in which a peptide containing the IFM amino acid sequence crucial for inactivation was found to compete with veratridine for binding to the channel (Ghatpande and Sikdar 1997).

Receptor site 3

Receptor site 3 binds several classes of peptide toxins that slow or block channel inactivation. These include the α-scorpion toxins from the Old World scorpions such as *Leiurus, Buthus* and *Androctonus*; sea anemone toxins from *Anthopleura* (anthopleurins) and *Anemonia* (ATX) and venom of Australian funnel-web spiders from the family *Atracinae* (atracotoxins). Photoaffinity labeling experiments pinpoint the location of this binding site to a region on the S5-S6 extracellular loop of domains I and IV (Tejedor and

Catterall 1988, Thomsen and Catterall 1989, Benziger *et al.* 1997). Site-directed mutagenesis suggests that a region on the short S3-S4 extracellular loop segment of domain IV is also critical (Rogers *et al.* 1996). This region of the molecule is believed to be involved in the coupling of activation to fast inactivation (Denac *et al.* 2000). These toxins do not act directly by blocking the internal inactivation gate at the intracellular loop connecting domains III and IV. Rather they uncouple inactivation from activation from the extracellular surface of the channel by binding to the activation gate in domain IV and trapping it in a position which is permissive to activation, but prevents the conformational change(s) required for fast inactivation (Rogers *et al.* 1996, Sheets *et al.* 1999).

This receptor site is allosterically linked to receptor site 2, and enhances persistent activation by toxins active at that site. This readily follows from their pharmacology; inhibition of inactivation by the site 3 toxins results in a greater number of channels in the conducting state, and site 2 toxins bind with greater affinity to open channels, stabilizing that conformation.

Receptor site 4

This site binds the β-scorpion toxins isolated from New World scorpions of the genus *Centruroides*. Unlike the α-scorpion toxins, the β-scorpion toxins have no effect on inactivation, but rather enhance activation of the channel. As a result, channels open at membrane potentials where they would normally exist in the resting state. The receptor site has been localized to the extracellular region of the S4 transmembrane segment of domain II and perhaps including the S3-S4 loop (Denac 2000). From this position, the toxins can stabilize and "trap" the voltage gate in its outward (conducting) conformation, thus enhancing channel activation. Although the receptor site for these toxins must be physically near that of the α-scorpion toxins, the lack of competition in binding studies demonstrates the receptor sites to be distinct. It is interesting that two classes of polypeptide toxins can bind at sites in such close proximity and yet one effects activation and the other inactivation by similar "trapping" effects on the S4 voltage gate. The fact that these receptor sites are remote from the site of the intracellular inactivation gate adds an additional layer of interest, and suggests the complexity of the activation/inactivation process.

Other Sodium Channel Receptor Sites

Other toxins bind specifically to the voltage-gated sodium channel. However, at present, there is no universally accepted numbering system for these, and they are less well understood than sites 1–5.

Pumiliotoxin B is an alkaloid neurotoxin found in the skin secretions of the poisonous frog *Dendrobates pumilio*. It has cardiotonic and myotonic activity, and it has been shown to stimulate Na^+ influx into synaptosomes that can be blocked by TTX. Based upon a lack of inhibition of STX or batrachotoxin binding, and synergistic effects on sodium influx with α- and β-scorpion toxins and brevetoxin, Gusovsky *et al.* (1988) suggested that pumiliotoxin-B binds to a distinct receptor site on the sodium channel. The exact location of this receptor on the channel protein and the nature of the allosteric interactions with other known sites have not yet been determined.

The pyrethroids are a large group of synthetic excitatory neurotoxins that were developed as insecticides. They are analogs of the natural pyrethrins found in the flowers of certain species of *Chrysanthemum*. Although they have biological effects on other channels as well, their main target appears to be the voltage-gated sodium channel (Denac *et al.* 2000 and references therein). Binding to the channel results in a prolongation of the sodium current that results in the development of a depolarizing afterpotential. Electrophysiological studies (De Weille *et al.* 1990, Narahashi 1992) suggest preferential binding to the activated channel state. Although blocked by TTX and allosterically coupled to sites 2, 3 and 5 (Bloomquist and Soderland 1988, Brown *et al.* 1988, Eells *et al.* 1993, Trainer *et al.* 1993, 1997), the pyrethroids are considered to act at a separate distinct site (Lombet *et al.* 1988). The location of this site is not currently known.

Local anesthetics such as lidocaine and procaine are small lipophilic molecules that act by blocking ion channels, including the sodium channel (Denac *et al.* 2000 and references therein). Several studies demonstrate an allosteric coupling of anesthetic binding to site 2 (Deffois *et al.* 1996, Postma and Catterall 1984, Linford *et al.* 1998). Interaction with other receptor sites does not appear to occur. The location of this site is believed to be inside the pore on the S6 segment of domain IV, and perhaps bridging to other domains as well (Catterall 2000, Denac 2000, and references therein). This site is near, but distinct from, receptor site 2, in the proposed vicinity of the activation gate.

The conotoxin TxVIA isolated from the venom of the cone snail *Conus textile* is highly toxic to mollusks, but only mildly toxic to vertebrates. However, it binds with high affinity to sodium channels in both molluscan and vertebrate tissues. It specifically inhibits sodium current inactivation in a similar manner to toxins that bind to site 3 (Hasson *et al.* 1993). However, based upon allosteric interactions with other receptor sites that differ dramatically from the site 3 toxins, Fainzilber *et al.* (1994) proposed a distinct receptor site. The location of this proposed site is not yet known. At least one research group (Cestele *et al.* 1995, Cestele and Catterall 2000) refer to this site as neurotoxin receptor site 6, although this has not been universally accepted.

Receptor site 5

This receptor site binds the brevetoxins and ciguatoxins. These toxins persistently activate channels by shifting the voltage-dependence of activation to more negative membrane potentials and inhibit fast inactivation (Wu and Narahashi 1988, Schreibmayer and Jeglitsch 1992, Baden *et al.* 1994, Jeglitsch *et al.* 1998). In addition, brevetoxins are known to stabilize more than one distinct conductance state of the sodium channel (Schreibmayer and Jeglitsch 1992, Gawley *et al.* 1995, Jeglitsch *et al.* 1998). Membrane depolarization, neurotransmitter release and all other known *in vivo* physiological effects (reviewed in Baden 1989) follow from these events.

Using the radiolabeled probe prepared by Baden *et al.* (1984), Poli *et al.* (1986) described the specific binding to sodium channels in rat brain synaptosomes. Binding was independent of membrane potential and occurred with high affinity and specificity to a single class of receptor sites. Rosenthal analysis and Hill plots suggested a single class of noninteracting binding sites. Natural brevetoxins, but not a synthetic nontoxic derivative, inhibited binding in a dose-dependent manner. Competition experiments with natural toxin probes specific for neurotoxin receptor sites 1–4 on the sodium channel indicated that brevetoxins bound to a previously undescribed site, denoted receptor site 5. Similar receptor sites were subsequently described in turtles (Edwards *et al.* 1990), fish (Stuart and Baden 1988, Edwards *et al.* 1990), and manatees (Trainer and Baden, 1999). In all of these models, Rosenthal analysis suggested a single class of receptor sites and the calculated binding affinities were similar (K_D 1-10 nM).

Over the range of 0-20 nM, only a single class of high-affinity binding sites was evident (Poli *et al.* 1985, Trainer *et al.* 1991). However, several studies (Trainer *et al.* 1991, Edwards *et al.* 1992, Whitney and Baden 1996, Dechraoui *et al.* 1999) later reported a second, low-affinity/high-capacity binding site for brevetoxins in rat brain tissues when label concentrations are increased to 100 nM or greater. In these studies, the high-affinity sites had affinities in the concentration range at which physiological effects are manifest. Maximum binding capacities calculated from Rosenthal analyses of binding data agreed well with those calculated for STX, suggesting brevetoxins bind in a 1:1 stoichiometry with the sodium channel. The low-affinity/high-capacity sites, however, demonstrated affinities two orders of magnitude lower, and the number of calculated binding sites greatly exceeded the density of sodium channels. This may be explained by either binding to proteins other than those associated with the channel (Trainer *et al.* 1991), binding to lipid components of the membrane (Edwards *et al.* 1992) or the presence of multiple receptor sites per channel (Whitney and Baden 1996). Although the physiological relevance of these sites is questionable, they may still be useful for probing channel structure or its membrane environment.

Subsequent work (Lombet *et al.* 1987, Holmes *et al.* 1991, Lewis *et al.* 1991) demonstrated that the ciguatoxin family of neurotoxins (produced by the dinoflagellate *Gambierdiscus toxicus* and transferred to humans who ingest tropical reef fishes) competitively inhibits brevetoxin binding to its receptor site in rat brain membranes. Based upon these data, plus similarities in structure and physiological effects *in vivo* and *in vitro,* the ciguatoxins are thought to share the brevetoxin receptor site on the voltage-sensitive sodium channel (Baden 1989 and references therein). At present, no other compounds, including endogenous molecules, are known to bind at this site.

Binding of brevetoxins to receptor site 5 has no effect on receptor sites 1 and 3, but markedly enhances binding and sodium influx mediated by ligands of receptor site 2 (Catterall and Risk 1981, Catterall and Gainer 1985, Poli *et al.* 1986). Poli *et al.* (1986) reported that the crude venom from *Centruroides sculpturatus* binding at site 4 has minimal effects on [^{3}H]PbTx-3 binding to rat brain synaptosomes. Similarly, Yuhi, *et al.,* (1994) reported that simultaneously applying veratridine and β-scorpion venom had no effect on [^{3}H]PbTx-3 binding to bovine adrenal chromaffin cells. However, Sharkey *et al.* (1987) reported enhancement of [^{125}I]CsTxII binding to receptor site 4 in rat brain synaptosomes by PbTx-2 and suggested an allosteric mechanism. Evidently, the allosteric linkage between these sites is unidirectional in synaptosomes, which are partially depolarized. It is likely bi-directional *in vivo.*

Note that allosteric interactions between brevetoxins and site 2 and 4 toxins differ among channel subtypes. For example, Cestele *et al.* (1995) noted differences in allostery between sites 2, 3 and 5 in rat brain and insect sodium channels. Wada *et al.* (1992) reported no allosteric interactions between site 5 and either site 2 or 3 in the bovine adrenal medulla.

Binding of a brevetoxin-linked photoaffinity probe to rat brain synaptosomes (Trainer *et al.* 1991) and purified reconstituted sodium channels from rat brain (Trainer *et al.* 1994) identified regions on S6 of domain I and S5 of domain IV as the likely location of site 5. From these data, Trainer *et al.* (1994) proposed a model whereby brevetoxins bind at the transmembrane interface between domains I and IV with the A-ring extending towards the intracellular opening and the K-ring near the extracellular surface of the pore. In this position, the photoreactive aryl azide moiety can react with amino acids on the extracellular side of the S5 of domain IV (Trainer *et al.* 1994). The extracellular segments of the S5 region of domain IV is also implicated in α-scorpion toxin binding, which may explain in part the allosteric interactions between the two sites. In addition, this region of domain IV is known to be the site of missense mutations in the congenital muscle diseases hyperkalemic periodic paralysis (HPP) and paramytonia congenita (PC). These diseases also result in modified channels with persistent depolarization and non-inactivating sodium conductance (Baden *et al.* 1994).

Gawley *et al.* (1992) first hypothesized that the two classes of brevetoxins

(PbTx-1-type and PbTx-2-type backbones) plus the structurally similar ciguatoxins share a common pharmacophore for receptor binding. Having noted that the K_D for each class correlated to the mouse potency data, they suggested that the flexibility of the polyether ladder backbones was critical and increased flexibility resulted in increased potency. They addressed this question elegantly, albeit indirectly, with an internal coordinate Monte Carlo search program to perform conformational analyses of the PbTx-2 (Gawley *et al.* 1992) and the PbTx-1 (Rein *et al.* 1994a) backbones. They found a range of conformations to be possible in solution. The four terminal rings (G-H-I-J of PbTx-1 and H-I-J-K of PbTx-2) are rigid and virtually identical. PbTx-2 has regions of flexibility around the D and H rings; PbTx-1 round the D, E, F, and G rings. Each toxin has a lactone function on the A ring. Energy minimization calculations suggested that "straight" conformations would dominate "bent" conformations. These energetically preferred conformations were then compared to structural parameters as well as binding and potency data derived from natural toxins and a series of synthetic derivatives (Rein *et al.* 1994b). From this, they concluded that the geometric requirements for binding are rather strict. They hypothesized the preferred structure for binding to the receptor site to be a cigar-shaped molecule approximately 30 Å long, which is bound primarily with hydrophobic and nonpolar solvation forces. This conclusion agreed well with the model of Trainer *et al.* (1994).

Further strengthening and refinement of this model came from experiments in which synthetic brevetoxin derivatives were evaluated and compared by *in vivo* toxicity, binding to the sodium channel and *in vitro* effects on single sodium channels. Baden *et al.* (1994) reduced the A-ring lactone of PbTx-3 with sodium borohydride and ethanol to form the saturated and unsaturated diols. These compounds were less potent by bioassay and sodium channel-binding experiments by three-to-four orders of magnitude, and they were similarly less potent in their ability to activate single sodium channels in nodose ganglia of newborn rats. From this, they concluded that the A-ring lactone is a functional requirement, probably by hydrogen bonding to some portion of the intracellular inactivation loop. Further, they postulated that the length of the brevetoxin molecule (26 Å) requires that the J-ring of PbTx-1 (or the K ring of PbTx-2) must reside 6-8Å inside the extracellular surface of the pore for the A ring to extend deep enough to affect the inactivation gate.

Gawley *et al.* (1995) tested this hypothesis further by synthesizing two additional derivatives: a truncated form of PbTx-2, which retained the functional groups of the native toxin but was 10 Å shorter, and a desoxy-derivative of PbTx-3, which lacked the A ring carbonyl oxygen. The truncated form was unable to competitively inhibit [^{3}H]PbTx-3 binding to rat brain membranes and was nontoxic to mosquito fish. In isolated single sodium channels, it did not prolong mean open times or inhibit inactivation, although it did shift the activation potential slightly. The desoxy derivative inhibited specific radiolabel binding and was ichthyotoxic to mosquito fish. In single sodium channels, it shifted the activation potential and inhibited inactivation, but did not induce longer mean open times. The A-ring diol did not inhibit specific radiolabel binding and was not toxic to mosquito fish. In single channel recordings, at high concentration (10 μM), it induced the channels to populate five subconductance states. Taken together, these experiments confirm that the full length of the molecule is necessary for all the observed physiological effects, and that the loss of the A-ring oxygen both reduces binding affinity and modifies the toxin's effects on inactivation.

Taking these experiments one step further, Purkerson-Parker *et al.* (2000) modified the K ring of PbTx-3 into the benzoyl and napthoyl derivatives. Each of these derivatives bound to the receptor site and competitively inhibited [^{3}H]PbTx-3 binding. The benzoyl derivative actually demonstrated an order of magnitude greater affinity for the site than for the native toxin, and activated channels. The α- and β-napthoyl derivatives, however, did not activate channels and were considered brevetoxin antagonists. Thus, the backbone structure is important for binding, while the side chain on the K-ring has a more functional role in interaction with the channel protein. The large napthoyl side chain evidently destabilizes the open configuration relative to nonconducting states.

Taken together, these experiments led to the current hypothesis of brevetoxin binding to neurotoxin receptor site 5 on the voltage-gated sodium channel (Figure 3). In this model, a nearly linear conformation of the toxin binds to the channel within the lipid bilayer at the juncture of domains I and IV of the α subunit. The head (A-ring) portion of the molecule forms hydrogen bonds with S6 of domain I, from where it physically obstructs the fast inactivation gate. The tail (J or K ring) extends to within 6-8 Å of the pore opening and interacts with amino acid residues on the extracellular end of S5 of domain IV. The length and conformation of the backbone, the carbonyl oxygen in the A-ring and the side chain on the K ring are all critical for binding and expression of the full range of pharmacological effects.

Toxicokinetics

The distribution, metabolism and excretion of brevetoxins in animals have been sorely neglected to date. In spite of the obvious relevance to both medical treatment of human intoxication and evaluation of seafood safety, little is known about the fate of brevetoxins as xenobiotics. Until recently, the working model of NSP was one in which the toxins were synthesized by the dinoflagellates, biomagnified in filter-feeding mollusks, and

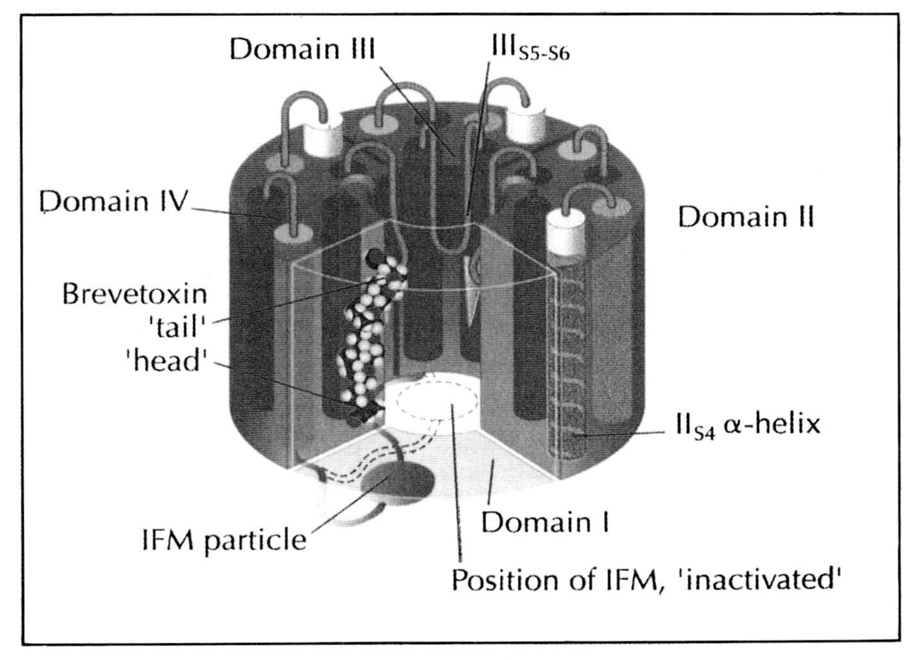

Figure 3. Hypothetical model of brevetoxin binding to neurotoxin receptor site 5 on the voltage-gated sodium channel. (Reprinted from Gawley *et al.* (1995) with permission from Elsevier Science.)

passed on in their native form to the seafood consumer. We now know this to be an overly simplistic model fraught with difficulties. In fact, most of the brevetoxins themselves are likely metabolic by-products of only two parent compounds, PbTx-1 and PbTx-2. Baden *et al.* (1989) reported that these two compounds account for 98% and 95%, respectively, of the total toxin profile in logarithmic phase cultures. During the stationary phase, however, the hydroxylated, acetylated and epoxide forms appear as the fractions of PbTx-1 and -2 decrease. These compounds demonstrate increased water-solubility and decreased potency compared to the parents. Although they may serve to protect the cells from self-intoxication by some unknown mechanism, as hypothesized by Baden *et al.* (1989), they may also simply reflect the fact that the parent molecules possess the proper functional group chemistries to be substrates of common metabolic enzyme systems. When other organisms are exposed to this mixture, either by diffusion across gill membranes (fish) or filter feeding (bivalve mollusks, fish), further metabolism occurs. If these organisms are in turn preyed upon by predators, the next trophic level is exposed to these secondary metabolites rather than the parent compounds. To adequately evaluate the exposure to, or presence of, brevetoxins in any organism, it is critical to understand its distribution, metabolism and excretion profile. This applies equally to monitoring seafood stocks for consumer safety and diagnosis of human victims of NSP.

Fish

Kennedy *et al.* (1992) performed the first toxicokinetic studies in fish. After injecting a bolus dose of [^{3}H]PbTx-3 into toadfish, they reported that radioactivity concentrated primarily in the muscle, liver and intestine. Elimination occurred slowly and primarily via the bile, although the kidneys and gills probably contributed as well. Only 24% of the radioactivity was eliminated in 4 days. Chromatographic analysis of blood and bile suggested the presence of metabolites more polar than the parent toxin.

These results were confirmed in toadfish by Washburn *et al.* (1994), who also examined whether brevetoxins could induce cytochrome P450 levels in immature redfish. After oral administration of very high toxin levels (25 μg/kg, but not 15 μg/kg), they found a two- to threefold increase in ethoxyresorufin *O*-deethylase (EROD) activity, but no induction of pentoxyresorufin *O*-depentylase (PROD) or glutathione *S*-transferase (GST) activities.

In a subsequent study (Washburn *et al.* 1996), who used even higher toxin concentrations (430 μg/kg, oral administration to striped bass), observed an identical two- to threefold increase in EROD levels, and minor (35-50%) induction of cytosolic GST isozymes. In contrast, β-naphthoflavone increased EROD levels 30-fold, but had no effect on GST isozymes.

While the EROD results are interesting, the relatively modest increases observed and the very high levels of toxin required for induction call the physiological relevance into question. This question was further clarified when it was shown that neither PbTx-2 nor PbTx-3 are ligands of the aryl hydrocarbon (aH) receptor. Instead, the epoxide form (PbTx-6) is an agonist of the aH receptor; in radioligand binding studies, it inhibited radiolabelled dioxin binding to the aH receptor 50%-75% at a concentration of 200 nM (Washburn *et al.* 1997). PbTx-6 is a probable metabolite of PbTx-2 via phase I cytochrome P450 activity, and is likely responsible for the EROD induction in fish.

These results are interesting and worthy of further investigation. However, the physiological relevance is yet to be determined. More work is urgently needed to identify the metabolites observed by Kennedy *et al.* (1992) so that the relative importance of PbTx-6 can be determined in relation to the other metabolites.

Shellfish

It is surprising, given that brevetoxins have been known for nearly 50 years to be vectored to humans by shellfish, that very little is known about molluscan metabolism of these toxins. After an outbreak of shellfish-associated intoxications in New Zealand in 1992, Ishida *et al.* (1995) isolated what they called a new brevetoxin (brevetoxin B1) from the cockle *Austrovenus stutchburyi*. This compound was essentially identical to PbTx-2 (brevetoxin B), except that the C-42 aldehyde moiety was replaced with a $-CO-NH-CH_2-CH_2-SO_3Na$ side chain. This was later followed by the isolation of brevetoxin B3, the cysteine sulfoxide derivative (Morohashi *et al.* 1995), and its oxidation product, brevetoxin B2 (Murata *et al.* 1998) from the green mussel *Perna canaliculus*. Brevetoxin B4 (Morohashi *et al.* 1999), the major toxin from the green mussel, was later isolated and identified as a mixture of N-myristoyl and N-palmitoyl derivatives of brevetoxin B2. All of these toxins are now considered to be metabolites of PbTx-2 or PbTx-3. Interestingly, B1 was not identified in green mussels, nor was B2, B3 or B4 found in cockles collected from the same area. This suggests that shellfish metabolism is at least somewhat species specific. Although these metabolites maintain various levels of toxicity to mice injected i.p., they are uniformly nontoxic to fish when applied to the water. This may be due to a species-specific alteration of their ability to modulate sodium channel function (as discussed in the section on pharmacology), or, more likely, it is the result of solubility changes that prevent the metabolites from crossing the gill membranes.

This species specificity of metabolism was confirmed by Dickey *et al.* (1999) and Poli *et al.* (2000) who isolated three brevetoxin metabolites from toxic clams (*Mercenaria* spp. and *Chione cancellata*) and a predatory whelk (*Busycon contrarium*) from Florida, as well as from toxic oysters (*Crassostrea virginica*) from Texas. The structures of these metabolites were not elucidated, but the chemical properties and masses identified by LC-MS analysis ($[M + H^+] = 1005, 1018$, and 1035) did not correspond to those metabolites isolated from New Zealand shellfish. In these two studies, the Florida shellfish produced the same metabolites as the Texas shellfish, albeit in different ratios. Even the whelk, which is carnivorous and likely accumulated brevetoxins by feeding on contaminated clams, yielded the same metabolites. Thus, while some species specificity undoubtedly exists, metabolic pathways common to many mollusks must clearly take part, at least in Atlantic species. Alternately, differences in metabolism may be related to differences in associated bacterial flora. Although this question has yet to be addressed, it could be an important new research direction.

Mammals

The first toxicokinetic investigation of brevetoxin in mammals was reported by Poli *et al.* (1990a, b). We used isolated hepatocytes, perfused livers and whole animals to investigate distribution, metabolism and elimination of [^{3}H]PbTx-3 in rats. After intravenous administration, brevetoxin rapidly cleared the bloodstream and distributed to various tissues. The highest concentration was found in the skeletal muscle, liver and intestinal tract. Heart, kidney, testes, lung, stomach, brain and spleen accumulated only small amounts. Elimination was nearly complete after 48 hours; after 6 days, 75% of the administered label was eliminated in the feces, and 14% in the urine. The liver was the major organ of metabolism and biliary excretion the major route of elimination. Chromatographic analysis of bile demonstrated several more polar peaks of radioactivity thought to be hepatic metabolites. This was confirmed by the demonstration of metabolites in the bile from isolated perfused livers and in the culture medium of isolated hepatocytes. The skeletal muscle was a storage depot, but not a site of metabolism. The identities of the metabolites were not determined.

These results were confirmed in later studies in rats by other routes of administration. Cattet and Geraci (1993) published a similar study using oral administration. They reported similar results, although they felt that the renal route of administration was more important after the first 48 hours. They did not analyze for metabolites. Bensen *et al*. (1999) used intratracheal instillation as their route of administration and again reported very similar distribution patterns. Consistent with their route of administration, they reported a greater percentage of radioactivity in the lungs after 7 days. They felt that the respiratory tract may capture a small percentage of an inhaled dose of brevetoxins and perhaps lead to accumulation upon repeated exposure. Again, they did not analyze for metabolites.

Taken together, these studies demonstrate a clear pattern of rapid distribution in rats, primarily to the skeletal muscle and liver, regardless of the administration route. The muscle mass is a storage depot from which toxin re-distributes slowly to the liver for hepatic metabolism and biliary excretion via the feces. The majority of elimination occurs in the first 48 hours, with fecal elimination dominating renal by at least 2:1. Later, renal elimination may equal or exceed fecal. Hepatic metabolism results in a number of more polar metabolites of unknown structure.

In 1996, an epizootic in manatees coincided with a bloom of *K. brevis* in south Florida and provided an opportunity to study the distribution of brevetoxin in another species after a natural exposure (Bossart *et al*. 1998). In this case, the exposure was likely chronic, occurring through daily inhalation of toxin-containing aerosols. There was positive immunohistochemical staining in lymphocytes and macrophages of the lung, liver and secondary lymphoid tissues. Lymphocytes and macrophages associated with inflammatory lesions of the nasal mucosa and meninges also stained positive. The authors hypothesized a two-part toxicosis scenario. First, after passing through the lungs and entering the general circulation, brevetoxins are phagocytized by macrophages and/or internalized by lymphocytes where they act as competitive inhibitors of the degradative enzymes of the cells, thus programming them for cell death. Secondly, internalization by macrophages initiates the release of a cascade of inflammatory mediators resulting in toxic shock and death. Either mechanism, or a combination of both, was thought to be the cause of death.

This scenario is supported circumstantially by the fact that immunostaining of brevetoxin correlated to the locations of highest lysosome concentrations. It further followed from a report by Sudarsanam *et al.* (1992), in which brevetoxins were shown to be inhibitors of cathepsin-L, a lysosomal proteinase thought to be important in cellular protein turnover. That this inhibition required μM concentrations of brevetoxins, and the neurotoxic effects of brevetoxins are manifest at nM concentrations in all species tested, must bring a certain reservation to this hypothesis. However, manatees may be more resistant to brevetoxins than other mammals, or toxin concentrations within macrophages and/or lymphocytes can locally exceed other tissues. Until these questions can be answered, this hypothesis must be considered along with traditional theories of neuromuscular paralysis. Additionally, more experiments in animal models given multiple aerosol challenges and measurement of serum concentrations of inflammatory mediators may help resolve to this issue.

Toxicokinetic and metabolism data on human intoxications are lacking. Poli *et al.* (2000) measured brevetoxin activity in the urine of two small children diagnosed with NSP in Florida. We found high levels of brevetoxin by radioimmunoassay several hours after ingestion of toxic shellfish, but no activity four days later. This agrees well with the rat data, although, because no intermediate urine samples were collected at the hospital, construction of elimination curves was impossible. HPLC-MS analysis of this urine detected low levels of both PbTx-3 and PbTx-2, but also peaks that were thought to be metabolites. Although the structures of the metabolites were not identified, they differed from the major shellfish metabolites.

Clearly, much work is needed to clarify issues regarding metabolism of brevetoxins in both shellfish and mammals. From the standpoint of seafood safety, it is critical to know which compounds are present in foodstuffs, and to tailor assay methods to the appropriate compounds. Similarly, we will better monitor human exposure and diagnose NSP victims when we learn what samples to collect and metabolites to target. For these reasons, detection of brevetoxins has received a great deal of attention.

Detection

As with many marine toxins, detection of brevetoxins was for many years dependent upon *in vivo* bioassays such as the mouse bioassay of McFarren *et al.* (1965) or the fish bioassay of Baden *et al.* (1979). This changed in 1984 when Baden *et al.* (1979) published the first paper describing the production of specific antibodies against the PbTx-2-type toxins. Since that time, detection has taken two parallel courses. The first course is immunological and is dependent upon antibody recognition of the polyether backbone structure. The second is pharmacological and is based upon binding of the toxin to its sodium channel receptor. These assays come in a variety of forms; some measure binding alone, others measure one or more sequelae. Both immunological and pharmacological methods can *detect* and *quantify* very well. However, neither can truly *identify* the toxins present. This requires analysis of the physical attributes of the molecule that can be accomplished only through traditional analytical chemical techniques such as mass spectrometry and nuclear magnetic resonance spectroscopy. While these methods can also be used for detection, they are usually time-consuming and require expensive instruments with highly trained operators. For these reasons, they are not routinely used to analyze large numbers of samples, and they will not be discussed here.

Immunological Methods

Baden *et al.* (1984) published the first report on the production of specific antibodies against the brevetoxins. The immunogen was a PbTx-3/bovine serum albumin (BSA) conjugate injected intramuscularly into a male goat. The resulting serum was specific for the PbTx-2-type backbone structure, but it was low in titer, non-neutralizing in the fish bioassay and characterization data were lacking. No binding affinity was measured, but the linear portion of the competition curve for their highest titer serum preparation suggested a detection limit of about 3-4 ng/ml. Although no validation data were presented, this was the first *in vitro* brevetoxin assay, and it improved sensitivity by at least two orders of magnitude over the mouse bioassay.

Poli and Hewetson (1992) produced polyclonal anti-PbTx serum in a goat by using an analogous immunogen. The resulting serum yielded antibodies of high affinity (K_D = 0.8 nM) and reasonably high capacity (7-9 μg of PbTx-3/ml serum). It was also neutralizing in rats (Poli *et al.* 1990a). From this serum we developed a highly specific and sensitive radioimmunoassay for the PbTx-2-type brevetoxins. The limit of accurate quantitation for this assay was 300 pg/ml. All of the PbTx-2-type brevetoxins were recognized equally; PbTx-1 was recognized at about 100-fold higher concentration. Other marine toxins (okadaic acid, pectenotoxin, yessotoxin, STX and palytoxin) did not cross-react. More importantly, we evaluated urine and shellfish extracts and optimized for these matrices.

This assay has been used to detect brevetoxins in shellfish extracts as well as brevetoxin metabolites in the urine of NSP victims (Poli 1996, Poli *et al.* 2000). Later, a similar antigen was used to produce antibodies in rabbits. This serum demonstrated much higher affinity (K_D = 26 pM), but considerably lower binding capacity (0.24 μg/ml).

Levine and Shimizu (1992) also elicited anti-PbTx antibodies in rabbits by conjugating PbTx-2 to BSA in the presence of $NaBH_4$. Similar to our rabbit antibody, they reported a much higher affinity (K_D = 8-12 pM), but lower capacity (approx 0.4 μg/ml) than the goat antiserum. As with the others, this serum recognized the PbTx-2-type backbone, but had little cross-reactivity with PbTx-1. They reported the ability to distinguish PbTx-2 and PbTx-3, although these data were not reproducible in collaborative work with our laboratory (unpublished data). Unfortunately, this serum was not developed and validated as an assay.

A common thread running through the data of each group is the specificity of their antibodies. When a PbTx-2-type brevetoxin is used as the hapten, cross-reactivity with the PbTx-1-type backbone is low. However, the specific binding epitopes on the PbTx-2-type backbone were unknown. To address this issue, brevetoxin derivatives were synthesized whereby different portions of the molecule were altered and the effects on binding by the different antibodies were assessed (Melinek *et al.* 1994, Poli *et al.* 1995). Each investigation demonstrated that the "head" or A-ring portion of the molecule is a distinct epitope from the "bend" or H-ring portion. The K-ring portion may or may not be an additional epitope. This was difficult to assess because toxin was coupled to the protein carrier through that portion of the molecule. Interestingly, each animal tested produced antibodies specific to only one region of the molecule, and competition curves were sigmoidal, suggesting that even within polyclonal sera, there was only one major recognition class of antibodies active at the radiolabel concentration used in the assays. This information is important for assay development; isolation of sera with different antigen recognition properties lends flexibility to assay formats developed from those sera.

While radioimmunoassays are rapid, specific and reproducible, there are serious obstacles to regulatory field use. The most important of these is their dependence upon radioisotopes and the associated handling and disposal issues. In addition, they require expensive equipment to evaluate results. To overcome these obstacles, antibodies were further developed for use in a direct ELISA (Trainer and Baden 1991, Baden *et al.* 1995). In this format, toxin was adsorbed directly to the plate, then exposed to antibody. The amount of bound antibody, which is a direct measure of toxin present, was detected with an enzyme-labeled anti-goat antibody. Using affinity-purified antibody to decrease background binding, they reported a detection limit of 0.04 pmoles/well (358 pg/ml), similar to that of the radioimmunoassay. However, there was a significant decrease in sensitivity in the presence of shellfish extracts, and they felt that difficulties in extracting metabolites from fish tissue precluded it for this use. The advantages of this assay clearly outweigh the tradeoff in sensitivity because the 96-well microtiter plate format is more amenable to rapid-throughput testing of large numbers of samples. The reported sensitivity is still orders of magnitude greater than that required to detect brevetoxins at the regulatory action level of 80 μg/100 g of shellfish.

Meanwhile, a research group in New Zealand was developing an ELISA based upon a competition format (Garthwaite *et al.* 1996). In their assay, the plates were coated with specific antibody, then exposed to a mixture of sample (toxin) and enzyme-labeled brevetoxin. This set up a competition between the labeled and unlabeled toxin. The amount of labeled toxin bound by the coated antibody was inversely proportional to the amount of unlabeled toxin in the sample. The reported working range was 2.5-75 ng/ml. Although less sensitive than other assays, it was sufficient to detect brevetoxins well below regulatory limits, and it was instrumental in monitoring and evaluating shellfish stocks after the 1992-93 NSP event in New Zealand (Garthwaite *et al.* 1996).

Pharmacological Methods

Pharmacological methods are those based upon binding of toxin to its endogenous sodium channel receptor. The simplest of these is the direct receptor-binding assay (Poli *et al.* 1986, Poli 1996, Trainer and Poli 2000). Because it measured binding in three dimensions by using a molecule designed (or evolved) to bind specifically the analyte in question, it was essentially analogous to the radioimmunoassay. The major difference lay in the binding molecule. When the binding molecule is the endogenous receptor, binding provides a measure of the total *pharmacological activity* in the sample. Hence, all molecules active at the receptor site, regardless of structure, are detected. In contrast, antibody binding measures *structural elements,* thus differentiating the two backbone types or other types of toxins. In this case, the assay directly measured binding of radiolabeled brevetoxin to receptor sites in rat brain membranes. Set up in a competition format, it measured unlabeled toxin in samples by detecting the decrement in bound radioactivity resulting from competition with labeled toxin for the available receptor sites. The assay was simple, straightforward and reproducible; sensitivity was < 1 ng/ml in assay buffer. Matrix effects of shellfish extracts were species dependent.

Because of its analogy to radioimmunoassay, however, it suffered from essentially the same drawbacks. One additional drawback of the membrane assay was specific to the format. Separation of bound from free radioactivity was accomplished by either centrifugation or filtration, which added a time-consuming step and was a major obstacle to high-throughput use. This obstacle was eliminated when Van Dolah *et al.* (1994) reformatted the membrane assay into a 96-well microtiter plate format. Although slightly less sensitive, this format was effective and opened the door for regulatory use of the receptor binding assay as a screening tool. An additional advantage in a regulatory setting is that the same format and equipment — and in some cases the same tissue preparations — can be used as a common platform to develop assays for several different toxins of regulatory importance. The National Ocean Services Laboratory in Charleston, South Carolina has designed procedures using this technology to test for ciguatoxins, STX and domoic acid in addition to brevetoxins.

Trainer *et al.* (1995) reported detection of brevetoxins by purified sodium channels isolated from rat brain membranes and reconstituted into phospholipid vesicles. This assay is analogous to the membrane-binding assay. However, the time and labor costs associated with membrane isolation and reconstitution probably preclude it from ever being adopted as a routine assay.

A different approach to pharmacological assay for brevetoxins was developed by Manger *et al.* (1993, 1995). Rather than measuring direct binding of brevetoxin to brain membranes, they used cell viability as an assay endpoint. In this assay, the sodium channels in cultured neuroblastoma cells were activated by veratridine in the presence of ouabain to inhibit sodium efflux through the Na^+-K^+ ATPase. This led to sodium influx through the activated channels and reduced cell viability. Co-exposing cells to brevetoxin caused enhanced effects through allosteric interactions between the receptor sites. This dose-dependent enhancement of cell death was measured in 96-well microtiter plates by colorimetrically monitoring the cells' ability to metabolically reduce a tetrazolium compound to a colored product. Their reported detection limit for brevetoxins was 0.25 ng/10 μl sample in each well (25 ng/ml). Although less sensitive than other assays, several advantages were nonetheless apparent. The assay required only 4-6 hours to complete and the format was well suited to automation. However, the biggest advantage was one of flexibility. As mentioned above for the membrane assay, this format offers the flexibility of testing for other toxins as well. For example, with only minor modifications, one can use this assay to test for sodium channel blockers as well as activators.

However, the cell-based assay suffers from some serious drawbacks. First, it requires highly trained operators and a great deal of support in the form of cell culture with the associated details of feeding, splitting, and timing of the cell cultures. More importantly, analysis of common samples indicated poor agreement between cell culture assays and either radioimmunoassay, membrane assay, or mass spectrometry results (Dickey *et al.* 1996, Garthwaite *et al.* 1996). The reason for this poor agreement has been difficult to understand, although it may be related to the artifacts inherent within the system. The neuroblastoma cells used in the cytotoxicity assay are an immortal cell line and thus not truly representative of primary cells. Cell death is caused by Na^+ influx into a cell where efflux through the Na^+-K^+ ATPase has been blocked with oubain. Hence, the assay endpoint is also not representative of the normal cellular physiology secondary to channel activation. This combination of artifacts may somehow adversely affect the assay. Berman and Murray (1999) recently reported a secondary mechanism of action of brevetoxins in cultured cells. They demonstrated that brevetoxin neurotoxicity in primary cerebellar granule neurons is not a direct result of sodium channel binding and activation, but instead a secondary effect resulting from an increase in intracellular Ca^{++} secondary to Na^+ influx. The increased $[Ca^{++}]$ resulted in release of excitatory amino acids and subsequent activation of NMDA receptors leading to cell death. The lack of NMDA receptors in the Neuro2A cell line used in the cytotoxicity assay precludes this from being a contributory problem in that assay, but it underscores the potential problems involved in using immortal cell lines. Further clarification of these issues awaits the development of cell assays using primary neuronal cells and measuring endpoints of natural physiological processes. In addition, comparison of different neuronal cell types (i.e. excitatory vs inhibitory) may yield important information.

Further modification of the cell-based cytotoxicity assay led to the development of a reporter gene assay (Fairey *et al.* 1997, Fairey and Ramsdell 1999). Based upon earlier work (Peng *et al.* 1995), demonstrating that brevetoxins and ciguatoxins induce the immediate response gene *c-fos* in cultured cells, this assay used a *c-fos*-luciferase construct stably transfected into neuroblastoma cells as a "reporter" of brevetoxin interaction with cells. Similar assays were developed for domoic acid and the putative active component produced by *Pfiesteria* cultures using different cell lines (Fairey and Ramsdell 1999). All of these assays use light generation through the luciferin/luciferase pathway as the assay endpoint. Although the signal was modest — the IC_{50} occurred at approximately twice background — the authors demonstrated a saturable, dose-dependent response to brevetoxins and a detection limit about 3 ng/ml (Fairey and Ramsdell 1999).

The main drawback to this assay is the lack of specificity, either real or perceived. Although the authors demonstrated a certain level of selectivity based upon differential receptor expression among cell types, they never adequately addressed the issue of specificity within cell types. The action of *c-fos* as an immediate response gene is well-documented and unchallenged. However, by its very nature, it is a reporter of cellular stress rather than specific pharmacological events. Thus, it may respond to a myriad of compounds exclusive of brevetoxins that result in physiologic stress to the cells. This creates the potential for a wide range of matrix effects that require evaluation and characterization. Until these data are available, it is unlikely that this assay will be widely accepted.

Finally, brevetoxin has been detected by an *in vitro* neurophysiological model (Kerr *et al.* 1999). This assay measured changes in antidromic and orthodromic spike amplitudes, field EPSP and afferent action potential volley in rat hippocampal slices after toxin exposure. Using these parameters, these researchers found characteristic signatures for brevetoxins, STX and domoic acid that they considered diagnostic. Further, they demonstrated detection capability in shellfish extracts. Although the assay time was relatively rapid, the complexities and expense involved with animal use and tissue preparation combined with the necessity for expensive equipment and highly trained operators will likely limit the utility of assays of this nature to the research laboratory.

Summary: Immunological vs. Pharmacological Assays

Immunological and pharmacological assays both possess major advantages and disadvantages. As mentioned above, antibody-based assays detect structure only, and thus can differentiate the PbTx-1 and PbTx-2 backbones. This can be an advantage in the research laboratory, but perhaps a disadvantage if one wishes to screen for total toxicity, or to "double up" and use the same assay to detect ciguatoxin. The latter is possible with pharmacological assays, which detect composite biological activity, but cannot differentiate toxins of like pharmacology. Composite biological activity is useful for regulatory screening, but may be misleading. For example, in *K. brevis* cultures, PbTx-1 amounts to only 1-16% of the total toxin produced (Baden and Tomas 1989), but is approximately fivefold more potent than the predominant toxin PbTx-2 (Baden *et al.* 1988). Immunological analysis of this culture would not detect the PbTx-1 (based upon 1% cross-reactivity) and therefore underestimate the toxicity by at least 50%. Pharmacological analysis would detect the total potency but give no information on the toxin profile. One might expect total toxicity to be the more important issue for regulatory screening. In this scenario, it could be argued that pharmacological assays would make the better screening tests, with analytical methods held in reserve to elucidate individual components in those samples that screen positive. However, new information regarding shellfish metabolites suggest this may not be the case. The major metabolites isolated from toxic shellfish from the Gulf coast of the United States (Dickey *et al.* 1999, Poli *et al.* 2000) were recognized three- to ten-fold better by radioimmunoassay than by membrane binding while the parent compound was recognized equally. This suggests that metabolic conversion affects binding affinity, and pharmacological assays will underestimate total toxin concentration. While this might be acceptable if no further metabolic conversion occurs, if the adducts are cleaved back to the parent compounds *in vivo*, then toxicity will revert to that of the parent compounds and the membrane assay will have underestimated the threat. Although there are no data presently available on this question, it is notable that none of the shellfish metabolites was detected in the urine of two children suffering from NSP (Poli *et al.* 2000).

Clearly, each type of assay has its own strengths and weaknesses; the best assay is determined by the purpose for which it is intended. At the present time, one can argue that immunological assays offer the best hope for regulatory screening although more research into the pharmacology and toxicology of shellfish metabolites in humans is needed before this question can be answered with certainty.

References

Ahmed, M.S., O. Arakawa, and Y. Onoue. 1995. Toxicity of cultured *Chattonella marina*. Pages 499-504 in *Harmful Algal Blooms*, P. Lassus, G. Arzul, E. Erard, P. Gentien, and C. Marcaillou, eds., Lavoisier, Paris.

Asai, S., J.J. Krzanowski, W.H. Anderson, D.F. Martin, J.B. Polson, R.F. Lockey, S.C. Burkantz, and A. Szentivanyi. 1982. Effects of the toxin of red tide, *Ptychodiscus brevis*, on canine tracheal smooth muscle: A possible new asthma triggering mechanism. *J. Allergy. Clin. Immunol.* 69: 418-428.

Baden, D.G. 1983. Marine food-borne dinoflagellate toxins. *Intl. Rev. Cytol.* 82: 99-150.

Baden, D.G. 1989. Brevetoxins: unique polyether dinoflagellate toxins. *FASEB J.* 3: 1807-1817.

Baden, D.G., R. Melinek, V. Sechet, V.L. Trainer, D.R. Schultz, K. S. Rein, C.R. Tomas, J. Delgado, and L. Hale. 1995. Modified immunoassays for polyether toxins: implications of biological matrices, metabolic states, and epitope recognition. *J. AOAC Intl.* 78: 499-508.

Baden, D.G., T.J. Mende, and R.E. Block. 1979. Two similar toxins isolated from *Gymnodinium breve*. Pages 327-334 in *Toxic Dinoflagellate Blooms*, Taylor, D.L. and Seliger, H.H., eds., Elsevier, New York.

Baden, D.G., Mende, T.J., and L.E. Roszell. 1989. Detoxification mechanisms of Florida's red tide dinoflagellate *Ptychodiscus brevis*. Pages 391-394 in *Red Tides: Biology, Environmental Science, and Toxicology*, Okaichi, K., Anderson, D. M., and Nemoto, T., eds., Elsevier, New York.

Baden, D.G., T.J. Mende, J. Walling, and D.R. Schultz. 1984. Specific antibodies directed against toxins of *Ptychodiscus brevis* (Florida's red tide dinoflagellate). *Toxicon* 22: 783-789.

Baden, D.G., K.S. Rein, R.E. Gawley, G. Jeglitsch, and D.J. Adams. 1994. Is the A-ring lactone of brevetoxin PbTx-3 required for sodium channel orphan receptor binding and activity? *Natural Toxins* 2: 212-221.

Baden, D.G. and C.R. Tomas. 1989. Variations in major toxin composition for six clones of *Ptychodiscus brevis*. Pages 415-418 in *Red Tides: Biology, Environmental Science, and Toxicology*, Okaichi, K., Anderson, D.M., and Nemoto, T., eds., Elsevier, New York.

Benson, J.M., D.L. Tischler, and D.G. Baden. 1999. Uptake, tissue distribution, and excretion of brevetoxin 3 administered to rats by intratracheal instillation. *J. Toxicol. Env. Health* 56(part A): 345-355.

Benziger, G.R., C.L. Drum, L.Q. Chen, R.G. Kallen, and D.A. Hanck. 1997. Differences in the binding sites of two site-3 sodium channel toxins. *Pflügers Arch.* 434: 742-749.

Berman, F.W. and T.F. Murray. 1999. Brevetoxins cause acute excitotoxicity in primary cultures of rat cerebellar granule neurons. *J. Pharmacol. Exp. Ther.* 290: 439-444.

Bloomquist, J.R. and D.M. Soderland. 1988. Pyrethroid insecticides and DDT modify alkaloid-dependent sodium channel activation and its enhancement by sea anemone toxin. *Mol. Pharmacol.* 33: 543-550.

Bonhaus, D.W., R.C. Herman, C.M. Brown, Z. Cao, L.F. Chang, D.N. Loury, P. Sze, L. Zhang, and J.C. Hunter. 1996. The β1 sodium channel subunit modifies the interactions of neurotoxins and local anesthetics with the rat brain IIA α sodium channel in isolated membranes but not in intact cells. *Neuropharmacology* 35: 605-613.

Bossart, G.D., D.G. Baden, R.Y. Ewing, B. Roberts and S.D. Wright. 1998. Brevetoxicosis in manatees (*Trichechus manatus latirostris*) from the 1996 epizootic: gross, histologic, and immunohistochemical features. *Toxicol. Pathol.* 26: 276-282.

Brown, G.B., J.E. Gaupp, and R.W. Olsen. 1988. Pyrethroid insecticides: stereospecific allosteric interaction with the batrachotoxinin—A benzoate binding site of mammalian voltage-sensitive sodium channels. *Mol. Pharmacol.* 34: 54-59.

Catterall, W.A. 1980. Neurotoxins that act on voltage-sensitive sodium channels in excitable membranes. *Ann. Rev. Pharmacol. Toxicol.* 20: 15-43.

Catterall, W.A. 2000. From ionic currents to molecular mechanisms: The structure and function of voltage-gated sodium channels. *Neuron* 26:13-25.

Catterall, W.A. and M. Gainer. 1985. Interaction of brevetoxin A with a new receptor site on the sodium channel. *Toxicon* 23: 497-504.

Catterall, W.A., C.S. Morrow, J.W. Daly, and G.B. Brown. 1981. Binding of batrachotoxinin A 20-α-benzoate to a receptor site associated with sodium channels in synaptic nerve ending particles. *J. Biol. Chem.* 256: 11379-11387.

Catterall, W.A. and M.A. Risk. 1981. Toxin T_{46} from *Ptychodiscus brevis* (formerly (*Gymnodinium breve*) enhances activation of voltage-sensitive sodium channels by veratridine. *Mol. Pharmacol.* 19: 345-348.

Cattet, M. and J.R. Geraci. 1993. Distribution and elimination of ingested brevetoxin (PbTx-3) in rats. *Toxicon* 31: 1483-1486.

Cestele, S. and W.A. Catterall. 2000. Molecular mechanisms of neurotoxin action on voltage-gated sodium channels. *Biochimie* 82: 883-892.

Cestele, S., R.B. Khalifa, M. Pelhate, H. Rochat and D. Gordon. 1995. α-Scorpion toxins binding on rat brain and insect sodium channels reveal divergent allosteric modulations by brevetoxin and veratridine. *J. Biol. Chem.* 270: 15153-15161.

Daugbjerg, N., G. Hansen, J. Larsen, and O. Moestrup. 2000. Phylogeny of some of the major genera of dinoflagellates based on ultrastucture and partial LSU rDNA sequence data, including the erection of three new genera of unarmored dinoflagellates. *Phycologia* 34: 302-317.

Davis, C.G. 1947. *Gymnodinium breve*, n. Sp. A cause of discolored water and animal mortality in the Gulf of Mexico. *Bot. Gaz.* 109: 58-360.

De Weille, J.R., L.D. Brown, and T. Narahashi. 1990. Pyrethroid modifications of the activation and inactivation kinetics of the sodium channels in squid giant axon. *Brain Res.* 512: 26-32.

Dechraoui, M.-Y., J. Naar, S. Pauillac, and A.-M. Legrand. 1999. Ciguatoxins and brevetoxins, neurotoxic compounds active on sodium channels. *Toxicon* 37: 125-143.

Deffois, A., D. Fage, and C. Carter. 1996. Inhibition of synaptosomal veratridine-induced sodium influx by antidepressants and neuroleptics used in chronic pain. *Neurosci. Lett.* 220: 117-120.

Denac, H., M. Mevissen, and G. Scholtysik. 2000. Structure, function, and pharmacology of voltage-gated sodium channels. *Naun.-Schmied. Arch. Pharmacol.* 362: 453-479.

Dickey, R., E. Jester, R. Granade, D. Mowdy, C. Moncreiff, D. Rebarchik, M. Robl, S. Musser and M. Poli. 1999. Monitoring brevetoxins during a *Gymnodinium breve* red tide:

Comparison of sodium channel specific cytotoxicity assay and mouse bioassay for determination of neurotoxic shellfish toxins in shellfish extracts. *Nat. Toxins* 7: 157-165.

Dickey, R.W., R.L. Manger, M.A. Poli, F.M. Van Dolah, J.A. Delgado-Arias, S. Musser, D.G. Baden, J. Hungerford, K.S. Rein, S. Lee, T. Leighfield, and H.R. Granade. 1996. Assessment of methods for the determination of ciguatoxins. *Toxicon* 34: 306.

Dudley, S.C.J.R., H. Todt, G. Lipkind, and H.A. Fozzard. 1995. A μ-conotoxin-insensitive Na^+ channel mutant: possible localization of a binding site at the outer vestibule. *Biophys. J.* 69:1657-1665.

Edwards, R.A., A.M. Stuart and D.G. Baden. 1990. Brevetoxin binding in three phylogenetically diverse vertebrates. Pages 290-293 in *Toxic Marine Phytoplankton*, E. Granelli, B. Sundström, L. Edler, and D. Anderson, eds. Elsevier Science Publishing Company, New York.

Edwards, R.A., V.L. Trainer, and D.G. Baden. 1992. Brevetoxins bind to multiple classes in rat brain synaptosomes. *Mol. Brain Res.* 14: 64-70.

Eells, J.T., J.L. Rasmussen, P.A. Bandettini, and J.M. Propp. 1993. Differences in the neuroexcitatory actions of pyrethroid insecticides and sodium channel-specific neurotoxins in rat and trout brain synaptosomes. *Toxicol. Appl. Pharmacol.* 123: 107-119.

Fainzilber, M., O. Kofman, E. Zlotkin, and D. Gordon. 1994. A new neurotoxin receptor site on sodium channels is identified by a conotoxin that affects sodium channel inactivation in mollusks and acts as an antagonist in rat brain. *J. Biol. Chem.* 269: 2574-2580.

Fairey, E.R., S.G. Edmunds, and J.S. Ramsdell. 1997. A cell-based assay for brevetoxins, saxitoxins, and ciguatoxins using a stably expressed *c-fos*-luciferase reporter gene. *Anal. Biochem.* 251: 129-132.

Fairey, E.R. and J.S. Ramsdell. 1999. Reporter gene assays for algal-derived toxins. *Nat. Tox.* 7: 415-421.

Garthwaite, I., K.M. Ross, M. Poli, and N. Towers. 1996. Comparison of immunoassay, cellular, and classical mouse bioassay methods for detection of neurotoxic shellfish toxins. Pages 404-412 in *Immunoassays for Residue Analysis*, Beier, R. and Stanker, L., eds., American Chemical Society, Washington, D.C.

Gawley, R.E., K.S. Rein, G. Jeglitsch, D.J. Adams, E.A. Theodorakis, J. Tiebes, K.C. Nicolou, and D.G. Baden. 1995. The relationship of brevetoxin 'length' and A-ring functionality to binding and activity in neuronal sodium channels. *Chem. Biol.* 2: 533-541.

Gawley, R.E., K.S. Rein, M. Kinoshita, and D.G. Baden. 1992. Binding of brevetoxins and ciguatoxin to the voltage-sensitive sodium channel and conformational analysis of brevetoxin B. *Toxicon* 39: 780-785.

Geraci, J.R. 1989. Clinical investigation of the 1987-88 mass mortality of bottlenose dolphins along the U.S. central and south Atlantic coast. Final Report, U.S. Marine Mammal Commission, Washington D.C., 63 pp.

Ghatpande, A.S. and S.K. Sikdar. 1997. Competition for binding between veratridine and KIFMK: an open channel blocking peptide of the RIIA sodium channel. *J. Memb. Biol.* 160: 177-182.

Golik, J., J.C. James, and K. Nakanishi. 1982. The structure of brevetoxin C. *Tet. Lett.* 23: 2535-2538.

Gunter, G., R.H. Williams, C.C. Davis and, F.G.W. Smith. 1948. Catastrophic mass mortality of marine animals and coincident phytoplankton bloom on the west coast of Florida, November 1946 to August 1947. *Ecol. Monogr.* 18: 309-324.

Gusovsky, F., D.P. Rossignol, E.T. McNeal, and J.W. Daly. 1988. Pumiliotoxin B binds to a site on the voltage-dependent sodium channel that is allosterically coupled to other binding sites. *Proc. Natl. Acad. Sci. USA* 85: 1272-1276.

Hallegraeff, G.M., B.L. Munday, D.G. Baden, and P.L. Whitney. 1998. *Chattonella marina* raphidophyte bloom associated with mortality of cultured bluefin tuna (*Thunnus maccoyii*) in South Australia. Pages 93-96 in *Harmful Algae*, B. Reguera, J. Blanco, M.L.

Fernandez and T. Wyatt, eds., Xunta de Galicia and Intergovernmental Oceanographic Commission of UNESCO.

Holmes, M.J., R.J. Lewis, M.A. Poli, and N.C. Gillespie. 1991. Strain-depedent production of ciguatoxin precursors (gambiertoxins) by *Gambierdiscus toxicus* (Dinophyceae) in culture. *Toxicon* 29: 761-775.

Ishida, H., A. Nozawa, K. Totoribe, N. Muramatsu, H. Nukaya, K. Tsuji, K. Yamaguchi, T. Yasumoto, H. Kaspar, N. Berkett, and T. Kosuge. 1995. Brevetoxin B1, a new polyether marine toxin from the New Zealand shellfish *Austrovenus stutchburyi*. *Tet. Lett.* 36: 725-728.

Jeglitsch, G., K. Rein, D.G. Baden, and D.J. Adams. 1998. Brevetoxin-3 (PbTx-3) and its derivatives modulate single tetrodotoxin-sensitive sodium channels in rat sensory neurons. *J. Pharmacol. Exp. Ther.* 284: 516-525.

Kennedy, C.J., L.S. Schulman, D.G. Baden, and P.J. Walsh. 1992. Toxicokinetics of brevetoxin PbTx-3 in the gulf toadfish (*Opsanus beta*), following intravenous administration. *Aquat. Toxicol.* 22: 3-14.

Kerr, D.S., D.M. Briggs, and H.I. Saba. 1999. A neurophysiological method of rapid detection and analysis of marine algal toxins. *Toxicon* 37: 1803-1825.

Khan, S., M.S. Ahmed, O. Arakawa, and Y. Ohoue. 1995. Properties of neurotoxins separated from a harmful red tide organism *Chattonella marina*. *Israeli J. Aquacul.* 47: 137-141.

Khan, S., O. Arakawa, and Y. Onoue. 1996a. A toxicological study of the marine phytoflagellate, *Chattonella antiqua* (Raphidophyceae). *Phycologia* 35: 239-244.

Khan, S., O. Arakawa, and Y. Onoue. 1996b. Neurotoxin production by a chloromonad *Fibrocapsa japonica* (Raphidophyceae). *J. World Aquacul. Soc.* 27: 254-263.

Khan, S., O. Arakawa, and Y. Onoue. 1997. Neurotoxins in a toxic red tide of *Heterosigma akashiwo* (Raphidophyceae) in Kagoshima Bay, Japan. *Aquacul. Res.* 28: 9-14.

Landsberg, J.H. and K.A. Steidinger. 1998. A historical review of *Gymnodinium breve* red tides implicated in mass mortalities of the manatee (*Trichechus manatus latirostrus*) in Florida, USA. Pages 97-100 in *Harmful Algae*, Reguera, B., Blanco, J., Fernandez, M.L., and Wyatt, T., eds., Xunta de Galicia and Intergovernmental Oceanographic Commission of UNESCO.

Lasker, R. and F.G.W. Smith. 1954. Red tide. *U.S. Fish Wildl. Serv. Fish. Bull.* 55: 173-176.

Layne, J.N. 1965. Observations on marine mammals in Florida waters. *Bull. Fla. State Mus.* 9: 131-181.

Lewis, R.J., M. Sellin, M.A. Poli, R.S. Norton, J. MacLeod, and M.M. Sheil. 1991. Purification and characterization of ciguatoxins from moral eel (*Lycodontis javanicus*, Muraenidae). *Toxicon* 29: 1115-1127.

Linford, N.J., A.R. Cantrell, Q. Yusheng, T. Scheuer, and W.A. Catterall. 1998. Interaction of batrachotoxin with the local anesthetic receptor site in transmembrane segment IVS6 of the voltage-gated sodium channel. *Proc. Nat. Acad. Sci. USA* 95: 13947-13952.

Lipkind, G.M. and H.A. Fozzard, 1994. A structural model of the tetrodotoxin and saxitoxin binding site of the Na^+ channel. *Biophys. J.* 66: 1-13.

Lombet, A., C. Mourre, and M. Lazdunski. 1988. Interaction of insecticides of the pyrethroid family with specific binding sites on the voltage-dependent sodium channel from mammalian brain. *Brain Res.* 459: 44-53.

Manger, R.L., L.S. Leja, S.Y. Lee, J.M. Hungerford, and M.M. Wekell. 1993. Tetrazolium-based cell bioassay for neurotoxins active on voltage-sensitive sodium channels: Semiautomated assay for saxitoxins, brevetoxins, and ciguatoxins. *Anal. Biochem.* 214: 190-194.

Manger, R.L., L.S. Leja, S.Y. Lee, J.M. Hungerford, Y. Hokama, R.W. Dickey, H.R. Granade, R. Lewis, T. Yasumoto, and M.M. Wekell. 1995. Detection of sodium channel toxins: directed cytotoxicity assays of purified ciguatoxins, brevetoxins, saxitoxins, and seafood extracts. *J. AOAC Intl.* 78: 521-527.

Marban, E., T. Yamagishi, and G.F. Tomaselli. 1998. Structure and function of voltage-gated sodium channels. *J. Physiol.* 508(part 3): 647-657.

McFarren, E.F., H. Tanabe, F.J. Silva, W.B. Wilson, J.E. Campbell, and K.H. Lewis. 1965. The occurrence of a ciguatera-like poison in oysters, clams, and *Gymnodinium breve* cultures. *Toxicon* 3: 123.

Morohashi, A., M. Satake, K. Murata, H. Naoki, H. Kaspar, and T. Yasumoto. 1995. Brevetoxin B3, a new brevetoxin analog isolated from the greenshell mussel *Perna canaliculus* involved in neurotoxic shellfish poisoning in New Zealand. *Tet. Lett.* 36: 8995-8998.

Morohashi, A., M. Sataki, H. Naoki, H. Kaspar, Y. Oshima, and T. Yasumoto. 1999. Brevetoxin B4 isolated from greenshell mussels *Perna canaliculus*, the major toxin involved in neurotoxic shellfish poisoning in New Zealand. *Nat. Toxins* 7: 45-48.

Murata, K., M. Satake, H. Naoki, H. Kaspar, and T. Yasumoto. 1998. Isolation and structure of a new brevetoxin analog, brevetoxin B2, from greenshell mussels from New Zealand. *Tetrahedron* 54: 735-742.

Narahashi, T. 1992. Nerve membrane sodium channels as targets of insecticides. *Trends Pharmacol. Sci.* 13: 236-241.

O'Shea, T.J., G.B. Rathbun, R.K. Bonde, C.D. Buergelt, and D.K. O'Dell. 1991. An epizootic of Florida manatees associated with a dinoflagellate bloom. *Mar. Mammal Sci.* 7: 165-179.

Peng, Y.G., T.B. Taylor, R.E. Finch, P.D.R. Moeller, and J.S. Ramsdell. 1995. Neuroexcitatory actions of ciguatoxins on brain regions associated with thermoregulation. *NeuroReport* 6: 305-309.

Pierce, R.H. 1986. Red tide (*Ptychodiscus brevis*) toxin aerosols: A review. *Toxicon* 24: 955-965.

Pierce, R.H., M.S. Henry, L.S. Proffitt, and P.A. Hasbrouck. 1990. Red tide (brevetoxin) enrichment in marine aerosol. Pages 397-402 in *Toxic Marine Phytoplankton*, Graneli, E., Sundström, B., Edler, L., and Anderson, D.M., eds., Elsevier, New York.

Poli, M.A. 1985. *Characterization of the binding of Ptychodiscus brevis neurotoxin T17 to sodium channels in rat brain synaptosomes.* Doctoral Dissertation, University of Miami.

Poli, M.A. 1996. Three-dimensional binding assays for the detection of marine toxins. In: *Proceedings of the Workshop Conference on Seafood Intoxications: Pan American Implications of Natural Toxins in Seafood*, Baden, D.G., ed., University of Miami, Miami.

Poli, M.A. and J.F. Hewetson, 1992. Antibody production and development of a radioimmunoassay for the PbTx-2-type brevetoxins. Pages 115-127 in *Ciguatera*, Tosteson, T.R., ed., Polyscience Publications, Quebec.

Poli, M.A., T.J. Mende, and D.G. Baden. 1986. Brevetoxins, unique activators of voltage-dependent sodium channels, bind to specific sites in rat brain synaptosomes. *Mol. Pharmacol.* 30: 129-135.

Poli, M.A., S.M. Musser, R.W. Dickey, P. Eilers, and S. Hall. 2000. Neurotoxic shellfish poisoning and brevetoxin metabolites: a case study from Florida. *Toxicon* 38: 981-993.

Poli, M.A., C.B. Templeton, J.G. Pace, and H.B. Hines. 1990a. Detection, metabolism, and pathophysiology of brevetoxins. Pages 176-191 in *Marine Toxins: Origin, Structure, and Molecular Pharmacology*, Hall, S. and Strichartz, G., eds., American Chemical Society, Washington, D.C.

Poli, M.A., C.B. Templeton, W.L. Thompson, and J.F. Hewetson. 1990b. Distribution and elimination of brevetoxin PbTx-3 in rats. *Toxicon* 28: 903-910.

Postma, S.W. and W.A. Catterall. 1984. Inhibition of binding of [^{3}H]batrachotoxinin-A 20-alpha-benzoate to sodium channels by local anesthetics. *Mol. Pharmacol.* 25: 219-227.

Purkerson-Parker, S.L., L.A. Fieber, K.S. Rein, T. Podona, and D.G. Baden. 2000. Brevetoxin derivatives that inhibit toxin activity. *Chem. Biol.* 7: 385-393.

Rein, K.S., D.G. Baden, and R.E. Gawley. 1994a. Conformational analysis of the sodium channel modulator, brevetoxin A, comparison with brevetoxin B conformations, and a hypothesis about the common pharmacophore of the "site 5" toxins. *J. Org. Chem.* 59: 2101-2106.

Rein, K.S., B. Lynn, R.E. Gawley, and D.G. Baden. 1994b. Brevetoxin B: chemical modifications, binding, toxicity, and an unexpected conformational effect. *J. Org. Chem.* 59: 2107-2113.

Rogers, J.C., Y. Qu, T.N. Tanada, T. Scheuer, and W.A. Catterall. 1996. Molecular determinants of high affinity binding of α-scorpion toxin and sea anemone toxin in the S3-S4 extracellular loop in domain IV of the Na^+ channel α subunit. *J. Biol. Chem.* 271: 15950-15962.

Schreibmayer, W. and G. Jeglitsch. 1992. The sodium channel activator brevetoxin-3 uncovers a multiplicity of different open states of the cardiac sodium channel. *Biochim. Biophys. Acta* 1104: 233-242.

Sheets, M.F., J.W. Kyle, R.G. Kallen, and D.A. Hanck. 1999. The Na^+ channel voltage sensor associated with inactivation is localized to the external charged residues of domain IV, S4. *Biophys. J.* 77: 747-757.

Sheridan, R.E. and M. Adler. 1989. The actions of a red tide toxin from *Ptychodiscus brevis* on single sodium channels in mammalian neuroblastoma cells. *FEBS Lett.* 247: 448-452.

Starr, T.J. 1958. Notes on a toxin from *Gymnodinium breve*. *Texas Rep. Biol. Med* 16: 500-507.

Stefan, M.M., J.F. Potts, and WS. Agnew. 1994. The μ1 skeletal muscle sodium channel: mutation E403Q eliminates sensitivity to tetrodotoxin but not to μ-conotoxins GIIIA and GIIIB. *J. Mem. Biol.* 137: 1-8.

Steidinger, K.A. and D.G. Baden. 1984. Toxic marine dinoflagellates. Pages 201-261 in *Dinoflagellates*, D. Sfecter, ed., Academic Press, New York.

Steidinger, K.A., M.A. Burklew, and R.M. Ingle. 1973. The effects of *Gymnodinium breve* toxin on estuarine animals. Pages 179-202 in *Marine Pharmacognosy*, Martin, D.F. and Padilla, G.M., eds., Academic Press, New York.

Steidinger, K.A., P. Carlson, D. Baden, C. Rodriguez, and J. Seagle. 1998. Neurotoxic shellfish poisoning due to toxin retention in the clam *Chione cancellata*. Pages 457-458 in *Harmful Algae*, Reguera, B., Blanco, J., Fernandez, M.L. and Wyatt, T., eds., Xunta de Galicia and Intergovernmental Oceanographic Commission of UNESCO.

Strichartz, G., T. Rando, and G.K. Wang. 1987. An integrated view of the molecular toxinology of sodium channel gating in exciteable cells. *Ann. Rev. Neurosci.* 10: 237-67.

Stuart, A.M. and D.G. Baden. 1988. Florida red tide brevetoxins and binding in fish brain synaptosomes. *Aquat. Toxicol.* 13: 271-280.

Tejedor, F.J. and W.A. Catterall. 1988. Site of covalent attachment of alpha-scorpion toxin derivatives in domain I of the sodium channel alpha subunit. *Proc. Nat. Acad. Sci. USA* 85: 8742-8746.

Thomsen, W.J. and W.A. Catterall. 1989. Localization of the receptor site for alpha-scorpion toxins by antibody mapping: implications for sodium channel topography. *Proc. Nat. Acad. Sci. USA* 86: 10161-10165.

Trainer, V.L. and D.G. Baden. 1991. An enzyme immunoassay for the detection of Florida red tide brevetoxins. *Toxicon* 29: 1387-1394.

Trainer, V.L. and D.G. Baden. 1999. High affinity binding of red tide neurotoxins to marine mammal brain. *Aquatic Toxicol.* 46: 139-148.

Trainer, V.L., D.G. Baden, and W.A. Catterall. 1994. Identification of peptide components of the brevetoxin receptor site of rat brain sodium channels. *J. Biol. Chem* 269: 19904-19909.

Trainer, V.L., D.G. Baden, and W.A. Catterall. 1995. Detection of marine toxins using reconstituted sodium channels. *J. AOAC Intl.* 78: 570-573.

Trainer, V.L., G.B. Brown, and W.A. Catterall. 1996. Site of covalent labelling by a photoreactive batrachotoxin derivative near transmembrane segment IS6 of the sodium channel alpha subunit. *J. Biol. Chem.* 271: 11261-11267.

Trainer, V.L., J.C. McPhee, H. Boutelet Bochan, C. Baker, T. Scheuer, D. Babin, J.P. Demoute, D. Guedin, and W.A. Catterall. 1997. High affinity binding of pyrethroids to the alpha subunit of brain sodium channels. *Mol. Pharmacol.* 51: 651-657.

Trainer, V.L., E. Moreau, D. Guedin, D. Baden, and W.A. Catterall. 1993. Neurotoxin binding and allosteric modulation at receptor sites 2 and 5 on purified and reconstituted rat brain sodium channels. *J. Biol. Chem.* 268: 17114-17119.

Trainer, V.L. and M.A. Poli. 2000. Assays for dinoflagellate toxins, specifically brevetoxin, ciguatoxin, and saxitoxin. Pages 1-19 in *Animal Toxins. Tools in Cell Biology. A Laboratory Companion*, H. Rochat and M.-F. Martin-Euclaire, eds., Birkhauser, Berlin, Germany.

Trainer, V.L., W.J. Thomsen, W.A. Catterall, and D.G. Baden. 1991. Photoaffinity labelling of the brevetoxin receptor on sodium channels in rat brain synaptosomes. *Mol. Pharmacol.* 40: 988-994.

Van Dolah, F.M., E.L. Finley, B.L. Haynes, G.J. Doucette, P.D. Moeller, and J.S. Ramsdell. 1994. Development of rapid and sensitive high throughput pharmacologic assays for marine phycotoxins. *Nat. Tox.* 2: 189-196.

Wada, A., Y. Uezono, M. Arita, T. Yuhi, H. Kobayashi, N. Yanagihara, and F. Izumi. 1992. Cooperative modulation of voltage-dependent sodium channels by brevetoxin and classical neurotoxins in cultured bovine adrenal medullary cells. *J. Pharmacol. Exp. Ther.* 263: 1347-1351.

Wang, S.Y. and G.K. Wang. 1998. Point mutations in segment I-S6 render voltage-gated Na^+ channels resistant to batrachotoxin. *Proc. Nat. Acad. Sci. USA* 95: 2653-2658.

Washburn, B.S., D.G. Baden, N.J. Gassman, and P.J. Walsh. 1994. Brevetoxin: tissue distribution and effect on cytochrome P450 enzymes in fish. *Toxicon* 32: 799-805.

Washburn, B.S., K.S. Rein, D.G. Baden, P.J. Walsh, D.E. Hinton, K. Tullis, and M.S. Denison. 1997. Brevetoxin-6 (PbTx-6), a nonaromatic marine neurotoxin, is a ligand of the aryl hydrocarbon receptor. *Arch. Biochem. Biophys.* 343: 149-156.

Washburn, B.S., C.A. Vines, D.G. Baden, D.E. Hinton, and P.J. Walsh. 1996. Differential effects of brevetoxin and β-naphthoflavone on xenobiotic metabolizing enzymes in striped bass (*Morone saxatilis*). *Aquat. Toxicol.* 35: 1-10.

Whitney, P.L. and D.G. Baden. 1996. Complex association and dissociation kinetics of brevetoxin binding to voltage-sensitive rat brain sodium channels. *Nat. Toxins* 4: 261-270.

Wu, C.H. and T. Narahashi. 1988. Mechanism of action of novel marine neurotoxins on ion channels. *Ann. Rev. Pharmacol. Toxicol.* 28: 141-161.

Yuhi, T., A. Wada, R. Yamamoto, M. Urabe, H. Niina, F. Izumi, and T. Yanagita. 1994. Characterization of [^{3}H]brevetoxin binding to voltage-dependent sodium channels in adrenal medullary cells. *Naun-Schmied. Arch. Pharmacol.* 350: 209-212.

Azaspiracid, a New Marine Toxin Isolated from Mussels: Chemistry and Histopathology

Emiko Ito[1] *and Masayuki Satake*[2]

[1]Research Center for Pathogenic Fungi and Microbial Toxicoses, Chiba University, 1-8-1 Inohana, Chuo-ku, Chiba 260-8673, Japan
[2]Faculty of Agriculture, Tohoku University, Tsutsumidori-Amamiya, Aobaku, Sendai 981-8555, Japan

Introduction

A food poisoning event of unknown etiology resulting from the ingestion of mussels, *Mytilus edulis*, was first reported in The Netherlands in November 1995. The mussels were cultivated in Killary Harbour on the west coast of Ireland (McMahon and Silke 1996, Satake *et al.* 1998a). The symptoms observed in the patients included nausea, vomiting, severe diarrhea and stomach cramps, resembling those of diarrhetic shellfish poisoning (DSP). Subsequently, mussel samples from the production area were tested for the presence of DSP toxins, using both rat and mouse bioassays, and strongly positive results were obtained. However, mouse symptoms induced by intraperitoneal injection (i.p.) of acetone extracts of mussel hepatopancreas were distinctly different from those normally associated with DSP toxins, showing prominent neurological symptoms such as respiratory difficulties, spasms, paralysis of the limbs, and death within 20 min at higher doses. Analysis revealed that DSP toxins were present at very low levels and paralytic shellfish poisoning (PSP) toxins were undetectable. Furthermore, no known toxic phytoplanktons were observed in water samples from the production area. Thus the involvement of a new toxin(s) was strongly suggested. The toxicity persisted in Killary Harbour from November 1995 through May 1996. Similar mussel toxicity and associated human illness was recorded in the Arranmore Island region of Donegal, northwest Ireland, in October 1997~April 1998 (McMahon and Silke 1998). From the toxic samples, the major toxins azaspiracid (AZA) was isolated (Satake *et al.* 1998b), and subsequently four new analogs of azaspiracid, azaspiracid-2 and azaspiracid-3, azaspiracid-4 and -5 (Ofuji *et al.* 1999a, 2001), were isolated, and it was proposed that the poisoning be termed azaspiracid poisoning (AZP). AZA-toxic shellfish were confirmed in Italy (1997) and France (1998), and then in England and Norway (James *et al.* 2000). The background of the appearance these toxic accidents may be related to global warming.

Chemistry

A. Isolation

Azaspiracid (AZAI, Fig. 1) and its four analogs, azaspiracid-2 to azaspiracid-5 (AZA2-AZA5), were isolated from the whole meat of the mussels, *Mytilus edulis*, collected in Killary Harbor in February 1996 and Arranmore Island in November 1997 by using various chromatographies (Satake *et al.* 1998b, Ofuji *et al.* 1999a, 2001). The whole meat of mussels was extracted with acetone once and then twice with MeOH. The extracts were combined and evaporated. The residue thus obtained was partitioned between hexane and 80% MeOH. The toxins obtained in the MeOH phase were chromatographed on silica gel 60 (Merck, Darmstadt, Germany) with acetone and MeOH in this order. The toxins eluted in the second fraction were applied to gel permeation chromatography over Toyopearl HW-40

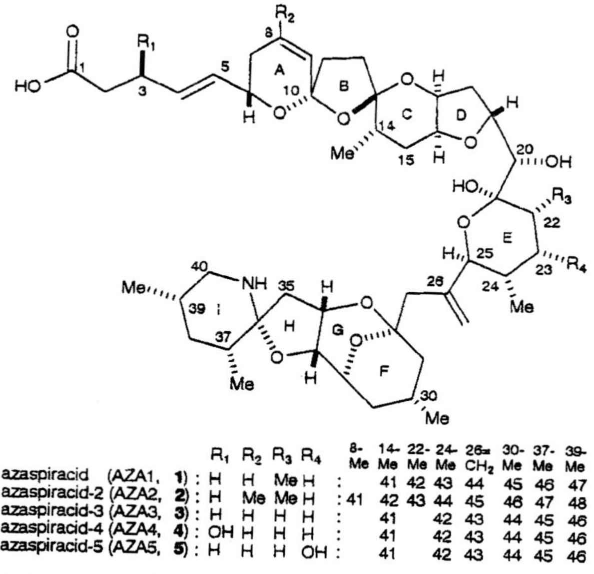

	R_1	R_2	R_3	R_4	8-Me	14-Me	22-Me	24-Me	26=CH_2	30-Me	37-Me	39-Me
azaspiracid (AZA1, **1**)	H	H	Me	H		41	42	43	44	45	46	47
azaspiracid-2 (AZA2, **2**)	H	Me	Me	H	41	42	43	44	45	46	47	48
azaspiracid-3 (AZA3, **3**)	H	H	H	H		41		42	43	44	45	46
azaspiracid-4 (AZA4, **4**)	OH	H	H	H		41		42	43	44	45	46
azaspiracid-5 (AZA5, **5**)	H	H	H	OH		41		42	43	44	45	46

Figure 1. Chemical structures of azaspiracid and its analogs.

super fine (Tosoh, Tokyo, Japan) with MeOH containing 1% AcOH. The toxic residue was dissolved in 82.5% MeOH containing 0.25% AcOH, charged onto a Develosil Lop ODS column (Nomura Chemical, Seto, Japan), and chromatographed with the same solvent. The toxins were next adsorbed on DEAE-Toyopearl from 80% MeOH solution and then eluted with 82.5% MeOH containing 0.25% AcOH. The toxic residue was dissolved in 80% MeOH and charged onto a CM-Toyopearl 650M column (Tosoh). The column was washed with 80% MeOH and then the toxin was eluted with 82.5% MeOH containing 0.25% AcOH. Finally, AZA1 and its analogs were purified on Asahipak ODP-50 (Showa Denko, Tokyo, Japan) with linear gradient elution from 50% MeOH containing 0.1% AcOH to MeOH containing 0.1% AcOH. Throughout the purification, elution of AZAs was checked by electron spray ionization mass spectrometry (ESI MS) and by mouse toxicity assays. On the Asahipak ODP-50 column, AZA1 was eluted at 38 min, while AZA2 and AZA3 were eluted at 39 and 35 min, respectively.

B. Characters of Azaspiracids

The structures of azaspiracids were determined by NMR and FAB CID MS/MS experiments. Azaspiracid ($C_{47}H_{71}NO_{12}$) is a colorless amorphous solid, with no UV absorption maxima above 210 nm. AZA is characterized by a trispiro assembly, an unusual azaspiro ring structure fused with a 2,9-dioxabicyclo[3.3.1] nonane ring, and carboxylic acid. AZA1 is stable in both mild acidic and alkaline conditions. AZA1 differs from any of the previously known nitrogen-containing toxins from shellfish or dinoflagellates, e.g., prorocentrolide, pinnatoxin, gymnodimine and the spirolides. The structures of analogs were determined by comparison of NMR and FAB CID MS/MS data with those of AZA1. Azaspiracid-2 ($C_{46}H_{69}NO_{12}$) was 8-methylazaspiracid, AZA3 ($C_{46}H_{69}NO_{12}$) was 22-demethyl-azaspiracid, AZA4 ($C_{46}H_{69}NO_{13}$) was 3-hydroxy 22-demethyl-azaspiracid, and AZA5 ($C_{46}H_{69}NO_{13}$) was 23-hydroxy-22-demethyl-azaspiracid. The relative stereostructures of two segments, rings A-E and rings F-1, were determined independently. The stereostructure at C3 in AZA4 was determined as ***R*** configuration, but the stereochemistry of ring A was not correlated with the absolute configuration at C3 in azaspiracid-4. Mouse (ddY, male 17g) lethality by i.p. injection of the toxins ranged from 0.1 to 0.5 mg/kg, except that AZA5 did not show mouse lethality even at 1 mg/kg (Ofuji *et al.* 1999a, 2001).

C. LC/MS Analysis of Azaspiracids

A liquid chromatography/mass spectrometry (LC/MS) method was developed for the sensitive and specific determination of azaspiracid and two of its analogs, AZA2 and AZA3 (Ofuji *et al.* 1999b). Azaspiracids contain nitrogen in the molecules, and they indicated an abundant $[M+H]^+$ ion on positive mode with electrospray ionization (ESI). Toxins were traced with protonated molecular ions at *m/z* 842.5 for azaspiracid-1, 856.5 for azaspiracid-2, and 828.5 for azaspiracid-3. Good separation of AZAs was achieved under these chromatographic conditions: column, Capcell Pak C18 UG-120 column (2 ´ 150 mm, Shiseido, Tokyo, Japan); solvent, MeOH/H_2O/AcOH (700:300:1); flow rate, 0.2 ml/min; column temperature; 35°C. The LC/MS method provided a detection limit of 50 pg for an azaspiracid standard. Thus, when compared with the mouse bioassay, which required 4 µg of AZA1 to kill a mouse, 80000 times higher sensitivity was achieved. With a simple cleanup procedure, LC/MS was applicable for detection of toxins from toxic mussels collected at Arranmore Island in 1997.

Histopathology of Azaspiracid-1 in Mice

A. Lethal Dose

When AZA1 was administered to ICR male mice orally with a gastric tube (p.o.), the lethality showed fluctuations between the 250 µg/kg~500 µg/kg levels. Among 3 age groups—4~5 weeks old, 6 weeks, and 5 months old—the youngest group was relatively resistant, that is, AZA1 even at 700 µg/kg did not kill some mice. In contrast, the oldest group of 5 months of age showed sensitivity to AZA1, and the lethal dose might well be lower than 250 µg/kg, but they were not tested at less than 250 µg/kg. The reason for these rather random lethality doses was still unclear. AZA1 killed mice during the first 2 days but not thereafter. Thus, in young mice, the p.o. lethal dose was suggested to be about 2.5 times that of the intraperitoneal route (i.p.) dose of 200 µg/kg.

B. Pathological Changes by Lethal Doses

Mice did not show behavioral or body conditional changes with AZA1 until sudden death, which occurred after 4 hr, even at a higher dose of 900 µg/kg. With a lethal dose of 500 µg/kg or more, some mice died or became weakened within 24 hr, but they did not show prominent body weight loss. Two prominent pathological changes appeared in the weakened mice, one in the liver and the other in the stomach (Table 1). The liver showed fatty change (data not shown), with the ratio of liver to body weight being 8.5% in the AZA1 group compared to 5.8% in control mice, and similar swelling was confirmed in a separate experiment (Table 2). The stomach showed dilatation because of contained juice. Microscopically, the following changes

Table 1. Pathological changes in organs of weakened mice by azaspiracid-1

Organ	Changes	Appearance (n=12)		%
Lung	Bleeding, congestion, inflammation of interstitium	11/12 9/12		91 75
Liver	Fatty change and swelling	12/12		100
Thymus	Necrotic lymphocytes	9/12 3/12	severe moderate	75 25
Spleen	Small number of lymphocytes in marginal zone and necrotic lymphocytes in red pulp	12/12		100
Stomach	Ulceration	12/12		100
Small intestine	Ulceration	12/12		100

Table 2. Weight of organs and the number of non-granulocytes in the spleen at 24 hr with AZA1 at 500 μg/kg

Group	Mouse	Body (g)	Liver (g)	Spleen	
				(g)	Non-granulocytes (´ 10^6)
Control	1	30.0	1.81	0.107	28
	2	30.6	2.03	0.139	20
	3	29.3	2.04	0.097	24
	Mean	29.9±0.5	1.96±0.10	0.114±0.018	24±3.2
AZA1	1	29.4	2.54	0.098	20
	2	19.4	2.98	0.098	14
	3	26.7	2.62	0.122	14
	Mean	28.5±1.3	2.71±0.10	0.106±0.010	16±2.8

Non-granulocytes: lymphocytes + monocytes + macrophages.

were confirmed: bleeding and inflammation in the alveolar wall of the lung, necrotic lymphocytes in the thymus and spleen, and ulceration in the stomach and small intestine.

C. Effect on Lymphocytes

As shown in Table 1, weakened mice had necrotic lymphocytes in the thymus and spleen. As the weight of the spleens decreased, lymphocytes were counted. Three 5-week-old mice were administered azaspiracid, 500 μg/kg. All three mice survived beyond 24 hr, and the number of lymphocytes was counted at that time. The number of non-granulocytes (lymphocytes + monocytes + macrophages) had decreased to 2/3 of the normal level (Table 2). Immunostaining methods demonstrated that both T cells (CD4 and CD8) and B cells in the spleen were damaged (data not shown).

D. Small Intestinal Changes Different from Those with Okadaic Acid

DSP mainly involves the small intestine, and the changes in the organ by AZA1 were compared to those by the representative DSP toxin, okadaic acid. Four-week-old mice (male ICR) were used in the experiment. AZA1 at 900 μg/kg caused desquamation of the surface epithelial cells as well as crypt cells (Fig. 2a). Similar changes were observed with AZA1 at 700 μg/kg, but they were less severe and crypt cells were preserved (Fig. 2b). However, at 600 μg/kg, the recovery did not start even after 24 hr (Fig. 2c).

Both okadaic acid and AZA1 induce ulceration at higher doses, and in these severe cases differences between them are no longer distinguishable. Thus, basic changes were induced by sublethal doses at 300 μg/kg of AZA1 and 130 μg/kg of okadaic acid, and the differences between them were compared. With AZA1 at 300 μg/kg, the initial changes in the villi appeared after 4 hr, where many vacuoles appeared in the surface epithelial cells, and lamina propria started degenerating, resulting in its disappearance in some villi (Fig. 3a). Surface mucosa did not reflect these changes, as the appearance was rather normal by SEM (scanning electron microscope). On the other hand, okadaic acid at 130 μg/kg caused edema between lamina propria and epithelial cells after 15 min, and desquamation of the epithelial cells developed within 30 min (Fig. 3b). Among the villi, some naked lamina proprias were exposed in the lumen after peeling of epithelial cells (SEM).

The differences in the intestine are summarized as follows:

1) Okadaic acid transiently injured the small intestine, and hypersecretion and erosion occurred briefly, ending within 1 hr. Recovery started soon, and complete recovery was seen by 48 hr (Ito and Terao, 1994). AZA1 at a sublethal dose induced changes after about 4 hr, even after 8 hr the degeneration was still progressing, and there were as yet no signs of recovery at 24 hr. This means that AZA1 had a lag time for initiating its activity, and that it was much more persistent than okadaic acid.
2) The initial change by okadaic acid was edema, and that by AZA1 was degeneration of the lamina propria. The reason for the late recovery were not only because of the basic injury to the lamina propria, but also because of multiple organ damages (Ito *et al.* 2000).

Conclusion

The new toxin AZA1 features different pathological changes from those caused by DSP toxin, namely that it causes multiple organ damage. Mice given AZA1 orally in our experiments did not show the neurological

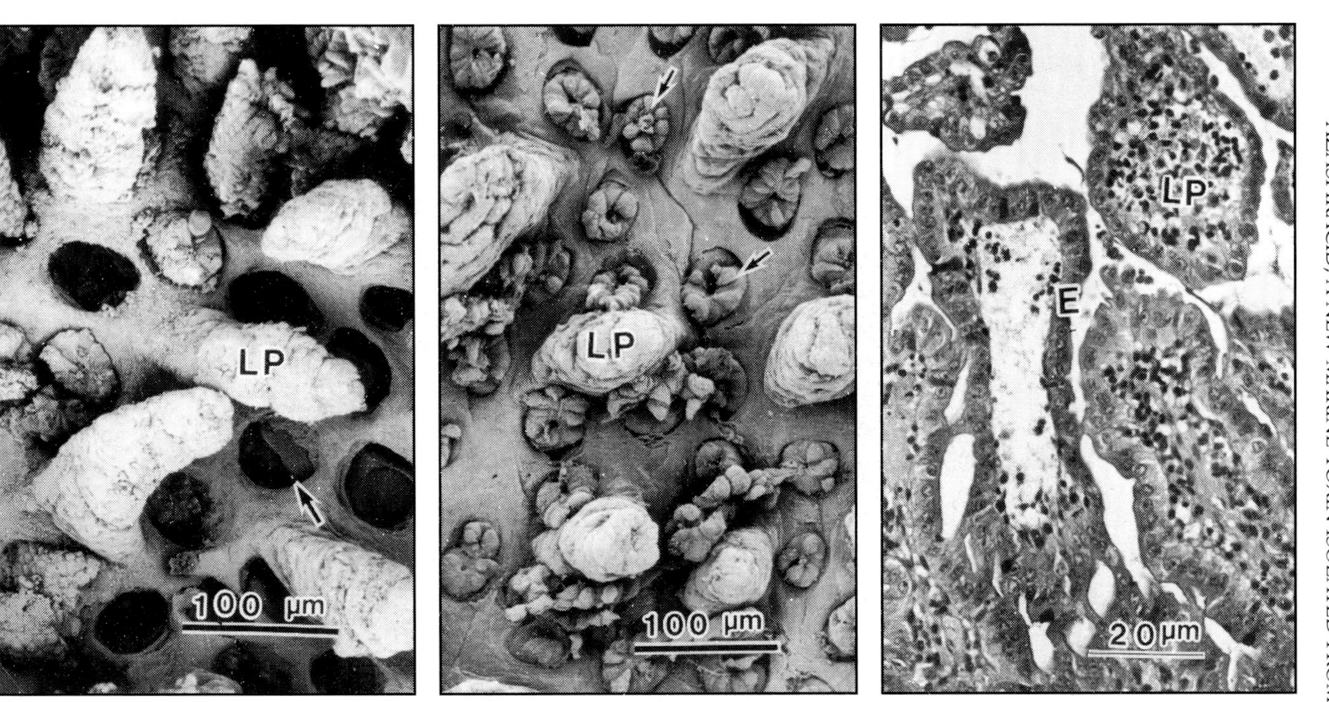

Figure 2. Small intestinal changes caused by p.o. lethal doses.
a. AZA1 900 µg/kg at 4 hr in weakened mouse. Not only desquamation of the surface epithelial cells but also gland cells in the crypt are destroyed. The crypts have become holes (arrow). LP: lamina propria. SEM.
b. AZA1 700 µg/kg at 8 hr. Surface epithelial cells are desquamated, but gland cells in the crypt are exposed in the lumen. LP: lamina propria: arrows: crypt cells. SEM.
c. AZA1 600 µg/kg at 24 hr, showing non-recovery. Some villi have lost the lamina propria and are empty inside, and some are inflamed by accumulation of lymphocytes. E: surface epithelial cells, LP: lamina propria: LM.

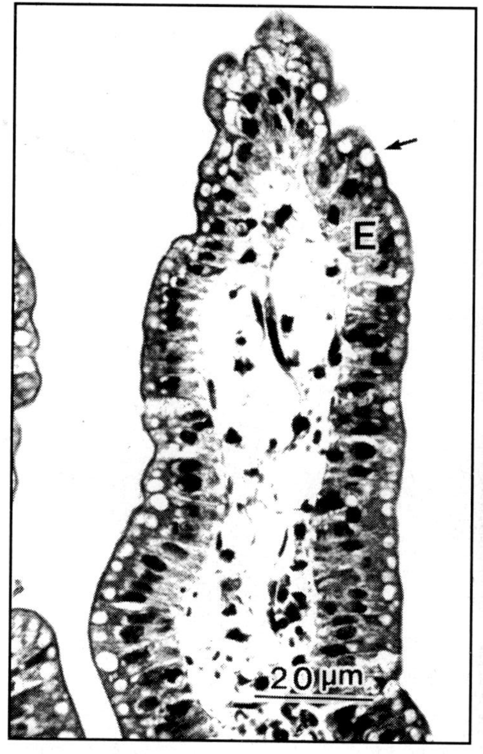

Figure 3. Comparison of basic changes between AZA1 and okadaic acid.

a. Essential changes observed by AZA1 at 7 hr with 300 µg/kg. Initial changes in the lamina propria started after 4 hr, and the lamina propria was gradually destroyed to finally disappear. Residual degenerating epithelial cells (E) show vacuolization (arrow), finally resulting in shortened villi. LM.
b. The initial change in okadaic acid after 30 min at 130 µg/kg consisted of edema between the lamina propria and surface epithelial cells causing desquamation of surface epithelial cells from the top. E: surface epithelial cells. SEM.

symptoms observed by acetone extraction from toxic mussels (McMahon and Silke 1996), and the abundant secretion in the small intestinal lumen was also absent, a finding confirmed by semi-crude AZA1 (KT3: Killary Toxin fraction 3) poisoning (Satake *et al.* 1998a, Ito *et al.* 1998). We studied the pathological changes by AZA1 only, but interaction of AZP with other AZAs, such as with AZA-2 and -3, must also occur; as well as with the recently isolated AZA-4 and -5 (Ofuji *et al.* 1999b, 2001). The observed multiple organ damage has unveiled the potentially high human health risks from this new toxin, and the urgency for additional toxicological studies to assess and delineate these risk is emphasized. Although the history of AZA is relatively new, and therefore there is an inadequacy in the amount of toxic samples available for study, further investigations, including the mode of action of this insidious toxin, are required without delay.

References

Ito, E., and K. Terao. 1994. Injury and recovery process of intestine caused by okadaic acid and related compounds. *Nat. Toxins* 2: 371-377.

Ito, E., K. Terao, T. Yasumoto and T. McMahon. 1998. Histopathological changes in mice caused by a new toxin from Irish mussels. Pages 588-589 in *Harmful Algae,* B. Reguera, J. Blanco, M.L. Fernández, and T. Wyatt, eds., IOC of UNESCO and Xunta de Galicia, Santiago de Compostera, Spain.

Ito, E., M. Satake, K. Ofuji, N. Kurita, T. McMahon, K.J. James, and T. Yasumoto. 2000. Multiple organ damage caused by a new toxin azaspiracid, isolated from mussels produced in Ireland. *Toxicon* 38: 917-930.

James, K. J., A. Furey, M. Satake, and T. Yasumoto. 2000. A new shellfish toxic syndrome in Europe. *Proceedings of International Conference on Harmful Algae (HAB 2000).* p. 25.

McMahon, T., and J. Silke, 1996. Winter toxicity of unknown aetiology in mussels. *Harmful Algae News,* 14: 2.

McMahon, T., and J. Silke. 1998. Re-occurrence of winter toxicity. *Harmful Algae News,* 17: 12.

Ofuji, K., M. Satake, T. McMahon, J. Silke, K.J. James, H. Naoki, Y. Oshima, and T. Yasumoto, 1999a. Two analogs of azaspiracid isolated from mussels, *Mytilus edulis,* involved in human intoxication in Ireland. *Nat. Toxins* 7: 99-102.

Ofuji, K., M. Satake, Y. Oshima, T. McMahon, K.J. James, and T. Yasumoto 1999b. A sensitive and specific determination method for azaspiracids, principal toxins of azaspiracid poisoning, by liquid chromatography mass spectrometry. *Nat. Toxins,* 7: 247-250.

Ofuji, K., M. Satake, T. McMahon, J. Silke, K.J. James, H. Naoki, Y. Oshima, and T. Yasumoto, 2001. Structure of azaspiracid analogs, azaspiracid-4 and azaspiracid-5, causative toxins of azaspiracid poisoning in Europe. *Biosci. Biotechnol. Biochim,* 65: 740-742.

Satake, M., K. Ofuji, K. James, A. Furey, and T. Yasumoto, 1998a. New toxic event caused by Irish mussels. Pages 468-469 in *Harmful Algae,* B. Reguera, J. Blanco, M.L. Fernández, and T. Wyatt, eds., IOC OF UNESCO and Xunta de Galicia, Santiago de Compostera, Spain.

Satake, M., K. Ofuji, H. Naoki, K. James, A. Furey, T. McMahon, J. Silke, and T. Yasumoto, 1998b. Azaspiracid, a new marine toxin having unique spiro ring assemblies, isolated from Irish mussels, *Mytilus edulis. J. Am. Chem. Soc.* 120: 9967-9968.

Enterotoxin of *Vibrio cholerae*

Yoshio Iijima[1] *and Takeshi Honda*[2]

[1]Laboratory of Bacteriology, Kobe Institute of Health, 4-6 Minatojima-nakamachi, Chuo-Ku, Kobe 650-0046, Japan
[2]Department of Bacterial Infections, Research Institute for Microbial Diseases, Osaka University, 3-1 Yamadaoka, Suita 565-0871, Japan

Introduction

Genome analyses showed that *Vibrio* species such as *V. cholerae*, *V. parahaemolyticus*, and *V. vulnificus* contain two circular chromosomes (Trucksis *et al.* 1998, Yamaichi *et al.* 1999). Subsequently, Heidelberg and colleagues (2000) reported the complete genome sequence of *V. cholerae* El Tor N16961 strain. The large chromosome encodes 2,770 open reading frames including essential cellular functions, toxins, surface antigens, and adhesions. The small chromosome carries 1,115 open reading frames including the virulence genes encoding a hemolysin and a metalloprotease. However, 46% of the open reading frames encode hypothetical proteins whose functions are currently unknown.

Although modern research has continued to accumulate information about this organism, human beings continue to suffer from cholera. The seventh pandemic by *V. cholerae* O1 biotype El Tor that began in Indonesia in 1961 spread to Asia, Africa, the Middle East, southern Europe, and Pacific islands, and in 1991 appeared in South America where cholera had not been reported previously. Before the subsidence of the pandemic, the cholera epidemic caused by a new serotype, namely *V. cholerae* O139 Bengal, which causes a clinically similar disease occurred in India and Bangladesh in late 1992 and early 1993. Cholera has frequently been a serious health problem among the people living in unhygienic circumstances, especially refugees. Cholera outbreaks recently occurred among refugees in Malawi from 1987 to 1993, in Somalia in 1993, in Congo (formerly Zaire) and Kenya in 1994, and again in Congo 1997 (Iijima *et al.* 1995, Matthys *et al.* 1998). In the case of the outbreak in Goma, Zaire, in July 1994, more than 45,000 refugees died during the first month (Goma Epidemiology Group 1995, Siddique *et al.* 1995).

Deaths due to cholera are mostly attributed to dehydration and loss of electrolytes by cholera toxin (formerly called choleragen), which is produced by *V. cholerae* O1 and O139. Cholera toxin activates adenylate cyclase of host cells, leading to intracellular cAMP accumulation, disturbance of ion transport (especially Cl^- secretion), and ultimately diarrhea. In this chapter, we describe the characteristics of and purification method for cholera toxin that is the most important pathogenic factor of toxigenic *V. cholerae*.

Cholera Enterotoxin

History

Koch (1884) hypothesized that a special "poison" could be responsible for the symptoms of cholera. After three quarters of a century, De (1959) demonstrated that the sterile culture filtrates of *V. cholerae* caused fluid to accumulate in ligated ileal loops of rabbits. In the same year, Dutta *et al.* (1959) showed that ingestion of sterile lysates of *V. cholerae* produced fatal diarrhea in infant rabbits. These studies indicated that severe diarrhea due to cholera was attributed to the toxin produced by *V. cholerae* rather than *V. cholerae* itself. Finkelstein and LoSpalluto (1969, 1970) succeeded in purifying cholera toxin. Availability of purified cholera toxin allowed elucidation of the characteristics of cholera toxin, including structure and mechanism of action.

Structure and Activity

Cholera toxin is one of the best-characterized bacterial toxins (reviewed by Spangler 1992, Burnette 1994, Kaper *et al.* 1994). Cholera toxin is the prototype of bacterial A-B toxins: A for "active" and B for "binding." Cholera toxin consists of one molecule of A subunit and five identical molecules of B subunits in the native form. The A subunit possesses a specific enzymatic activity and the B subunits bind to the ganglioside GM_1 receptor on the cell surface. The molecular weights of the A and B subunits are 27.2 kDa and 11.7 kDa, respectively. By proteolytic cleavage, the mature A subunit was shown to be composed of two disulfide-linked peptides A_1 and A_2. The triangular A subunit is connected to the ring-shaped pentamer formed by the five B subunits (Zhang *et al.* 1995).

After cholera toxin binds to GM_1 ganglioside via the B subunits, the A subunit enters the cell, perhaps by endocytosis. Reduction of the disulfide bond between A_1 and A_2 seems to be necessary for penetration of the A_1 subunit into the cell. After penetration, the A_1 subunit ADP-ribosylates the a subunit of a G protein which is a positive regulator of adenylate cyclase, thereby increasing the concentration of intracellular cAMP. Increased cAMP activates protein kinase A, leading to activation of cAMP-dependent chloride channels, which are known as cystic fibrosis transmembrane conductance regulators (CFTR), and to subsequent fluid secretion (Gabriel *et al.* 1994).

Cholera Toxin Gene

The large chromosome of *V. cholerae* contains *ctxA* and *ctxB* encoding the A and B subunits of cholera toxin, respectively. The terminal of the *ctxA* gene overlaps with the initiation of the *ctxB* gene (Gennaro and Greenaway 1983, Lockmann and Kaper 1983, Mekalanos *et al.* 1983). The last four nucleotides of the *ctxA* gene (ATGA) include a stop codon (TGA) as well as the initiation codon of the *ctxB* gene (ATG). The *ctxA* and *ctxB* genes are expressed as a polycistronic mRNA and translated into two precursor peptides (Figure 1). Both signal peptides are cleaved off and the A subunit is nicked by proteolysis, resulting in the mature disulfide-linked peptides A_1 and A_2.

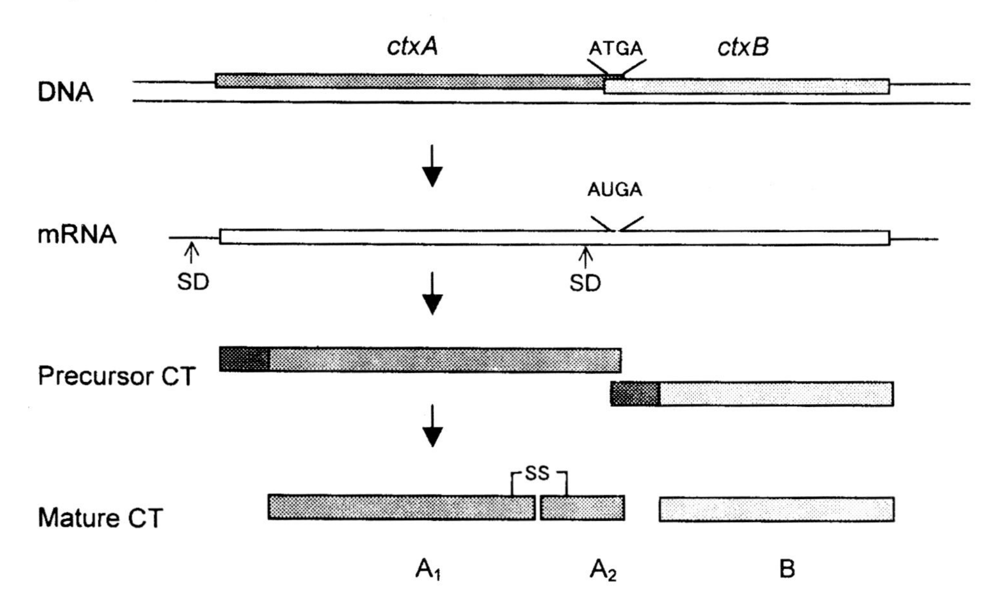

Figure 1. Cholera toxin genes to cholera toxin. *ctxA* and *ctxB* genes located in the large chromosome of *V. cholerae* are expressed as a polycistronic mRNA. Overlapped *ctxA* and *ctxB* genes are translated into precursor peptides. Both of the signal peptides (darkest area) of A and B subunits are cleaved off by a signal peptidase while excreted from the organism. The A subunit is proteolytically cleaved, but the A_1 and A_2 peptides are still linked by a disulfide bond before internalization. SD, Shine-Dalgarno sequence; CT, cholera toxin.

Related Toxin: Heat-Labile Enterotoxin of *Escherichia coli*

Diarrheagenic *Escherichia coli* is classified into several groups including enterotoxigenic *E. coli* that produces heat-stable enterotoxin and/or heat-labile enterotoxin. *E. coli* heat-labile enterotoxin is similar to cholera toxin with approximately 80% homology of their amino acid sequences (Figure 2).

E. coli heat-labile enterotoxin also resembles cholera toxin in crystal structure (Sixma *et al.* 1991, 1992, Zhang *et al.* 1995). Shiga toxins produced by *Shigella dysenteriae* and enterohemorrhagic *E. coli* also have 1A:5B structures similar to cholera toxin and *E. coli* heat-labile toxin (Fraster *et al.* 1994). Furthermore, the B subunits of Shiga toxins are highly similar to those of cholera toxin and *E. coli* heat-labile toxin (Sixma *et al.* 1993). Both of the B subunits of cholera toxin and *E. coli* heat-labile enterotoxin recognize GM_1 ganglioside, but the modes of binding are slightly different (Fukuta *et al.* 1988, Sixma *et al.* 1992). In contrast, the B subunits of Shiga toxins bind to globotriaosyl-ceramide, Gb3 (Hilaire *et al.* 1994).

No differences of enzymatic activities are found between cholera toxin and *E. coli* heat-labile enterotoxin, whereas the structures of the A_2 subunits are different. The A_2 subunit of cholera toxin remains helical throughout its course, but that of *E. coli* heat-labile enterotoxin has an extended conformation (Zhang *et al.* 1995). The enzymatic activity of Shiga toxin, which shows N-glycosidase activity, is completely different from those of cholera toxin and *E. coli* heat-labile enterotoxin.

Purification of Cholera Toxin

Not only purified cholera toxin but FITC-, biotin- and peroxidase-labeled cholera toxin are now commercially available from several sources. However, most commercially available cholera toxins are derived from the *V. cholerae* biotype classical 569B strain or its hypertoxigenic mutants. If researchers desire to study cholera toxin derived from a specific strain or mutant strains, they need to purify it by themselves. Since several methods for purification of cholera toxin have been reported (Finkelstein and LoSpalluto 1969, Duhamel *et al.* 1970, Finkelstein and LoSpalluto 1970, Heckly and Wolochow 1970, Richardson and Noftle 1970, Spyrides and Feely 1970, Mekalanos *et al.* 1978, Tayot *et al.* 1981, Mekalanos 1988, Dubey *et al.* 1990, Uesaka *et al.* 1994), we describe a rapid and simple method giving high yields as an example.

Cholera toxin purifications have been attempted using various media such as Syncase, TCY, TAY, CAYE, CYE, AKI, and yeast extract-peptone water (Finkelstein *et al.* 1966, Richardson 1969, Kusama and Craig 1970, Mekalanos *et al.* 1977, Iwanaga and Yamamoto 1985, Iwanaga and Kuyyakanond 1987). Cholera toxin production is markedly affected by the medium. For example, cholera toxin productions of some El Tor strains in

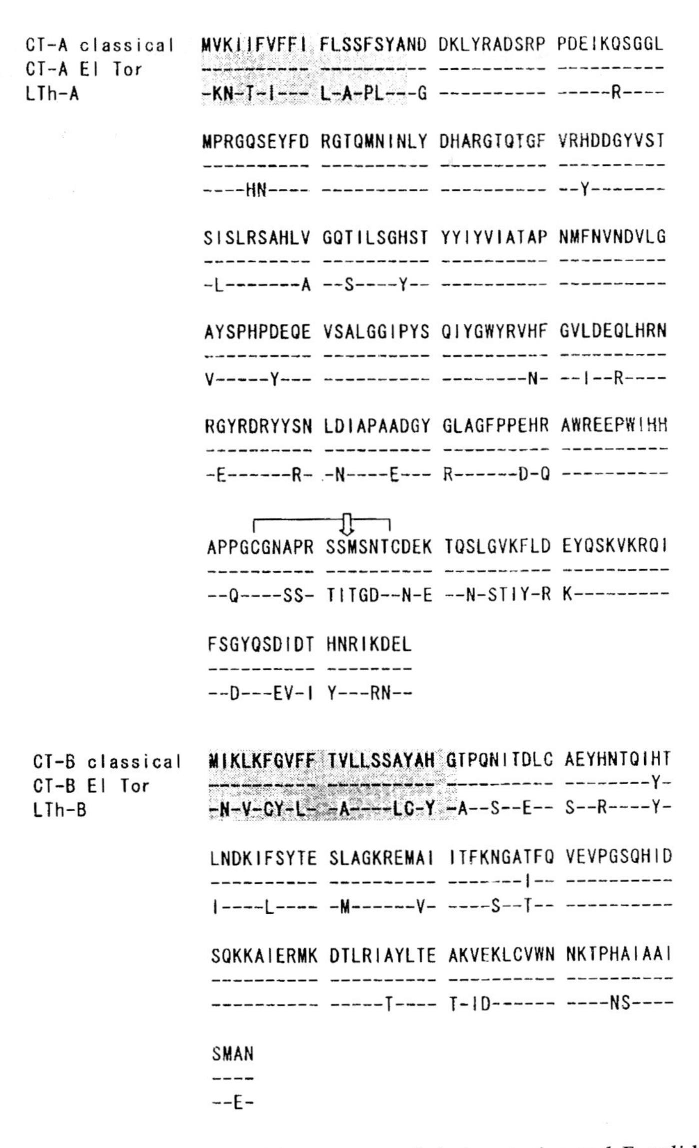

Figure 2. Alignment of amino acid sequences of cholera toxins and *E. coli* heat-labile enterotoxin. The dark areas indicate signal peptides. The arrow indicates the cleavage site of A subunit. The bridge between cysteines (C) indicates the disulfide bond between A_1 and A_2 peptides. The amino acid sequences of A subunits of *V. cholerae* classical strain 569B and El Tor strain 2125 are completely identical, but those of B subunits are different in two amino acids. The sequences of *E. coli* heat-labile enterotoxin are derived from *E. coli* strain H10407. The amino acid sequences of A and B subunits of *V. cholerae* O139 Bengal strain 854 is completely identical to those of El Tor strain 2125.

yeast extract-peptone water were more than 4,000-fold higher than those in Syncase media (Iwanaga and Kuyyakanond 1987). Furthermore, cholera toxin production is affected by the culture conditions, including volume of medium, incubation period, aeration, and temperature. Thus, preliminary examination is necessary before starting cholera toxin purification. One of the recommended media is yeast extract-peptone water consisting of 1.5% Bacto-Peptone (Difco, Detroit, MI), 0.4% yeast extract, and 0.5% NaCl, because most strains produce large amounts of cholera toxin in this medium (Iwanaga and Kuyyakanond 1987). The bacterial phase as well as aeration is also important for large production of cholera toxin. In the best conditions, most strains produce about 15 µg/ml of cholera toxin.

When a large amount of cholera toxin is obtained in the culture, the cells are removed by centrifugation at 10,000 ´ g for 40 min. Sodium hexametaphosphate is added to the culture supernatant to a final concentration of 2.5 g/l and the pH is adjusted to 4.5 with concentrated HCl. After stirring for 2 h at room temperature, the mixture is centrifuged at 10,000 ´ g for 15 min and the precipitate is dissolved in 1/20 volume of 100 mM sodium phosphate buffer (pH 8.0). The concentrate is dialyzed against 10 mM sodium phosphate buffer (pH 7.0) more than two times. Insoluble material containing lipopolysaccharide is removed by centrifugation at 10,000 ´ g for 1 h and filtration (0.22 µm pore size). Crude cholera toxin can also be prepared either by ammonium sulfate precipitation (90% saturation) or adsorption to ammonium hydroxide (Finkelstein and LoSpalluto 1970, Spyrides and Feely 1970). In general, however, maximal recovery of cholera toxin is obtained by using sodium hexametaphosphate.

After obtaining the crude cholera toxin, several procedures have been described, but affinity chromatography is more convenient than other chromatographies such as ion-exchange and gel filtration. While the affinity immunosorbent technique using anti-cholera toxin immunoglobulin can be used (Dafni *et al.* 1978), it is difficult to obtain the antibody. Affinity chromatography on GM_1 ganglioside is also possible because cholera toxin has a characteristic of binding to GM_1 ganglioside (Tayot *et al.* 1981, Dubey *et al.* 1990). In brief, monosialoganglioside-GM_1 (Sigma-Aldrich Co., G7641, St. Louis, Mo.) is coupled to Epoxy-activated Sepharose 6B (Amersham Pharmacia Biotech UK Ltd, Buckinghamshire, England). Crude cholera toxin is applied to a GM_1-Sepharose column equilibrated with 10 mM phosphate buffer containing 0.1% NaCl (pH 6.8) and washed with the buffer until no protein is detected. Cholera toxin is then eluted with 50 mM sodium citrate buffer (pH 2.8). The eluate is neutralized with 0.1 M NaOH and concentrated with an Ultrafilter membrane. After measuring the concentration of the protein, the purity of cholera toxin should be checked by SDS-PAGE. To obtain higher purification, high performance liquid chromatography (HPLC) using a Mono Q column gives good results.

Detection of Cholera Toxin

Cholera toxin can be assayed by either biological or immunological detection methods (reviewed by Nair and Takeda 1994). In the late 1950s and 1960s, some biological assays were developed, accompanied with purification and characterization of cholera toxin. In the 1970s, several immunological detection methods were established because of the availability of antibodies against cholera toxin as well as development of an enzyme linked immunosorbent assay (ELISA).

Biological Assays

Rabbit ileal loop assay. As mentioned above, the rabbit ileal loop assay was the first model to detect cholera toxin by eliciting fluid secretion into the small intestine (De 1959). Cholera toxin or other diarrheagenic substances were surgically inoculated into the rabbit small intestine that was ligated at distances from 5 to 10 cm. An appropriate time after the injection (18 hr is the most popular), cholera toxin or other diarrheagenic substances can be detected because of the fluid accumulation into the lumen. Although the assay is time-consuming and complicated, it has been extensively used as a model demonstrating the biological activity of cholera toxin or other diarrheagenic substances. The minimal dose of cholera toxin eliciting fluid accumulation was 0.5-10 μg (Coleman *et al.* 1968, Duhamel *et al.* 1970). The assay is also applicable for toxin neutralization studies.

Infant rabbit/mouse model. The infant rabbit model was originally developed to study the efficacy of chemotherapeutic agents against *V. cholerae* (Dutta and Habbu 1955). Small amounts of inocula containing cholera or cholera toxin were orally challenged into infant rabbits using a cannula. Cholera toxin can be semi-quantitatively detected by the wetness of the ventral surface. Like the infant rabbit model, infant mice can be also used to detect cholera toxin quantitatively by means of the fluid accumulation ratio, which is both time and dose dependent. Fluid accumulation by oral challenge with cholera toxin into infant mice occurred 6 to 8 hr after inoculation and reached a maximum at 10 hr. The minimal dose for a positive reaction is 0.5 μg of purified cholera toxin (Baselski *et al.* 1977). Neutralization of the toxin with a specific antibody can be tested in these assays.

Rabbit skin permeability test. Cholera toxin is detectable by intradermal injection into rabbits or guinea pigs. Cholera toxin increases the permeability of the small blood vessels of the skin, resulting in induration of the skin. Craig (1965) demonstrated a skin permeability factor in both of the filtrates of stool of cholera patients and cultures of *V. cholerae*. Some years later, the permeability factor was proven to be identical to cholera toxin by its purification (Finkelstein and LoSpalluto 1969). The skin permeability test has a wide range detection of cholera toxin, from 1 ng to 10 μg. The sensitivity of the skin permeability test was 500- to 1000-fold higher than with the ileal loop test and infant rabbit model (Duhamel *et al.* 1970).

Tissue culture assay. Cholera toxin can be detected by morphological changes of tissue culture cells. Donta *et al.* (1973) found that cholera toxin induced the cellular shape of Y1 mouse adrenal tumor cells from flat to round. Guerrant and colleagues (1974) demonstrated that cholera toxin evoked morphological alteration of the Chinese hamster ovary (CHO) cells from oval to spindle. The minimal dose inducing morphological change of CHO cells was 10 pg of purified cholera toxin. The morphological cellular changes due to cholera toxin occur by adenylate cyclase stimulation (Kanter *et al.* 1974).

Immunological Assays

The passive hemagglutination test was the first immunological assay to detect antibody in cholera patients using purified cholera toxin (Finkelstein and Peterson 1970). In 1970s, a variety of ELISAs was developed. Sandwich ELISA is one of the most popular methods. For example, cholera toxin is sandwiched between antibody adsorbed to a plastic surface and rabbit antibody. Rabbit antibody is detected by an enzyme-labeled goat/sheep antibody using color development. Alkaline phosphatase and horseradish peroxidase are the most commonly used enzymes for ELISAs. Like the sandwich ELISA, the detection method of sandwiching cholera toxin between immobilized GM_1 ganglioside and antibody was developed and designated as GM_1-ELISA (Svennerholm and Holmgren 1978). Radioimmunoassays have also been developed (Greenberg *et al.* 1977, Hejtmancik *et al.* 1977, Ceska *et al.* 1978). In the 1980s, using of the principle of cell hybridization (Köhler and Milstein 1975), monoclonal antibodies against cholera toxin were produced and immunological detection methods using them were developed (Remmers *et al.* 1982, Lindholm *et al.* 1983, Svennerholm *et al.* 1986).

Among the various immunological assays, the reversed passive latex agglutination (RPLA) test is the most convenient because of its simplicity. The procedures for this RPLA test are only dilution and mixing. Unlike an ELISA, neither washing nor addition of reagents is necessary for the RPLA test. In 1983, Ito *et al.* (1983) developed an RPLA with a sensitivity of 31 pg of cholera toxin per ml. Currently, RPLA kits are commercially available from some companies such as Oxoid (Hampshire, England) and Denka Seiken (Tokyo, Japan). The sensitivity of the kits is 0.5-2 ng of cholera toxin per ml. They are recommended for clinical diagnosis as well as laboratory use.

Use of Cholera Toxin

Cholera toxin has been used as a reagent that specifically causes the activation of adenylate cyclase leading to the accumulation of cytoplasmic cAMP (Spangler 1992). Cholera toxin is also used as an antigen for a component of cholera vaccine (Clemens *et al.* 1986).

Cholera toxin and *E. coli* heat-labile enterotoxin are currently used as adjuvants for mucosal immunity (reviewed by Rappuoli *et al.* 1999, Williams *et al.* 1999). Oral or nasal administration of protein antigen elicits little immunological reaction, which perhaps evolved as a means to avoid detrimental immune responses to innocuous dietary and airborne environmental antigens. Cholera toxin and *E. coli* heat-labile enterotoxin, however, elicit strongly antibody responses when administrated either orally or nasally (Elson *et al.* 1995). Antigens coadministrated with one of them also induce strong antibody responses. Thus, cholera toxin and *E. coli* heat-labile enterotoxin can possibly be used as adjuvants for mucosal immunization.

References

Baselski, V., R. Briggs, and C. Parker. 1977. Intestinal fluid accumulation induced by oral challenge with *Vibrio cholerae* or cholera toxin in infant mice. *Infect. Immun.* 15: 704-712.

Burnette, W.N. 1994. AB_5 ADP-ribosylating toxins: comparative anatomy and physiology. *Structure* 2: 151-158.

Ceska, M., F. Effenberger, and E. Grossmüller. 1978. Highly sensitive solid-phase radioimmunoassay for determination of low amounts of cholera toxin antibodies. *J. Clin. Microbiol.* 7: 209-213.

Clemens, J.D., D.A. Sack, J.R. Harris, J. Chakraborty, M.R. Khan, B.F Stanton, B.A. Kay, M.U. Khan, M. Yunus, W. Atkinson, A.-M. Svennerholm, and J. Holmgren. 1986. Field trial of oral cholera vaccine in Bangladesh. *Lancet* ii: 124-147.

Coleman, W.H., J. Kaur, M.E. Iwert, G.J. Kasai, and W. Burrows. 1968. Cholera toxin: purification and preliminary characterization of ileal loop reactive type 2 toxin. *J. Bacteriol.* 96: 1137-1143.

Craig, J.P. 1965. A permeability factor (toxin) found in cholera stools and culture filtrates and its neutralization by convalescent cholera sera. *Nature* 207: 614-616.

Dafni, Z., R.B. Sack, and J.P. Craig. 1978. Purification of heat-labile enterotoxin from four *Escherichia coli* strains by affinity immunoadsorbent: evidence for similar subunit structure. *Infect. Immun.* 22: 852-860.

De, S.N. 1959. Enterotoxicity of bacteria-free culture filtrate of *Vibrio cholerae*. *Nature* 183: 1533-1534.

Donta, S.T., M. King, and K. Sloper. 1973. Induction of steroidogenesis in tissue culture by cholera toxin. *Nat. New Biol.* 243: 246-247.

Dubey R.S., M. Lindblad, J. Holmgren. 1990. Purification of El Tor cholera enterotoxins and comparisons with classical toxin. *J. Gen. Microbiol.* 136: 1839-1847.

Duhamel, R.C., P. Talbot, and G.F. Grady. 1970. Production, purification, and assay of cholera enterotoxin. *J. Infect. Dis.* 121(Suppl): S85-91.

Dutta, N.K. and M.K. Habbu. 1955. Experimental cholera in infant rabbits: a method for chemotherapeutic investigation. *Br. J. Pharmacol. Chemother.* 10: 153-159.

Dutta, N.K., M.W. Panse, and D.R. Kulkarni. 1959. Role of cholera toxin in experimental cholera. *J. Bacteriol.* 78: 594-595.

Elson, C.O., S.P. Holland, M.T. Dertzbaugh, C.F. Guff, and A.O. Anderson. 1995. Morphologic and functional alterations of mucosal cells by cholera toxin and its B subunit. *J. Immunol.* 154: 132-140.

Finkelstein, R.A., P. Atthasampunna, M. Chulasamaya, and P. Charunmethee. 1966. Pathogenesis of experimental cholera: biological activities of purified procholeragen A. *J. Immunol.* 96: 440-449.

Finkelstein, R.A. and J.J. LoSpalluto. 1969. Pathogenesis of experimental cholera: preparation and isolation of choleragen and choleragenoid. *J. Exp. Med.* 130: 185-202.

Finkelstein, R.A. and J.J. LoSpalluto. 1970. Production, purification and assay of cholera toxin. Production of highly purified choleragen and choleragenoid. *J. Infect. Dis.* 121(Suppl): S63-72.

Finkelstein, R.A., and J.W. Peterson. 1970. In vitro detection of antibody to cholera enterotoxin in cholera patients and laboratory animals. *Infect. Immun.* 1: 21-29.

Fraster, M.E., M.M. Chernaia, Y.V. Kozlov, and M.N.G. James. 1994. Crystal structure of the holotoxin from *Shigella dysenteriae* at 2.5 Å resolution. *Nature Struc. Biol.* 1: 59-64.

Fukuta, S., E.M. Twiddy, J.L. Magnani, V. Ginsburg, and R.K. Holmes. 1988. Comparison of the carbohydrate binding specificities of cholera toxin and *Escherichia coli* heat-labile enterotoxins LTH-1, LT-1a, and LT-1b. *Infect. Immun.* 56: 1748-1753.

Gabriel, S.E., K.N. Brigman, B.H. Koller, R.C. Boucher, M.J. Stutts. 1994. Cystic fibrosis heterozygote resistance to cholera toxin in the cystic fibrosis mouse model. *Science* 266: 107-109.

Gennaro, M. and P.J. Greenaway. 1983. Nucleotide sequence within the cholera toxin operon. Nucleic Acids Res. 12: 3855-3861.

Goma Epidemiology Group. 1995. Public health impact of Rwandan refugee crisis: what happened in Goma, Zaire, in July, 1994? *Lancet* 345: 339-344.

Greenberg, H.B., D.A. Sack, W. Rodringuez, R.B. Sack, R.G. Wyatt, A.R. Kalica, R.L. Horswood, R.M. Chanock, and A.Z. Kapikian. 1977. Microtiter solid-phase radioimmunoassay for detection of *Escherichia coli* heat-labile enterotoxin. *Infect. Immun.* 17: 541-545.

Guerrant, R.L., L.L. Brunton, T.C. Schnaitman, L.I. Rebhun, and A.G. Gilman. 1974. Cyclic adenosine monophosphate and alteration of Chinese hamster ovary cell morphology: a rapid sensitive *in vitro* assay for the enterotoxin of *Vibrio cholerae* and *Escherichia coli. Infect. Immun.* 10: 320-327.

Heckly, R.J. and H. Wolochow. 1970. Characterization and purification of cholera toxin. *J. Infect. Dis.* 121(Suppl): S80-84.

Heidelberg, J.F., J.A. Eisen, W.C. Nelson, R.A. Clayton, M.L. Gwinn, R.J. Dodson, D.H. Haft, E.K. Hickey, J.D. Peterson, L. Umayam, S.R. Gill, K.E. Nelson, T.D. Read, H. Tettelin, D. Richardson, M.D. Ermolaeva, J. Vamathevan, S. Bass, H. Qin, I. Dragoi, P. Sellers, L. McDonald, T. Utterback, R.D. Fleishmann, W.C. Nierman, O. White, S.L. Salzberg, H.O. Smith, R.R. Colwell, J.J. Mekalanos, J.C. Venter, and C.M. Fraser. 2000. DNA sequence of both chromosomes of the cholera pathogen *Vibrio cholerae. Nature* 406: 477-483.

Hejtmancik, K.E., J.W. Peterson, D.E. Markel, and A. Kurosky. 1977. Radioimmunoassay for the antigenic determinants of cholera toxin and its components. *Infect. Immun.* 17: 621-628.

Hilaire, P.M.S, M.K. Boyd, and E.J. Toone. 1994. Interaction of the Shiga-like toxin 1B-subunit with its carbohydrate receptor. *Biochemistry* 33: 14452-14463.

Iijima, Y., J.O. Oundo, K. Taga, S.M. Saidi, and T. Honda. 1995. Simultaneous outbreak due to *Vibrio cholerae* and *Shigella dysenteriae* in Kenya. *Lancet* 345: 69-70.

Ito, T., S. Kuwahara, and T. Yokota. 1983. Automatic and manual latex agglutination tests for measurement of cholera toxin and heat-labile enterotoxin of *Escherichia coli. J. Clin. Microbiol.* 17: 7-12.

Iwanaga, M. and K. Yamamoto. 1985. New medium for the production of cholera toxin by *Vibrio cholerae* O1 biotype El Tor. *J. Clin. Microbiol.* 22: 405-408.

Iwanaga, M. and T. Kuyyakanond. 1987. Large production of cholera toxin by *Vibrio cholerae* O1 in yeast extract peptone water. *J. Clin. Microbiol.* 25: 2314-2316.

Kanter, H.S., S.P. Tao, and S.L. Gorbach. 1974. Stimulation of intestinal adenyl cyclase by *Escherichia coli* enterotoxin: comparison of strains from an infant and an adult with diarrhea. *J. Infect. Dis.* 129: 1-9.

Kaper, J.B., A. Fasano, and M. Trucksis. 1994. Toxins of *Vibrio cholerae*. Pages 145-176 in *Vibrio cholerae and Cholera: Molecular to Global Perspectives*. I.K. Wachsmuth, P.A. Blake, and Ø. Olsvik, eds., American Society for Microbiology, Washington, D.C.

Koch, R. 1884. An address on cholera and its bacillus. *Br. Med. J.* 2: 403-407.

Köhler, G. and C. Milstein. 1975. Continuous cultures of fused cells secreting antibody of predefined specificity. *Nature* 256: 495-497.

Kusama, H. and J.P. Craig. 1970. Production of biologically active substances by two strains of *Vibrio cholerae*. *Infect. Immun.* 1: 80-87.

Lindholm, L., J. Holmgren, M. Wilström, U. Karlsson, K. Anderson, and N. Lycke. 1983. Monoclonal antibodies to cholera toxin with specificial references to cross-reactions with *Escherichia coli* heat-labile enterotoxin. *Infect. Immun.* 40: 570-576.

Lockmann, H. and J.B. Kaper. 1983. Nucleotide sequence analysis of the A2 and B subunits of *Vibrio cholerae* enterotoxin. *J. Biol. Chem.* 258: 13722-13726.

Matthys, F., S. Malé, and Z. Labdi. 1998. Cholera outbreak among Rwandan refugees—Democratic Republic of Congo, April 1997. *MMWR* 47: 389-391.

Mekalanos, J.J., R.J. Collier, and W.R. Romig. 1977. Simple method for purifying choleragenoid, the natural toxoid of *Vibrio cholerae*. *Infect. Immun.* 16: 789-795.

Mekalanos, J.J., R.J. Collier, and W.R. Romig. 1978. Purification of cholera toxin and its subunits: new methods of preparation and the use of hypertoxinogenic mutants. *Infect. Immun.* 20: 552-558.

Mekalanos, J.J., D.J. Swartz, G.D. Pearson, N. Harford, F. Groyne, and M. de Wilde. 1983. Cholera toxin genes: nucleotides sequence, deletion analysis and vaccine development. *Nature* 306: 551-557.

Mekalanos, J.J. 1988. Production and purification of cholera toxin. Pages 169-175 in *Methods in Enzymology. Vol. 165: Microbial Toxins: Tools in Enzymology*. S. Harshman, ed. Academic Press Inc., San Diego.

Nair, G.B. and Y. Takeda. 1994. Detection of toxins of *Vibrio cholerae* O1 and non-O1. Pages 53-67 in *Vibrio cholerae and Cholera: Molecular to Global Perspectives*. I.K. Wachsmuth, P.A. Blake, and Ø. Olsvik, eds., American Society for Microbiology, Washington, D.C.

Rappuoli, C.M., M. Pizza, G. Douce, and D. Dougan. 1999. Structure and mucosal adjuvanticity of cholera and *Escherichia coli* heat labile enterotoxins. *Immunol. Today* 20: 493-500.

Remmers, E.F., R.R. Colwell, and R.A Goldsby. 1982. Production and characterization of monoclonal antibodies to cholera toxin. *Infect. Immun.* 37: 70-76.

Richardson S.H. 1969. Factors influencing *in vitro* skin permeability factor production by *Vibrio cholerae*. *J. Bacteriol.* 100: 27-34.

Richardson, S.H. and K.A. Noftle. 1970. Purification and properties of permeability factor/cholera enterotoxin from complex and synthetic media. *J. Infect. Dis.* 121(Suppl): S73-79.

Siddique, A.K., A. Salam, M.S. Islam, K. Akram, R.N. Majumdar, K. Zaman, N. Fronczak, and S. Laston. 1995. Why treatment centres failed to prevent cholera deaths among Rwandan refugees in Goma, Zaire. *Lancet* 345: 359-361.

Sixma, T.K., S.E. Pronk, K.H. Kalk, E.S. Wartna, B.A.M. van Zanten, B. Witholt, and W.G.J. Hol. 1991. Crystal structure of a cholera toxin-related heat-labile enterotoxin from *E. coli*. *Nature* 351: 371-377.

Sixma, T.K., S.E. Pronk, K.H. Kalk, B.A.M. van Zanten, A.M. Berghuls, and W.G.J. Hol. 1992. Lactose binding to heat-labile enterotoxin revealed by X-ray crystallography. *Nature* 355: 561-564.

Sixma, T.K., P.E. Stein, W.G.J. Hol., and R.J. Read. 1993. Comparison of the B-pentamers of heat-labile enterotoxin and verotoxin-1: two structure with remarkable similarity and dissimilarity. *Biochemisty* 32: 191-198.

Spangler, B.D. 1992. Structure and function of cholera toxin and related *Escherichia coli* heat-labile enterotoxin. *Microbiol. Rev.* 56: 622-647.

Spyrides, G.J. and J.C. Feely. 1970. Concentration and purification of cholera exotoxin by absorption on aluminum compound gels. *J. Infect. Dis.* 121(Suppl): S96-100.

Svennerholm, A.-M. and J. Holmgren. 1978. Identification of *Escherichia coli* heat-labile enterotoxin by means of a ganglioside immunosorbent assay (GM_1-ELISA) procedure. *Curr. Microbiol.* 1: 19-23.

Svennerholm, A.-M., M. Wikström, M. Lindblad, and J. Holmgren. 1986. Monoclonal antibodies to *Escherichia coli* heat-labile enterotoxin: neutralizing activity and differentiation of human and porcine LTs and cholera toxin. *Med. Biol.* 64: 23-30.

Tayot, J.L., J. Holmgren, L. Svennerholm, M. Lindblad, and M. Tardy. 1981. Receptor-specific large-scale purification of cholera toxin on silica beads derivatized with lysoGM_1 ganglioside. *Eur. J. Biochem.* 113: 249-258.

Trucksis, M., J. Michalski, Y.K. Deng, and J.B. Kaper. 1998. The *Vibrio cholerae* genome contains two unique circular chromosomes. *Proc. Natl. Acad. Sci. USA.* 95: 14464-14469.

Uesaka, Y., Y. Otsuka, Z. Lin, S. Yamasaki, J. Yamaoka, H. Kurazono, and Y. Takeda. 1994. Simple method of purification of *Escherichia coli* heat-labile enterotoxin and cholera toxin using immobilized galactose. *Microb. Pathog.* 16: 71-76.

Williams, N.A., T.R. Hirst, and T.O. Nashar. 1999. Immune modulation by the cholera-like enterotoxins: from adjuvant to therapeutic. *Immunol. Today* 20: 95-101.

Yamaichi, Y., T. Iida, K.-S. Park, K. Yamamoto, and T. Honda. 1999. Physical and genetic map of the genome of *Vibrio parahaemolyticus*: presence of two chromosomes in *Vibrio* species. *Mol. Microbiol.* 31: 1513-1521.

Zhang, R.-G., D.L. Scott, M.L. Westbrook, S.N. Nance, B.D. Spangler, G.G. Shipley, and E.M. Westbrook. 1995. The three-dimensional crystal structure of cholera toxin. *J. Mol. Biol.* 251: 563-573.

Exophysiology and Biosynthesis of Marine Biotoxins (TTX and PSP)

Deng-Fwu Hwang and Ya-Hui Lu

Department of Food Science, National Taiwan Ocean University, Keelung, Taiwan, R.O.C.

Introduction

It is well known that marine biotoxins, including tetrodotoxin (TTX) (Figure 1) and paralytic shellfish poisons (PSP) (Figure 2), are widely distributed in marine animals, especially TTX (Hwang and Tsai 1999). Therefore, food poisoning incidents due to ingesting these marine biotoxins sporadically occur in the world. TTX poisoning has been recognized for more than two thousand years. Japanese historical records show that the consumption of certain species of puffer resulted in paralytic intoxication (Hashimoto 1979, MHW 1997). This problem continues in modern times in various Asian countries, especially Japan, where puffers are still regarded as a delicacy. We also reported several incidents of TTX intoxication in Taiwan (Hwang *et al.* 1989b, 1995a,b, Lin *et al.* 1999). Clinical symptoms of TTX intoxication include numbness, paralysis, and in some instances death.

PSP are still most frequently encountered as the components responsible for the human disease named paralytic shellfish poisoning. They are accumulated from toxic dinoflagellates by filter-feeders and then passed through the food chain or web to humans. PSP are a large family of natural saxitoxin derivatives from a variety of apparent source organisms. Due to their high potency, they are a public concern in the world and may have a significant influence on ecological relationships. There are also several incidents of PSP intoxication in Taiwan (Hwang *et al.* 1987, 1992a, 1995b,c).

This chapter first reviews the present understanding of the distribution of TTX and PSP. Then, the susceptibility and preference of marine animals to biotoxins are discussed. Some biological significance of TTX and PSP in animals is described. Furthermore, we discuss the biosynthesis of TTX and PSP, and environmental and nutritional factors that affect the toxin-producing

Derivatives	MW (Da)	Lethal	Dose (ip)	TLC (Rf)*
Tetrodotoxin (TTX)	320	LD_{50}	8 μg/kg	0.67
4,9-anhydro-TTX	302	-		-
4-*epi*-TTX	320	-		-
6- *epi*-TTX	320	LD_{50}	60 μg/kg	0.49
4,9-anhydro-6-*epi*-TTX	302	-		-
11-deoxy-TTX	304	LD_{50}	71 μg/kg	0.89
11-deoxy-4-*epi*-TTX	304	-		-
4,9-anhydro-11-deoxy-TTX	286	-		-
11-*nor*-TTX-6-(r)-ol	390	LD_{99}	70 μg/kg	-
11-oxy-TTX	336	LD_{99}	120 μg/kg	0.42
Tetrodonic acid (TDA)	310	LD_{50}	30 mg/kg	0.50

*pyridine: ethyl acetate: acetic acid: H_2O (15:5:3:4)

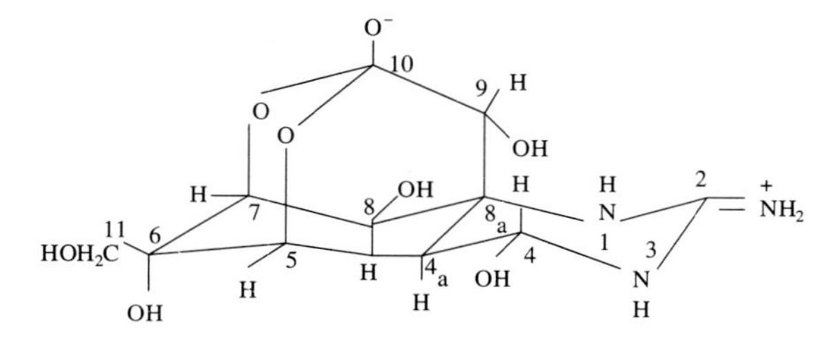

Figure 1. The chemical properties of tetrodotoxin derivatives.

ability of toxic dinoflagellates and bacteria. In the last section, the original source of TTX and PSP is discussed.

Distribution of TTX and PSP

Numerous species of edible shellfish and finfish are known to contain biotoxins (TTX and PSP) during all or part of their life cycle. TTX and PSP have been isolated from species of fish, starfish, crab, octopus, frog, newt, salamander, goby, gastropod, mollusk, flatworm, annelid, zooplankton, and alga (Mosher *et al.* 1964, Sheumack *et al.* 1978, Noguchi and Hashimoto 1973, Noguchi *et al.* 1981, 1982, 1983, 1984a,b, 1986a, Hwang *et al.* 1988, 1990a, 1991, 1992b, Maruyama *et al.* 1984, Jeon 1985, Koyama *et al.* 1983a, Hwang and Tsai 1999, Lin *et al.* 2000). The level of TTX in fish and other marine organisms varies among species, individual, and tissues of animals (Hwang *et al.* 1988). For examples, most puffers mainly contain toxin in liver and gonads, while coral reef puffers mainly contain TTX in skin (Hwang *et al.* 1992c). Terrestrial animals have also been shown to contain TTX, including frogs, newts, and their eggs (Mosher *et al.* 1964, Kim *et al.* 1975). The distribution of TTX in natural biology is shown in Table 1.

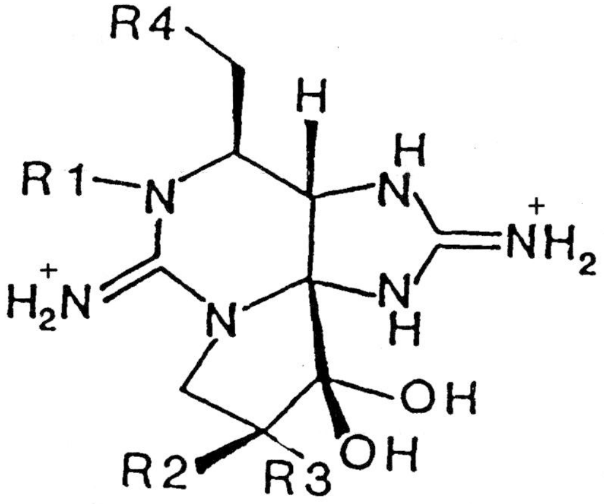

	R1	R2	R3	R4	Specific toxicity (MU/µmol)
Carbamate PSP group					
STX	H	H	H	$OCONH_2$	2483
hy STX	H	H	H	OCONHOH	1740
neo STX	OH	H	H	$OCONH_2$	2295
hyneo STX	OH	H	H	OCONHOH	1490
GTX1	OH	H	OSO_3^-	$OCONH_2$	2468
GTX2	H	H	OSO_3^-	$OCONH_2$	892
GTX3	H	OSO_3^-	H	$OCONH_2$	1584
GTX4	OH	OSO_3^-	H	$OCONH_2$	1803
N-sulfocarbamoyl PSP group					
GTX5	H	H	H	$OCONHSO_3^-$	160
GTX6	OH	H	H	$OCONHSO_3^-$	180
C1 (epi-GTX8)	H	H	OSO_3^-	$OCONHSO_3^-$	16
C2 (GTX8)	H	OSO_3^-	H	$OCONHSO_3^-$	239
C3	OH	H	OSO_3^-	$OCONHSO_3^-$	3
C4	OH	OSO_3^-	H	$OCONHSO_3^-$	143
Decarbamoyl PSP group					
DcSTX	H	H	H	OH	1274
dcneoSTX	OH	H	H	OH	30
dcGTX1	OH	H	OSO_3^-	OH	1200
dcGTX2	H	H	OSO_3^-	OH	382
dcGTX3	H	OSO_3^-	H	OH	935
dcGTX4	OH	OSO_3^-	H	OH	900
dcoxydcSTX	H	H	H	H	
dcoxydcGTX2	H	OSO_3^-	H	H	
dcoxydcGTX3	H	H	OSO_3^-	H	

Figure 2. Structure of PSP.

Table 1. Distribution of tetrodotoxin in marine organisms

	Category	Genus	Species
Chordata			
Amphibians	Salamandridae	*Taricha* sp., *Triturus* sp.	*Taricha torosa, T. granulosa, Cynops pyrrhogaster, C. ensicauda*
	Atelopodidae	*Atelopus* sp.	*Atelopus chiriquiensis, Colostethus inguinalis*
Osteichthyes	Tetraodotidae		Almost all species
	Gobiidae		*Yoneichthys nebulosus, Acenerogobius caninus*
	Scaridae		*Scarus ovifrous, S. gibbus*
	Pomacanthidae		*Pomacanthus semicirulatus*
Echinodermata			
Asteroidea	Astropectinidae		*Asteropecten polyacanthus, A. scoparius, A. latespinosis*
Chaetognatha	Sagittadae	*Parasagitta* sp.	*Parasagitta elegans*
Arthropoda			
Crustacea	Xanthidae		*Zosimus aeneus, Lophozozymus pictor, Atergatus floridus, Denania reynaudi, Atergatopsis germaini*
Merostomata	Xiphosuridae		*Carcinoscorpius rotundicauda*
Mollusca			
Cephalopoda	Octopodidae		*Octopus maculosa*
Gastropoda	Naticidae		*Natica tumidus, N. lineata, N. vitellus, Polinices didyna*
	Cymatiidae		*Charonia sauliae, Nonoplex echo*
	Bursidae		*Tutufa lissostona*
	Muricidae		*Rapana rapiformis, R. venosa*
	Buccinidae		*Babylonia japonica, B. formosae*
	Melongenidae		*Hemifusus ternatanus*
	Nassariidae		*Niotha clathrata, Nassarius condoidalis, Zeuxis siquijorensis, Z. scalaris, Z. sufftatus*
Platyhelminthes	Planoceridae		*Planocera multitenaculata, P. reticulata*
Rhynchocoela	Lineidae		*Lineus fuscoviridis, Tubulanus punctatus*
Annelida	Annedidae		*Pseudopotamilla occelata*
Marine bacteria			Some species: *Vibrio, Aeromonas, Pseudomonas, etc.*

On the other hand, PSP have been purified from many marine species, including fish, crabs, shellfish, annelids, and algae (Sato *et al.* 1997, Koyama *et al.* 1981, 1983, Hwang and Tsai 1999). PSP are mainly accumulated in the digestive gland of mussels, clams and mollusks (Onoue *et al.* 1981, Hwang *et al.* 1987), in the appendages of crabs (Koyama *et al.* 1981, 1983), in the muscle of Spanish abalone (Nagashima *et al.* 1995), and in the siphon of Alaska butter clam (Schantz and Magnusson 1964). Reports indicate that levels of PSP in dinoflagellates vary between species and strains (Bates 1998, Hwang and Lu 2000). For example, the predominant toxin components of

Alexandrium minutum obtained from different locations were as follows: neosaxitoxin (neoSTX) and STX in Plenty Bay of New Zealand (Chang *et al.* 1997), gonyautoxin (GTX) 1 and 4 in Whangarei of New Zealand (Cembella *et al.* 1987), Australia (Oshima *et al.* 1989, Hallegraeff *et al.* 1991), Portugal (Cembella *et al.* 1987) and Spain (Franco *et al.* 1994). However, the toxin profile of wild strain *A. minutum* (GTX 2 and 3) collected from Taiwan was different from that of the isolated strain (GTX 1 and 4) (Hwang and Lu 2000). Other toxic phytoplanktonic organisms are *A. catenella, A. tamarense, A. acatenella, A. polyedra, Gymnodinium catenatum, Pyrodinium bahamense, P. bahamense* var. *compressa, P. phoneus* and *Aphanizomenon flos-aquae* (Noguchi 1982). The composition of PSP in dinoflagellates and shellfish is shown in Table 2.

Recently, some papers reported that certain marine animals simultaneously contained TTX, PSP and other toxins (Arakawa 1988, Hwang *et al.* 1994a, 1996, Tsai *et al.* 1995, 1997). The toxin profiles in marine animals are quite different depending on regionality, season, and species. For example, toxic crabs have caused a number of human fatalities in various parts of the world, especially in the Western Pacific region (Halstead 1988, Hashimoto 1979). The toxic principle in the toxic crab was first supposed to consist mostly of saxitoxin (STX), a representative member of PSP (Noguchi *et al.* 1969). Koyama *et al.* (1981) and Yasumoto *et al.* (1981), however, reexamined the toxin composition of the three toxic crab species from Ishigaki Island, Okinawa, and demonstrated that each species also contained a considerable amount of neosaxitoxin (neoSTX) in addition to STX, along with small amounts of gonyautoxins (GTXs). Noguchi *et al.* (1983) examined the toxin composition of *Atergatus floridus* specimens from Miura Peninsula, Kanagawa, Japan, and found that about 90% of the total toxicity was accounted for by puffer toxin, TTX, and the remaining portion by PSP. But it was mainly composed of PSP along with minor TTX in specimens inhabiting Ryukyu (Noguchi *et al.*, 1986a). Recently, Arakawa *et al.* (1994) reported that specimens of *A. floridus* collected from Ishigashi Island, Ryukyu, contained mainly TTX and related compounds. Shiomi *et al.* (1982) pointed out that the toxin was mainly PSP and a minor amount of TTX in specimens of *A. floridus* inhabiting Chiba, Japan. Arakawa (1988) also reported that *Zosimus aeneus* contained mainly PSP and a minor amount of TTX in specimens collected from the Philippines. Yasumoto *et al.* (1986a) described that *Lophozozymus pictor* and *Demania alcala* collected from the Philippines contained palytoxin. The toxin of *D. reynaudi* specimens in the Philippines was reported to be a palytoxin-like compound (Alcala *et al.* 1988). Lau *et al.* (1995) also reported that specimens of *L. pictor* collected from Singapore contained the isomer of palytoxin, while Lloewellyn and Endean (1989) reported that the toxin of *L. pictor* collected from Australia contained GTX 2. Recently, we found that five xanthid crabs *Zosimus aeneus, L. pictor, Ategatopsis germaini, Atergatus floridus*, and *D. reynaudi* contained

Table 2. Composition of paralytic shellfish poison in dinoflagellates and shellfish

Species	Place	PSP component									
		STX	nSTX	GTX_1	GTX_2	GTX_3	GTX_4	GTX_5	GTX_6	dcSTX	others
Alexandrium (Protogonyaulax) catenella	Japan, Mia	+		+++	++	+	+	++			
"	U.S., Washington	++	++	++	++	++	+	+			+
Alexandrium tamarense	U.S., Massachusetts	++	+++	++	+++	+	+	+		-	
"	Japan, Iwate	-	+	+++	+	+	-	-	-		-
Alexandrium minutum	New Zealand, Bay of Plenty	+	++	+	+	+	+				
"	New Zealand, Tasman Bay	+	+	++	+	+	++	-	-	-	-
"	New Zealand, Whangarei	+	-	+++	+	+	+++	-	-	-	-
"	Australia, Port Riner	-	-	+++	+	+	+++	-	-	-	-
"	Taiwan	-	-	++	++	++	++	-	-	-	-
"	Portugal, Laguna Obidos	-	-	+++	+	+	+++	-	-	-	-
"	Spain, Ria de Vigo	-	-	+++	+	+	+++	-	-	-	-
"	France, Morlay Bay	-	-	-	++	++	-	-	-		++
Mytilus edulis	Japan, Mia	+		+++	++	+	+	++	+		
"	Spain	+++	+	+	++	+	+	++	+		
"	U.S., Alaska	+		+++	++	+	+	++	+		
Ruditapes philillinarum	Japan, Mia	+		+++	++	+	+	++	+		
Mya arenaria	U.S., Massachusetts	++	+	++	+++	+	+				
Placopecten magellanicus	Canada	+	+++	++	++	++	+	+			+
Saxidomus giganteus	U.S., Alaska	+++	+								
Turbo marmorata	Okinawa	+++	+	-	+	-	-	-	-		
Tridacna crocea	Palau islands	+++	+	-	-	-	-	+	-	++	
Pyrodinium bahamense var. *compressa*	Palau islands	++	+++	-	-	-	-	+	+	+	
Soletellina diphos	Taiwan	+	+	++	++	+++	+++				
Meretrix lamarckii	Japan, Kashima	-	-	++	+	++	++	+	+		
Crassostrea gigas	Japan, Nagasaki	-			+			+	+		++
Aulacomya ater	Chile	+	+	++	++	+	+	+		+	
Chlamys farreri	Japan, Nagasaki	+	+		+	+		+	+	+	
Pecten albicans	Japan, Nagasaki		+		++	+			+		+++

various toxin profiles in Taiwan (Hwang and Tsai 1999). The toxicity and toxin composition of toxic crabs are reviewed in Table 3.

Susceptibility of Marine Animals to Marine Biotoxins

TTX and PSP are well known as strong non-proteineous neurotoxins to

Table 3. Toxic composition and toxicity in toxic crabs

Species	Toxin	Approximate toxicity (MU/g)	Place
Xiphosuridae	PSP	40	Thailand
Carcinoscorpius rotundicauda	TTX	10	Thailand
Coenobitidae	Unknown	1	South Pacific Ocean
Birgus latro			Ryukyu
Xanthidae			
Zosimus aeneus	STX, neoSTX, dcSTX, GTXs	2000	Ryukyu
	STX, neoSTX, GTXs	20	Australia
	STX, neoSTX, GTXs, TTX	220	Philippines
	PSP, TTX	20	Philippines
	STX, neoSTX, GTXs	9	Fuji
	PSP	50	Palau Island
	TTX, GTX1-4 (minor)	7	Taiwan
Atergatus floridus	STX, neoSTX	1400	Ryukyu
	TTX and related compounds	20	Ishigashi Island
	TTX, STX (minor)	50	Miura Peninsula, Japan
	GTXs	180	Chiba, Japan
	STX, neoSTX, GTX2 (minor)	63	Fuji
	TTX	10	Australia
	TTX, GTX1-4 (minor)	8	Taiwan
Platypodia granulosa	70% STX, 30% unknown toxin	400	Ryukyu
Lophozozymus pictor	Palytoxin	1400	Philippines
	GTX2	5	Australia
	Isomer of palytoxin	6000	Singapore
Ategatopsis germaini	TTX, GTX1,3 (minor)	5	Taiwan
	STX, GTX, TTX (minor)	40	Taiwan
Demania spp.			
D. alcala	Palytoxin	3000	Philippines
D. reynaudi	Palytoxin-like	8	Philippines
	TTX (major), GTX2-4, neoSTX	3	Taiwan
D. toxica	Unknown	100	Philippines
Eriphia sebana	STX, neoSTX, GTX1,2	9	Australia
E. scabricula	STX, neoSTX, GTXs	16	Ryukyu
Leptodius sanguineus	PSP	2	Australia
Neoxanthias impressus	PSP	4	Ryukyu
Pilumnus vespertilio	STX, neoSTX, GTXs	3	Ryukyu
Portunidae	PSP or GTX1	4	Australia, Ryukyu
Thalamita sp.			
Grapsidae	PSP	1	Australia, Ryukyu
Grapsus albolineatus			

(Hwang and Tsai, 1999)

block the sodium channels of excitable membranes (Kao 1966). Incidents of fish mortality have been occasionally reported (Nishio 1982). It indicates that fish usually have low resistance to TTX and PSP. Saito *et al.* (1985a) reported that toxic puffer species possess higher resistance to TTX than do non-toxic puffers. They further found that common marine and freshwater fish have low resistance to TTX. Koyama et al. (1983b) also reported that toxic xanthid crabs possess higher resistance to TTX and PSP. Yamamori *et al.* (1988) pointed out that the gustatory responses to TTX and PSP in fish were much more sensitive so fish might commonly avoid the hazard of these toxins. In addition, our results showed that fish and crustaceans were more sensitive to both toxins, but most mollusks were much less susceptible (Hwang *et al.* 1990b). This may mean that mollusks are more likely to accumulate these toxins, and these toxins will also cause fish and crustaceans kills. Recently, we found the purple clam *Soletellina diphos*, which has been exposed to the species *A. minutum* in Taiwan (Hwang *et al.* 1987, 1992a), showed more resistibility to this toxic dinoflagellate than the hard clam *Meretrix lusoria* and oyster *Crassostrea gigas*. Judging from these data, it is possible that the resistance to TTX and PSP is a property of individual nerve fibers.

Biological Significance of TTX and PSP in Marine Biology

Marine biotoxins are thought to have a physiological significance for the host organisms. Sheumack *et al.* (1978) first reported that the blue octopus (*Octopus maculosa*) contains TTX in its posterior salivary gland and can kill an adult human with a single bite. Therefore, TTX is recognized as an attacking agent in catching prey. Saito *et al.* (1985b) indicated that toxic puffers could release appreciable amounts of TTX from the skin on mild stimuli such as handling. It is supposed that puffers release TTX as a kind of defense agent when they encounter enemies, or, in fewer cases, as a paralyzing agent when they attack prey. Kodama *et al.* (1985) also reported that puffers could release a lot of TTX when they were stimulated by electric shock. Following this, we found that the gastropods *Natica lineata* and *Niotha clathrata* secrete TTX in response to external stimulation, such as removal from seawater (Hwang *et al.* 1990c, 1992b). Saito *et al.* (1985b) found that toxic crabs release TTX and PSP from their shell when handled. Yamamori *et al.* (1988) found that non-toxic fish are able to detect low concentrations of TTX and PSP when they bite toxic animals, presumably a protective response. Even when the secreted toxin is of low concentration in the water, it may play an important role in defense or when attacking their predators or prey. Furthermore, Matsumura (1995) found that TTX was mostly distributed in the surface of puffer eggs, and might act as a pheromone to attract the male puffer.

Biosynthesis and Source of Marine Biotoxins in Marine Biology

Elucidation of the biosynthetic pathways of marine biotoxins in marine biology is an important step for understanding the toxigenesis of the toxic organisms. Such knowledge is still lacking for TTX (Shimizu and Kobayashi 1983). Shimizu *et al.* (1990) found that the PSP toxin is formed from arginine, acetate, serine, glycine and methionine by feeding ^{14}C-, ^{13}C- and ^{15}N-labeled putative precursors to cultures of the cyanobacterium *A. flos-aquae.* The research clearly demonstrates that PSP can be produced by a prokaryote. However, the enzyme system involved in each biosynthetic step needs further study.

The primary sources of PSP are generally recognized to be toxic dinoflagellates including *A. catenella, A. excavata, A. minutum, A. tamarense, A. polyedra, G. catenatum, P. bahamense, P. bahamense* var. *compressa* and *P. phoneus* (Proctor *et al.* 1975, Harada *et al.* 1982, Noguchi 1982). It is generally known that toxin production by toxic dinoflagellates varies with time during their growth. The toxicity of toxic dinoflagellates usually increases with lengthening of the culture time and then decreases. Boyer *et al.* (1987) and Anderson *et al.* (1990) pointed out that the maximum amount of toxin produced from toxic dinoflagellates was in the mid-exponential phase. However, White (1978) and Oshima and Yasumoto (1989a) reported it was in the stationary phase. Furthermore, several papers described that toxic dinoflagellates had higher cell toxicity when grown under conditions of low temperature, high salinity and low phosphate concentration (Proctor *et al.* 1975, Boyer *et al.* 1987, Ogata *et al.* 1987). Meanwhile, some papers indicated that the amount of toxin produced from cultured strains was higher than in wild strains (Kodama *et al.* 1982; Oshima *et al.* 1987, 1989b, Maranda *et al.* 1985, White 1986). In addition, Flynn *et al.* (1994) and Gledhill *et al.* (1997) reported that the growth and toxin production of dinoflagellates were influenced by nutritional factors. Therriault *et al.* (1985) and Sekiguchi *et al.* (1989) indicated that humic acid might be a nutrient of dinoflagellates. We found that the optimal environmental conditions for cell growth and toxin production of *A. minutum* were as follows: temperature 25°C, pH 7.5, light strength 120 $\mu Em^{-2}s^{-1}$ and salinity 15 ppt. The optimal level of nutrients supplemented in the 50% natural seawater medium was as follows: phosphate 0.002%, nitrate 0.01%, cupric ion 5.0 ppb, ferric ion 270 ppb and humic acid free. Both cell toxicity and total toxicity reached the maximum level at the post-stationary growth phase and decreased quickly. The toxic components of *A. minutum* were found to be GTX 1-4 only. Among these four toxin components, toxins 1 and 4 were the predominant components throughout the growth curve when the cells were grown in optimal environmental and nutritional conditions. But toxins GTX 2 and 3 increased when the cells were cultured in high salinity medium (Hwang and Lu 2000). Hence, the relationship between cellular toxicity and environmental factors revealed a complex relationship of growth rate with cell toxin profile and ambient nutrient levels. However, the biosynthesis of PSP in these toxic dinoflagellates is not confirmed.

Although it is generally considered that PSP are associated with dinoflagellates, it has been suggested that heterotrophic bacteria are responsible for toxin production in these organisms (Kodama 1989, 1990, Kodama *et al.* 1988, 1989, 1990, Ogata *et al.* 1990, Gallacher *et al.* 1997). The bacteria *Moraxella* sp., *Vibrio*-like sp. and *Bacillus* sp. have been reported to produce PSP. The levels produced from bacteria reported so far are quite low. On the contrary, several papers reported that the symbiotic bacteria could not produce PSP, but might influence the growth of toxic dinoflagellates (Dimanlig and Taylor 1985, Lu and Hwang 2000; Ishimaru 1985). Thus, while dinoflagellates are clearly the proximate source of PSP that reach high levels in filter-feeders, they may not be the sole or ultimate source of PSP in the marine environment. The real mechanism of PSP production in toxic dinoflagellates and/or bacteria is worthy of study. The PSP-producing plasmid is supposed to be able to transfer easily between these toxic dinoflagellates and bacteria. The proposed mechanism of PSP accumulation in marine animals is shown in Figure 3.

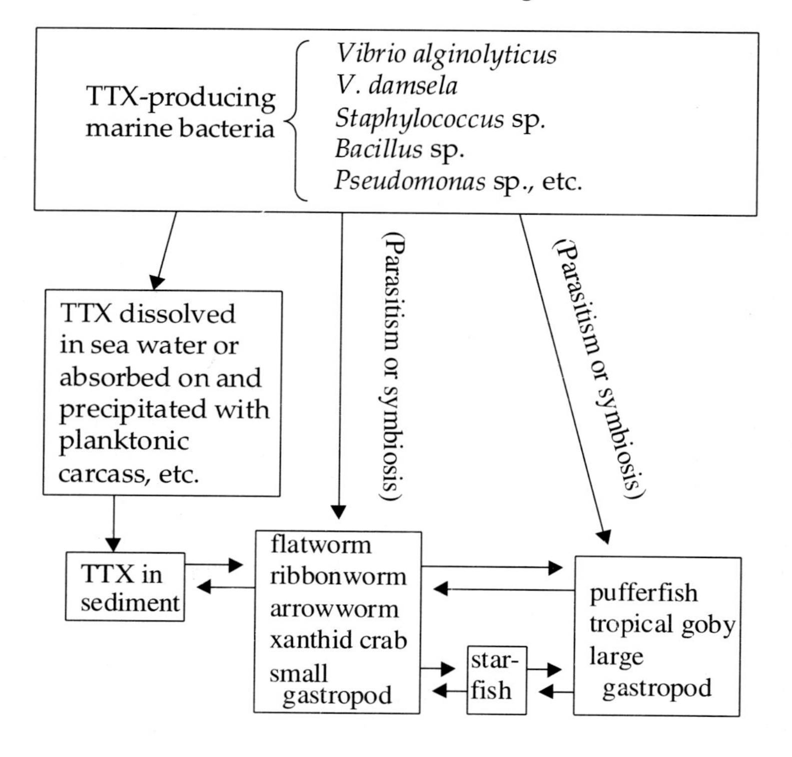

Figure 3. Proposed mechanism of TTX accumulation in marine animals.

For many years the source of TTX has also been a controversial subject. The discovery that TTX was in tissues of animals that were phylogenetically distinct resulted in speculation that TTX is bioaccumulated and/or originates from symbiotic microorganisms, such as bacteria. Since 1986, some papers have reported that many bacteria including *Pseudomonas, Aeormonas, Vibrio, Photobacterium, Plesiomonas, Bacillus* and *Escherichia* isolated from puffer, crab, octopus, gastropod and marine environment produce TTX (Noguchi *et al.* 1986b, 1987, Simidu *et al.* 1987, Kogue *et al.* 1988, Do *et al.* 1991, Hwang *et al.* 1989b, 1994b, Cheng *et al.* 1995), especially *Vibrio alginolyticus* and *V. parahaemolyticus.* However, TTX production by these bacteria has not yet been validated because TTX and its related substances are described as difficult to detect by using high performance liquid chromatography and gas chromatography-mass spectroscopy, and show no activity in the mouse bioassay. These experiments were conducted under aerobic conditions. Recently, we found that *V. alginolyticus* and *V. parahaemolyticus* might produce higher amounts of TTX and its related substances when they were cultured under facultative anaerobic conditions (Lin 1999).

The fact that the shellfish accumulate PSP from toxic dinoflagellates is well known; but the mechanisms involved in transfer of bacterial TTX and PSP to animal tissues such as liver, skin, intestine and gonad are unknown. TTX and PSP may originate from bacteria in the skin, intestines, or other internal tissues. Indeed, evidence indicates that *Vibrio* sp. may be normal flora of fish tissues (Tamplin 1990). Diet may also be a potential source of toxins, since other animals in the food chain are known to contain TTX (Noguchi *et al.* 1982, Lin and Hwang 2001). This has particular relevance since marine sediments were found to contain TTX, potentially affecting benthic-feeding animals. The toxic starfish *Asteropecten polyacanthus* is known to be eaten by the trumpet shell *Charonia sauliae* that accumulates large quantities of TTX (Noguchi *et al.* 1982, 1984b). Likewise, we found that cultured puffer *Takifugu rubripes* accumulated TTX from the toxic flatworm *Stylochus orientalis* (Lin *et al.* 1998), and the toxic starfish *A. scoparius* accumulates high amounts of TTX from the toxic gastropod *Umborium suturale* (Lin and Hwang 2001). Hence, the food chain may be the major way of accumulating TTX in TTX-bearing animals. The proposed mechanism of TTX accumulation is shown in Figure 4.

Summary

Tetrodotoxin (TTX) and paralytic shellfish poisons (PSP) are well known as potent neurotoxic marine biotoxins and are widely distributed in marine animals. TTX- and/or PSP-bearing animals possess high resistance to these

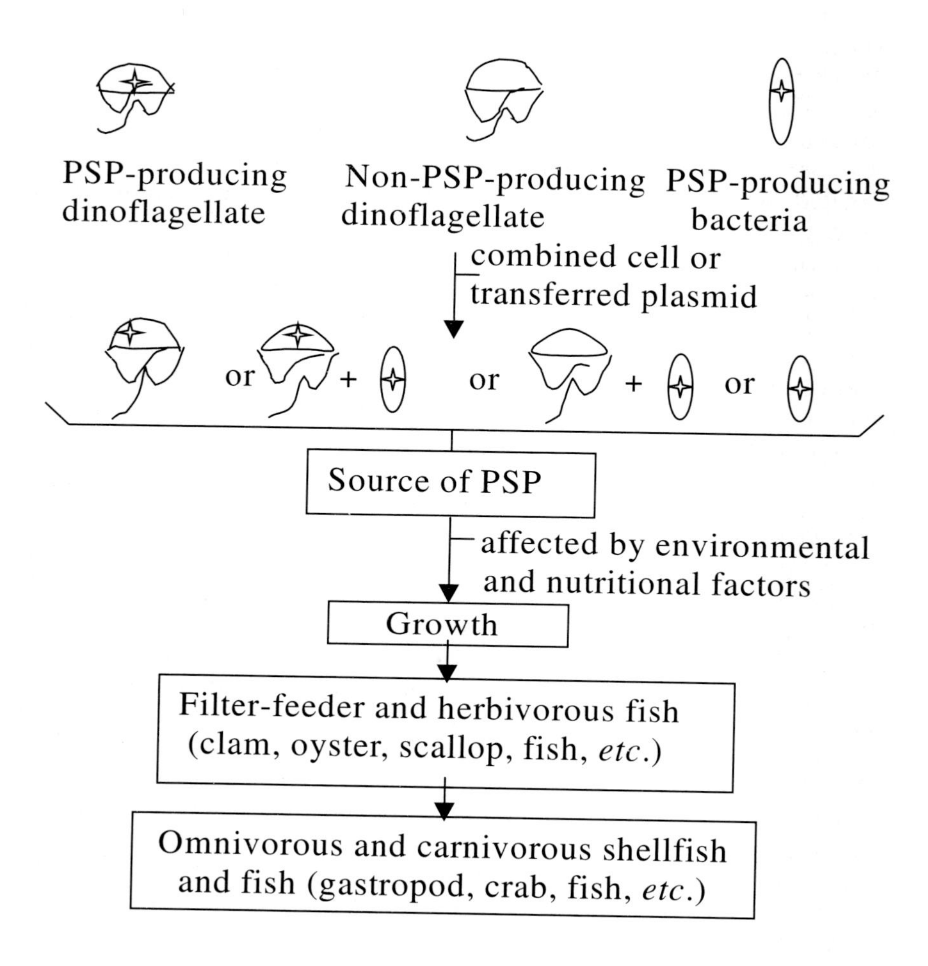

Figure 4. Proposed mechanism of PSP accumulation in marine animals.

toxins. These animals accumulate biotoxins mainly from the food chain and use them as defense and attack agents against their predators and prey. TTX and PSP originate from toxigenic dinoflagellates and/or bacteria. Among them, the component saxitoxin of PSP in the toxic cyanobacterium *Aphanizomenon flos-aquae* is biosynthesized from arginine, acetate, serine, glycine and methionine. However, the biosynthesis way of PSP and/or TTX in bacteria and dinoflagellate is still unknown. Food chain may play an important role for accumulating TTX and/or PSP in toxic animals.

References

Alcala, A.C., L.C. Alcala, J.S. Garth, D. Yasumura, and T. Yasumoto. 1988. Human fatality due to ingestion of the crab *Demania reynaudi* that contained a palytoxin-like toxin. *Toxicon* 26: 105-107.

Anderson, D.M., D.M. Kulis, J.J., Sullivan, and S. Hall. 1990. Toxin composition variations in one isolate of the dinoflagellate *Alexandrium fundyense*. *Toxicon* 28: 885-893.

Arakawa, O. 1988. *Studies on Paralytic Toxin in Crabs of the Family Xanthidae*. Ph.D. thesis, Tokyo University.

Arakawa, O., T. Noguchi, Y. Shida, and Y. Onoue. 1994. Occurrence of 11-oxotetrodotoxin and 11-nortetrodotoxin-6(R)-ol in a xanthid crab *Atergatis floridus* collected at Kojima, Ishigaki Island. *Fish. Sci.* 60: 769-771.

Bates, S.S. 1998. Ecophysiology and metabolism of ASP toxin production. Pages 405-426 in *Physiological Ecology of Harmful Algal Blooms*, D.M. Anderson *et al.*, eds, NATO-Advanced Study Institute Series, Vol. 41, Springer-Verlag, Heidelberg.

Boyer, G.L., J.J. Sullivan, P.J. Harrison, and F.J.R. Taylor. 1987. Effect of nutrient limitation on toxin production and composition in marine dinoflagellate *Protogonyaulax tamarensis*. *Mar. Biol.* 96: 123-128.

Cembella, A.D., J.J. Sullivan, G.L. Boyer, F.J.R. Taylor, and R.J. Anderson. 1987. Variation in paralytic shellfish toxin composition within the *Protogonyaulax tamaresis/catenella* species complex. *Biochem. System. Ecol.* 15: 171-186.

Chang, F.H., D.M. Anderson, D.M. Kulls, and D.G. Till. 1997. Toxin production of *Alexandrium minutum* (Dinophyceae) from the bay of Plenty, New Zealand. *Toxicon* 35: 393-409.

Cheng, C.A., D.F. Hwang, Y.H. Tsai, H.C. Chen, S.S. Jeng, T. Noguchi, K. Ohwada, and K. Hashimoto. 1995. Microflora and tetrodotoxin-producing bacteria in a gastropod, *Niotha clathrata*. *Fd. Chem. Toxic.* 33: 929-934.

Dimanlig, M.N. and F.J.R. Taylor. 1985. Extracellular bacteria and toxin production in *Protogonyaulax* species. Pages 103-108 in *Toxic Dinoflagellates*, Y. Fukuyo, ed., Koseisha-Koseikaku, Tokyo.

Do, H.K., K. Kogure, C. Imada, T. Noguchi, K. Ohwada, and U. Simidu. 1991. Tetrodotoxin production of actinomycetes isolated from marine sediment. *J. Clin. Bacteriol.* 70: 464-468.

Flynn, K., J.M. Franco, P. Fernandez, B, Reguera, M. Zapata, G. Wood, and K.J. Flynn. 1994. Changes in toxin content, biomass and pigments of the dinoflagellate *Alexandrium minutum* during nitrogen refeeding and growth into nitrogen or phosphorus stress. *Mar. Ecol. Prog. Ser.* 111: 99-109.

Franco, J.M., P. Fernandez, and B. Reguera. 1994. Toxin profiles of natural population and cultures of *Alexandrium minutum* Halim from Galician (Spain) coastal waters. *J. Appl. Phycol.* 6: 275-279.

Gallacher, S., K.J. Flynn, J.M. Franco, E.E. Brueggemann, and H.B. Hines. 1997. Evidence for production of paralytic shellfish toxins by bacteria associated with *Alexandrium* spp. (Dinophyta) in culture. *Appl. Envir. Microbiol.* 63: 239-245.

Gledhill, M., M. Nimmo, and S.J. Hill. 1997. The toxicity of copper (II) species to marine algae, with particular reference to macroalgae. *J. Phycol.* 33: 2-11.

Hallegraeff, G.M., C. J. Bolch, S.I. Blackburn, and Y. Oshima. 1991. Species of the toxigenic dinoflagellate genus *Alexandrium* in southeastern Australian waters. *Bot. Mar.* 34: 575-587.

Halstead, B.W. 1988. *Poisonous and Venomous Marine Animals of the World*. The Darwin Press, Inc., Princeton.

Harada, Y., Y. Oshima, H. Kamiya, and Y. Yasumoto. 1982. Confirmation of paralytic shellfish toxins in the dinoflagellate *Pyrodinium bahamense* var. *compressa* and bivalves in Palau. *Bull. Japan Soc. Sci. Fish.* 48: 821-825.

Hashimoto, Y. 1979. *Marine Toxins and Other Bioactive Metabolites*. Japan Scientific Societies Press, Tokyo.

Hwang, D.F. and Y.H. Tsai. 1999. Toxins in toxic Taiwanese crabs. *Food Rev. Int.* 15: 145-162.

Hwang, D.F. and Y.H. Lu. 2000. Influence of environmental and nutritional factors on growth, toxicity, and toxin profile of dinoflagellate *Alexandrium minutum*. *Toxicon* 38: 1491-1503.

Hwang, D.F., T. Noguchi, Y. Nagashima, I.C. Liao, and K. Hashimoto. 1987. Occurrence of paralytic shellfish poison in the purple clam *Soletellina diphos* (bivalve). *Bull. Japan Soc. Sci. Fish.* 53: 623-626.

Hwang, D.F., T. Noguchi, O. Arakawa, T. Abe, and K. Hashimoto. 1988. Toxicological studies on several species of puffer in Taiwan. *Bull. Japan Soc. Sci. Fish.* 54: 2001-2008.

Hwang, D.F., W.C. Wang, H.M. Chung, and S.S. Jeng. 1989a. First identification of acute tetrodotoxin-associated food poisoning in Taiwan. *J. Formosan Med. Assoc.* 88: 289-291.

Hwang, D.F., O. Arakawa, T. Saito, T. Noguchi, U. Simidu, K. Tsukamoto, Y. Shida and K. Hashimoto. 1989b. Tetrodotoxin producing bacteria from the blue-ringed *Octopus maculosus*. *Mar. Biol.* 100: 327-332.

Hwang, D.F., C.H. Chueh, and S.S. Jeng. 1990a. Occurrence of tetrodotoxin in the gastropod mollusk *Natica lineata* (lined moon shell). *Toxicon* 28: 21-27.

Hwang, D.F., C.H. Chueh, and S.S. Jeng. 1990b. Susceptibility of fish, crustacean and mollusk to tetrodotoxin and paralytic shellfish poison. *Bull. Japan Soc. Sci. Fish.* 56: 337-343.

Hwang, D.F., C.H. Chueh, and S.S. Jeng. 1990c. Tetrodotoxin secretion from the lined moon shell *Natica lineata* in response to external stimulation. *Toxicon* 28: 1133-1136.

Hwang, D.F., S.C. Lu, and S.S. Jeng. 1991. Occurrence of tetrodotoxin in the gastropods *Rapana rapiformis* and *R. venosa*. *Mar. Biol.* 111: 65-69.

Hwang, D.F., Y.H. Tsai, C.A. Cheng, and S.S. Jeng. 1992a. Comparison of paralytic toxins in aquaculture of purple clam in Taiwan. *Toxicon* 30: 669-672.

Hwang, D.F., L.C. Lin, and S.S. Jeng. 1992b. Variation and secretion of toxins in gastropod mollusk *Niotha clathrata*. *Toxicon* 30: 1189-1194.

Hwang, D.F., C.Y. Kao, H.C. Yang, S.S. Jeng, T. Noguchi, and K. Hashimoto. 1992c. Toxicity of puffer in Taiwan. *Bull. Japan Soc. Sci. Fish.* 58: 1541-1547.

Hwang, D.F., C.A. Cheng, and S.S. Jeng. 1994a. Gonyautoxin-3 as a minor toxin in the gastropod *Niotha clarthrata* in Taiwan. *Toxicon* 32: 1573-1579.

Hwang, D.F., C.A. Cheng, H.C. Chen, S.S. Jeng, T. Noguchi, K. Ohwada, and K. Hashimoto. 1994b. Microflora and tetrodotoxin-producing bacteria in the lined moon shell *Natica lineata*. *Fish. Sci.* 60: 567-571.

Hwang, D.F., C.A. Cheng, Y.H. Tsai, and S.S. Jeng. 1995a. Tetrodotoxin associated food poisoning due to unknown fish in Taiwan between 1988-1994. *J. Natural Toxins* 4: 165-171.

Hwang, D.F., C. A. Cheng, Y.H. Tsai, D.Y.C. Shih, H.C. Ko, R.Z. Yang, and S.S. Jeng. 1995b. Identification of tetrodotoxin and paralytic shellfish toxins in marine gastropods implicated in food poisoning. *Fish. Sci.* 61: 675-679.

Hwang, D.F., Y.H. Tsai, T. Noguchi, K. Hashimoto, I.C. Liao, and S.S. Jeng. 1995c. Two food poisoning incidents due to ingesting the purple clam occurred in Taiwan. *J. Natural Toxins* 4: 173-179.

Hwang, D.F., Y.H. Tsai, T. Chai, and S.S. Jeng. 1996. Occurrence of tetrodotoxin and paralytic shellfish poison in Taiwan crab *Zosimus aeneus*. *Fish. Sci.* 62: 500-501.

Ishimaru, T. 1985. Biology: Growth and environmental factors. pages 40-46 in *Toxic Dinoflagellate,* Y. Fukuyo, ed., Koseisha-Koseikaku, Tokyo.

Jeon, J.K. 1985. *Studies on Paralytic Toxins in Several Marine Invertebrates.* Ph.D. thesis, Tokyo University.

Kao, C.Y. 1996. Tetrodotoxin, saxitoxin and their significance in the study of excitation phenomena. *Pharmacol. Rev.* 18: 997-1049.

Kim, Y.H., G.B. Brown, H.S. Mosher, and F.A. Fuhrman. 1975. Tetrodotoxin: Occurrence in *Atelopid* frogs of Costa Rica. *Science* 189: 151-152.

Kodama, M. 1989. Possible association of paralytic shellfish toxins-producing bacteria with bivalve toxicity. Pages 391-398 in *Mycotoxins and Phycotoxins*, S. Natori *et al.*, eds., Elsevier Publishing Co., Ltd., Tokyo.

Kodama, M. 1990. Possible links between bacteria and toxin production in algal blooms. Pages 52-61 in *Toxic Marine Phytoplankton*, E. Graneli *et al.*, eds., Elsevier Science Publishing Co., Inc., New York.

Kodama, M., Y. Fukuyo, T. Ogata, T. Garashi, H. Kamiya, and F. Mastuura. 1982. Comparison of toxicities of *Protogonyaulax* cells of various sizes. *Bull. Japan Soc. Sci. Fish.* 48: 567-571.

Kodama, M., T. Ogata, and S. Sato. 1985. External secretion of tetrodotoxin from puffer fishes stimulated by electric shock. *Mar. Biol.* 87: 199-202.

Kodama, M., T. Ogata, and S. Sato. 1988. Bacterial production of saxitoxin. *Agric. Biol. Chem.* 52: 1075-1077.

Kodama, M., T. Ogata, and S. Sato. 1989. Saxitoxin-producing bacterium isolated from *Protogonyaulax tamarensis*. Pages 363-366 in *Red Tides: Biology, Environmental Science and Toxicity*, T. Okaichi *et al.*, eds., Elsevier Science Publishing Co., Ltd., New York.

Kodama, M., T. Ogata, S. Sakamoto, S. Sato, and T. Honda. 1990. Production of paralytic shellfish toxins by a bacterium *Moraxella* sp. isolated from *Protogonyaulax tamarensis*. *Toxicon* 28: 707-714.

Kogue, K., H.K. Do, E.V. Thuesen, K. Nanba, K. Ohwada, and U. Simidu. 1988. Accumulation of tetrodotoxin in marine sediment. *Mar. Ecol. Prog. Ser.* 45: 303-305.

Koyama, K., T. Noguchi, Y. Ueda, and K. Hashimoto. 1981. Occurrence of neosaxitoxin and other paralytic shellfish poisons in toxic crabs belonging to the family Xanthidae. *Bull. Japan Soc. Sci. Fish.* 47: 965.

Koyama, K., T. Noguchi, A. Uzu, and K. Hashimoto. 1983. Local variation of toxicity and toxin composition in a xanthid crab *Atergatis floridus*. *Bull. Japan Soc. Sci. Fish.* 49: 1883-1886.

Lau, C.O., C.H. Tan, H.E. Khoo, R. Yuen, R.J. Lewis, G.P. Corpuz, and G.S. Bignami. 1995. *Lophozozymus pictor* toxin: A fluorescent structural isomer of palytoxin. *Toxicon* 33: 1373-1377.

Lin, S.J. 1999. *Studies on Toxins and Their Sources of Torafugu, Goby and Starfish in Taiwan*. Ph.D. thesis, National Taiwan Ocean University, Keelung. pp. 208-228.

Lin, S.J. and D.F. Hwang. 2001. Possible source of tetrodotoxin in the starfish *Astropecten scoparius*. *Toxicon* 39: 573-579.

Lin, S.J., T, Chai, S.S. Jeng, and D.F. Hwang. 1998. Toxicity of the puffer *Takifugu rubripes* cultured in northern Taiwan. *Fish. Sci.* 64: 766-770.

Lin, S.J., J.B. Chen, K.T. Hsu, and D.F. Hwang. 1999. Acute goby poisoning in southern Taiwan. *J. Natural Toxins* 8: 141-147.

Lin, S.J., D.F. Hwang, K.T. Shao, and S.S. Jeng. 2000. Toxicity of Taiwanese gobies. *Fish. Sci.* 66: 547-557.

Llowellyn, I.E. and R. Endean. 1989. Toxicity and paralytic shellfish toxin profiles of the xanthid crabs, *Lophozozymus pictor* and *Zosimus aeneus*, collected from some Australian coral reefs. *Toxicon* 27: 590-600.

Lu, Y.H., T. Chai, and D.F. Hwang. 2000. Isolation of bacteria from toxic dinoflagellate *Alexandrium minutum* and their effects on algae toxicity. *J. Natural Toxins* 9: 409-417.

Maranda, L., D.M. Anderson, and Y. Shimizu. 1985. Comparison of toxicity between populations of *Gonyaulax tamarensis* of eastern north American waters. *Estuarine, Coastal and Shell Sci.* 21: 401-410.

Maruyama, J., T. Noguchi, J.K. Jeon, T. Harada, and K. Hashimoto. 1984. Occurrence of tetrodotoxin in the starfish *Astropecten latespinosus*. *Experientia* 40: 1395-1396.

Matsumura, K. 1995. Tetrodotoxin as a pheromone. *Nature* 378: 563-564.

MHW (Ministry of Health and Wealth, Japan). 1997. Pufferfishes Available in Japan—An Illustrated Guide to Their Identification, 2nd Edition, MHW, Tokyo.

Mosher, H.S., F.A. Fuhrman, H.D. Bruchward, and H.G. Fisher. 1964. Tarichatixin-tetrodotoxin: A potent neurotoxin. Science 144: 1100-1101.

Nagashima, Y., O. Arakawa, K. Shiomi, and T. Noguchi. 1995. Paralytic shellfish poisons of ormer, *Haliotis tuberculata*, from Spain. *J. Food Hyg. Soc. Jap.* 36: 627-631.

Nishio, S. 1992. Ichthyotoxicity. Pages 50-61 in *Toxic Phytoplankton—Occurrence, Mode of Action, and Toxins*, T. Okaichi *et al.*, eds., Koseisha-Koseikaku, Tokyo.

Noguchi, T, S. Konosu and Y. Hashimoto. 1969. Identify of the crab toxin with saxitoxin. *Toxicon* 7: 325-326.

Noguchi, T. 1982. Chemistry of paralytic shellfish poison. Pages 88-101 in *Toxic Phytoplankton—Occurrence, Mode of Action, and Toxins*, T. Okaichi *et al.*, eds., Koseisha-Koseikaku, Tokyo.

Noguchi, T. and Y. Hashimoto. 1973. Isolation of tetrodotoxin from a goby *Gobius criniger*. *Toxicon* 11: 305-308.

Noguchi, T., J. Maruyama, Y. Ueda, K. Hashimoto, and T. Harada. 1981. Occurrence of tetrodotoxin of tetrodotoxin in the Japanese ivory shell *Babylonia japonica*. *Bull. Japan Soc. Sci. Fish.* 47: 909-913.

Noguchi, T., H. Narita, J. Maruyama, and K. Hashimoto. 1982. Tetrodotoxin in the starfish *Astropecten polyacanthus*, in association with toxification of a trumpet shell, "boshubora" *Charonia sauliae*. *Bull. Japan Soc. Sci. Fish.* 48: 1173-1177.

Noguchi, T., A. Uzu, K. Koyama, J. Maruyama, Y. Nagashima, and K. Hashimoto. 1983. Occurrence of tetrodotoxin as the major toxin in a xanthid crab *Atergatis floridus*. *Bull. Japan Soc. Sci. Fish.* 49: 1887-1892.

Noguchi, T., J. Maruyama, H. Narita, and K. Hashimoto. 1984a. Occurrence of tetrodotoxin in the gastropod mollusk *Tutufa lissostoma* (frog shell). *Toxicon* 22: 219-226.

Noguchi, T., H. Narita, J. Maruyama, and K. Hashimoto. 1984b. Tetrodotoxin in the starfish *Astropecten polyacanthus*, in association with toxification of a trumpet shell, "boshubora" *Charonia sauliae*. *Bull. Japan Soc. Sci. Fish.* 48: 1173-1178.

Noguchi, T., O. Arakawa, K. Koyama, and K. Hashimoto. 1986a. Local differences in toxin composition of a xanthid *Atergatis floridus* inhabiting Ishigaki Island, Okinawa. *Toxicon* 24: 705-711.

Noguchi, T., J.K. Jeon, O. Arakawa, H. Sugita, Y. Deguchi, Y. Shida, and K. Hashimoto. 1986b. Occurrence of tetrodotoxin and anhydortetrodotoxin in *Vibrio* sp. isolated from the intestines of a xanthid crab, *Atergatis floridus*. *J. Biochem.* 99: 311-314.

Noguchi, T., D.F. Hwang, O. Arakawa, H. Sugita, Y. Deguchi, Y. Shida and K. Hashimoto. 1987. *Vibrio alginolyticus*, a tetrodotoxin-producing bacterium, in the intestines of the fish *Fugu vermicularis vermicularis*. *Mar. Biol.* 94: 625-630.

Ogata, T., T. Ishimaru, and M. Kodama. 1987. Effect of water temperature and light intensity on growth and light intensity on growth rate and toxicity change in *Protogonyaulax tamaresis*. *Mar. Biol.* 95: 217-220.

Ogata, T., M. Kodama, K. Komaru, S. Sakamoto, S. Sato, and S. Simidu. 1990. Production of paralytic shellfish toxins by bacteria isolated from toxic dinoflagellates. Pages 311-315 in *Toxic Marine Phytoplankton*, E. Granili *et al.*, eds., Elsevier Science Publishing Co., Inc., New York.

Onoue, Y., T. Noguchi, J. Maruyama, Y. Ueda, K. Hashimoto, and T. Ikeda. 1981. Comparison of PSP compositions between toxic oysters and *Protogonyaulax catenella* from Senzaki Bay, Yamaguchi Prefecture. *Bull. Japan Soc. Sci. Fish.* 47: 1347-1350.

Oshima, Y., M. Hasegawa, T. Yasumoto, G. Hallegraeff, and S.I. Blackburn. 1987. Dinoflagellate *Gonyaulax catenatum* as the source of paralytic shellfish toxins in Tasmanian shellfish. *Toxicon* 25: 1105-1111.

Oshima, Y. and T. Yasumoto. 1989a. Analysis of toxins in cultured *Gonyaulax excavata* cells originating in Ofunato Bay, Japan. Pages 309-312 in *Red Tide: Biology, Environmental Science and Toxicology*, T. Okaichi *et al.*, eds., Elsevier Science Publishing Co., Inc., New York.

Oshima, Y., M. Hirota, Y. Yasumoto, G.M. Hallegraeff, S.I. Blackburn, and D.A. Steffensen. 1989b. Production of paralytic shellfish toxins by the dinoflagellate *Alexandrium minutum* Halim from Australia. *Bull. Japan Soc. Sci. Fish.* 55: 925.

Proctor, N.H., S.L. Chan, and A.J. Trevor. 1975. Production of saxitoxin by cultures of *Gonyaulax catenella*. *Toxicon* 13: 1-9.

Saito, T., T. Noguchi, T. Takeuchi, S. Kamimura, and K. Hashimoto. 1985a. Resistibility of toxic and nontoxic pufferfish against TTX. *Bull. Japan Soc. Sci. Fish.* 51: 257-260.

Saito, T., T. Noguchi, T. Harada, O, Murata, and K. Hashimoto. 1985b. Tetrodotoxin as a biological defense agent for puffers. *Bull. Japan Soc. Sci. Fish.* 51: 1175-1180.

Sato, S., M. Kodama, T. Ogata, K. Saitanu, K. Hirayama, and K. Kakinuma. 1997. Saxitoxin as a toxic principle of a freshwater puffer, *Tetraodon fangi*, in Thailand. *Toxicon* 35: 137-140.

Schantz, E.J. and H.W. Magnusson. 1964. Observations on the origin of the paralytic poison in Alaska butter clam. *J. Protozool.* 11: 242-246.

Sekiguchi, K., N. Inoguchi, M. Shimizu, S. Saito, S. Watanabe, T. Ogata, M. Kodama, and Y. Fukuyo. 1989. Occurrence of *Protogonyaulax tamarensis* and shellfish toxicity in Ofunato Bay from 1980-1986. Pages 399-402 in *Red Tides: Biology, Environmental Science and Toxicity*, T. Okaichi *et al.*, eds., Elsevier Science Publishing Co., Inc., New York.

Sheumack, D.D., M.E.H. Howden, I. Spence, and J. Ouinn. Maculutoxin: 1978. A neurotoxin from the venom glands of the octopus *Hapalochaena maculosia* identified as tetrodotoxin. *Science* 199: 187-189.

Shimizu, Y. and M. Kobayashi. 1983. Apparent lack of tetrodotoxin biosynthesis in captured *Taricha torosa* and *Taricha granulosa*. *Chem. Pharm. Bull.* 31: 3625-3631.

Shimizu, Y., S. Gupta, A. Krishna, and V. Prasad. 1990. Biosynthesis of dinoflagellate toxins. Pages 62-71 in *Toxin Marine Phytoplankton*, E. Graneli *et al.*, eds., Elsevier Science Publishing Co., Inc., New York.

Shiomi, K., H. Inaoka, H. Yamanaka, and T. Kikuchi. 1982. Occurrence of a large amount of gonyautoxins in a xanthid crab *Atergatis floridus* from Chiba. *Bull. Japan Soc. Sci. Fish.* 48: 1407-1410.

Simidu, U., T. Noguchi, D.F. Hwang, Y. Shida, and K. Hashimoto. 1987. Marine bacteria which produce tetrodotoxin. *Appl. Environ. Microbiol.* 53: 1714-1715.

Tamplin, M. L. 1990. A bacterial source of tetrodotoxins and saxitoxins. Pages 79-86 in *Marine Toxins*, S. Hall and G. Strichartz, eds., American Chemical Society, Washington, DC.

Therriault, J.C., J. Painchaud, and F. Levasseur. 1985. Factors controlling the occurrence of *Protogonyaulax tamarensis* and shellfish toxicity in the St. Lawrence Estuary: Freshwater runoff and the stability of the water column. Pages 141-146 in *Toxic Dinoflagellate*, Y. Fukuyo, ed., Koseisha-Koseikaku, Tokyo.

Tsai, Y.H., D.F. Hwang, T. Chai, and S.S. Jeng. 1995. Occurrence of tetrodotoxin and paralytic shellfish poison in Taiwanese crab *Lophozozymus pictor*. *Toxicon* 33: 1669-1673.

Tsai, Y.H., D.F. Hwang, T. Chai, and S.S. Jeng. 1997. Toxicity and toxic components of two xanthid crabs *Atergatis floridus* and *Demania reynaudi* in Taiwan. *Toxicon* 35: 1327-1335.

White, A.W. 1978. Salinity effects on growth and toxin content of *Gonyaulax excavata*, a marine dinoflagellate causing paralytic shellfish poisoning. *J. Phycol.* 14: 475-479.

White, A.W. 1986. High toxin content in the dinoflagellate *Gonyaulax excavata* in nature. *Toxicon* 24: 605-610.

Yamamori, K., M, Nakamura, T. Matsui, and T. Hara. 1988. Gustatory responses to tetrodotoxin and saxitoxin in fish: A possible mechanism for avoiding marine toxins. *Can. J. Fish. Aquat. Sci.* 45: 2182-2186.

Yasumoto, T., Y. Oshima, M. Hosaka, and M. Miyakoshi. 1981. Analysis of paralytic shellfish toxins of xanthid crabs in Okinawa. *Bull. Japan Soc. Sci. Fish.* 47: 957-959.

Yasumoto, T., D. Yasumura, Y. Ohizumi, M. Takahashi, A.C. Alcala, and L.C. Alcala. 1986. Palytoxin in two species of xanthid crab from the Philippines. *Agric. Biol. Chem.* 50: 163-167.

Effects of 4-Aminopyridine in the Treatment of Saxitoxin Poisoning

Fat-Chun Tony Chang, Bernard Benton and David Spriggs*

Pharmacology Division
U.S. Army Medical Research Institute of Chemical Defense
Aberdeen Proving Ground, MD 21010-5400 U.S.A.

and

*U.S. Army Edgewood Chemical Biological Center
Aberdeen Proving Ground, MD 21010-5400 U.S.A.

Introduction

Saxitoxin (STX) is one of the deadliest naturally occurring substances known. In guinea pigs, the lethal dose of STX (im) is only 4 μg/kg (unpublished observation). STX is elaborated by marine dinoflagellates of the genus *Gonyaulax* (Taylor and Seliger 1979, Ikawa *et al.* 1982). Under favorable climatic conditions, dinoflagellate blooms can occur several times a year along the coastal waters world-wide. Filter-feeders such as Alaska butter clams (*Saxidomus*), cherrystone clams, sea scallops, and mussels are among some of the known marine organisms through which STX infiltrates the human food web (Taylor and Seliger 1979). To the extent that these marine organisms are routinely harvested for human consumption, the likelihood of accidental poisoning by STX-containing shellfish could be a considerable public health concern (Hughes 1979, Valenti *et al.* 1979).

The toxic effects of STX are mediated primarily through the blockade of voltage-dependent sodium channels (Kao 1972, Narahashi 1972). This blocking action can eliminate, or significantly alter, the excitable cell's ability to initiate action potentials. Sodium channels are among the most important defining characteristics of membrane excitability. Since many vital functions, including cardiorespiratory rhythms, are driven and modulated by sodium channel-bearing excitable cells, the physiological effects following intoxication by STX would be serious indeed.

In cases of accidental STX poisoning, the time to onset of symptoms depends on the amount of toxin ingested and could vary from half an hour to several hours (Hughes 1979, Rodrigue *et al.* 1990, Sutherland 1983, Valenti *et al.* 1979). Symptoms associated with STX poisoning include dyspnea, dysphagia, paresthesia, dystonia, hot-cold reversal, nausea, gastro-intestinal disturbance, and a state of generalized malaise. These symptoms often persist, albeit with abating severity, for many days. Death can ensue within 1–2 hr when large amounts of STX are accidentally ingested. The most life-threatening element of accidental STX poisoning is a state of combined vascular hypotension and ventilatory insufficiency attributable to the blockade of diaphragmatic neurotransmission (see Benton *et al.* 1994, 1997, Borison and McCarthy 1977, Borison *et al.* 1980, Chang *et al.* 1990, 1993, 1996, 1997, Evans 1972, Jaggard and Evans 1975, Kao 1972, Kao *et al.* 1967, Murtha 1960). There is no specific antidote for STX. To this date, gastric lavage and prolonged periods of respiratory support are still the only available means to clinically manage accidentally poisoned victims. Findings from animal research indicated that immunopharmacological agents (i.e., STX-specific antibodies and toxin-binding proteins) could be a promising alternative to gastric lavage and ventilation support (Benton *et al.* 1994, Davio 1985, Fukiya and Matsumura 1992, Kaufman *et al.* 1991, Rivera *et al.* 1995, Smith *et al.* 1989). However, in view of the cost involved in the design and development of these proteins, and of a less than profit-producing market potential, the inclusion of immunopharmacological products in the armamentarium of emergency medicine does not appear to be forthcoming. For these reasons, we began to search for pharmacological compounds known to be effective for certain indications that may also be used to antagonize STX toxicity.

CNS and Cardiorespiratory Effects of 4-AP

Following a comprehensive evaluation of STX-induced pathophysiology across a variety of cardiorespiratory system components (Chang *et al.* 1990, 1993), we have identified diaphragmatic neurotransmission blockade as being the most life-threatening element of STX toxicity. Based on this information, a systematic attempt was made to identify pharmacological agents known to enhance neuromuscular transmission and cardiovascular performance. One of the pharmacological compounds we studied was 4-aminopyridine (4-AP; Benton *et al.* 1997, Chang *et al.* 1990, 1993, 1996, 1997, for reviews, see Bowman and Savage 1981, Glover 1982). 4-AP is a potent potassium channel blocker (Pelhate and Pichon 1974, Yeh *et al.* 1976) and is best known for its ability in enhancing the impulse-evoked acetylcholine

(ACh) release from presynaptic motor nerve terminals in phrenic-diaphragm preparations and other nerve-muscle model systems (Lundh 1978, Miller *et al.* 1979, Molgó *et al.* 1977). In clinical experimental therapeutics studies, multiple sclerosis (Van Diemen *et al.* 1993), Lambert-Eaton syndrome (Lundh *et al.* 1977) and myasthenia gravis (Lundh *et al.* 1979, Murray and Newsom-Davis 1981) are a few of the disorders in which patients have shown varying degrees of improvement after treatment with 4-AP.

Figure 1 is a schematic illustration of enhanced neurotransmitter release in response to 4-AP. The mechanism through which 4-AP facilitates neuromuscular transmission is believed to be a corollary of potassium (K^+) channel blockade. The K^+ conductance block causes a delayed repolarization of the presynaptic neuron which promotes profound alterations in the voltage-

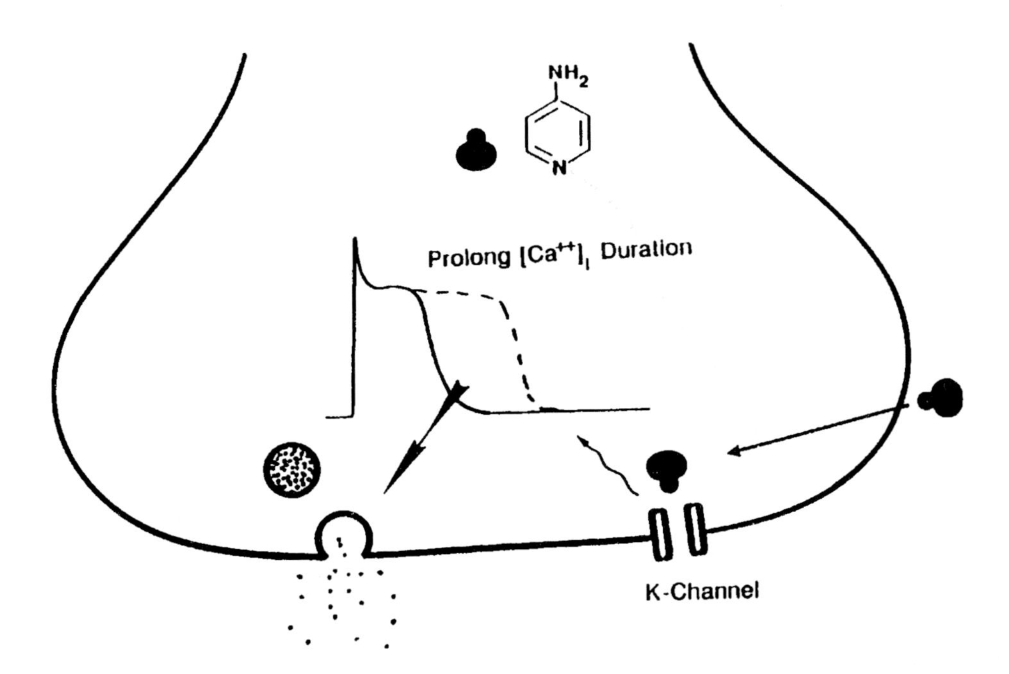

Figure 1. The mechanism through which 4-AP facilitates neuromuscular transmission is generally believed to be a corollary of potassium channel conductance blockade. This is depicted by the synaptic terminal cartoon in this figure. In brief description, 4-AP blocks the potassium channels on the cytosolic side of the excitable membrane and causes a delay in the repolarization of the presynaptic cell. This in turn prolongs the duration of calcium activation which allows a greater than normal calcium influx into the presynaptic terminals. To the extent that neurotransmitter release is a calcium-dependent phenomenon, the prolonged calcium activation would ultimately cause an enhanced level of ACh release from the motor nerve terminals.

dependence of calcium channels and their gating kinetics. These alterations in turn prolong the duration of intracellular calcium activation and allow a greater than normal calcium influx to ultimately cause an enhanced level of calcium-dependent neurotransmitter release from the presynaptic terminals.

In addition to enhancing neuromuscular transmission, 4-AP has also been shown to exert a powerful respiratory stimulant effect (Benton *et al.* 1997, Chang *et al.* 1996, 1997, Folgering *et al.* 1979), a pressor effect (Chang *et al.* 1996, Glover 1981, Sobek 1970, Yanagisawa and Taira 1979), cortical excitant effect (Chang *et al.* 1996, 1997) and, at a higher dose level, a seizurogenic/ convulsant effect (Chesnut and Swann 1990, Galvan *et al.* 1982, Rutecki *et al.* 1987, Szente and Baranyi 1989). Thus, within the framework of our search, it appears that 4-AP has several important pharmacological attributes that could be exploited in our design and evaluation of a treatment strategy against STX intoxication.

Figure 2 is an electrophysiological depiction of changes in cardiorespiratory activity profile in response to 2 mg/kg (im) 4-AP (Chang *et al.* 1997). The electrophysiograms were derived from an unanesthetized, freely behaving guinea pig electrophysiological model system (see Figure 2 legend for details). The most notable change in Figure 2 was the respiratory stimulant effect of 4-AP as indicated by an increase in diaphragmatic burst frequencies (i.e., respiratory rates) at 15 min and 30 min Post 4-AP stages. Respiratory rates typically began to return to control level at 2–3 hours post 4-AP. Also noteworthy was a sustained increase in the neck EMG amplitudes attributable to a tonic increase in ACh release at the neuromuscular junction in response to 4-AP. Figure 3 is a quantitative description of cardiorespiratory responses to 4-AP (1 and 2 mg/kg) measured over a period of 4 hours. Particularly noteworthy was the dose-dependent increase in respiratory frequency following 4-AP. The increase in respiratory rate following 1 or 2 mg/kg 4-AP could be maintained for at least 1.5 hours before returning to a level comparable to that of control. The magnitude of increase in heart rate after either 1 or 2 mg/kg 4-AP, by comparison, was less striking and exhibited no obvious signs of dose-dependency. After an initial increase during the first 5 min, the heart rate would return to a level only slightly above control for as long as 4 hours (see Fig. 3).

Antagonism of Cardiorespiratory Infirmity and Lethality

In a comprehensive electrophysiological survey, we have reported the severity as well as the temporal sequence of onset of STX-induced pathophysiology using an urethane-anesthetized guinea pig model system (Chang *et al.* 1993). This model system allows concurrent monitoring of a wide variety of cardiorespiratory and CNS activities. To examine respiratory system responses

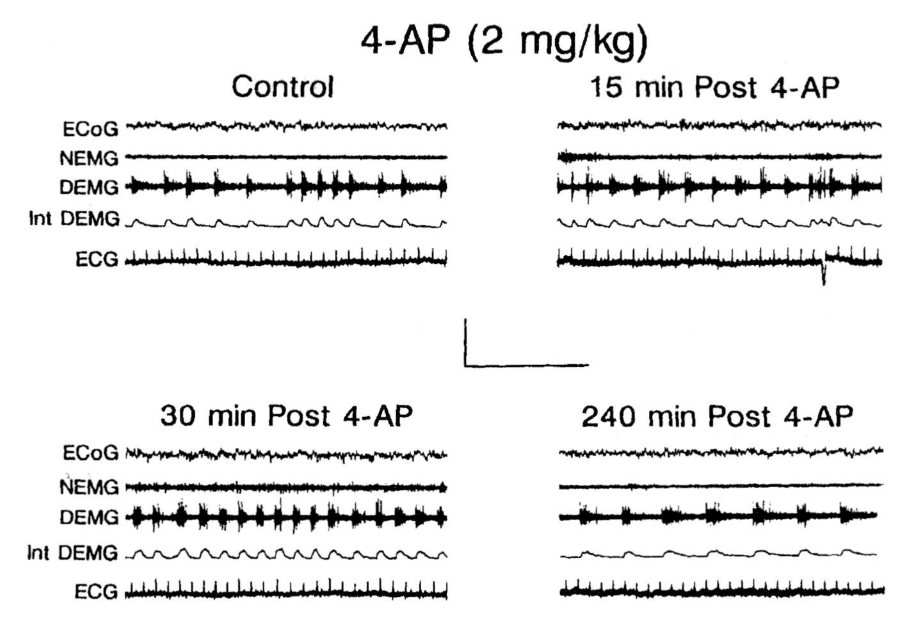

Figure 2. Electrophysiograms depicting changes in cardiorespiratory and electrocorticographic (ECoG) activities in response to 4-AP (2 mg/kg, im) across control condition and 15, 30 and 240 min after administration (Chang *et al.* 1997). The electrophysiograms were derived from an unanesthetized, freely behaving guinea pig electrophysiological model system. In this model system, the animals were surgically instrumented for concurrent recordings of electrocorticogram (**ECoG**), diaphragmatic electromyogram (**DEMG**), neck muscle electromyogram (**NEMG**) and Lead II electrocardiogram (**ECG**). Note the increase in DEMG burst frequencies shown in panels labeled 15 and 30 min Post 4-AP. Also noteworthy was a tonic augmentation in NEMG signal amplitudes at 15 and 30 min Post 4-AP stages indicating a tonic increase in ACh at the neuromuscular junctions. Voltage calibrations: ECoG, 210 µ V; DEMG, 2.60 mV. Time calibration: 5 sec.

to intoxication and 4-AP therapy, diaphragmatic electromyogram (DEMG; Chang and Harper 1989), tracheal end-tidal CO_2 (ET-CO_2), medullary inspiratory (Nucleus para-Ambiguus) and expiratory (Bötzinger Complex) neuronal activities (see Chang *et al.* 1988 for technical details) were concurrently monitored. Electrocardiogram (ECG, Lead II; from which heart rate can be derived) and carotid arterial blood pressure (BP) were recorded to assess the nature and extent of cardiovascular activity changes. Electromyographic recording from the neck muscle (NEMG; from dorsal portion of cervical trapezius) was also performed to measure the magnitude of 4-AP induced changes in skeletal muscle activities and the potential of muscle fasciculation resulting from enhanced ACh release. Finally, the extent of cortical excitant effects, and in particular, the seizurogenic properties of 4-AP reported earlier (Chesnut and Swann 1990, Galvan *et al.* 1982, Rutecki *et al.* 1987) were monitored with electrocorticogram (ECoG).

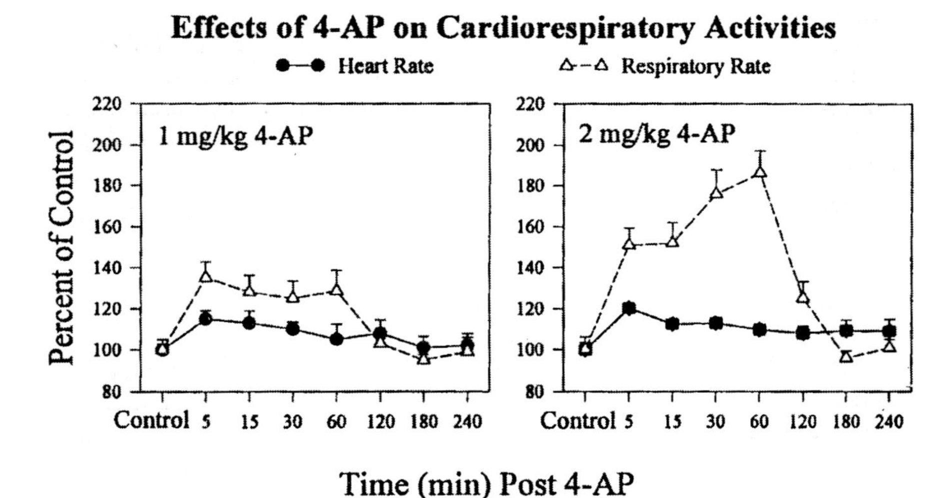

Figure 3. Changes (% Control) in mean heart rate (filled circle/solid line) and mean respiratory rate (open triangle/dashed line) in response to 1 mg/kg (left panel) and 2 mg/kg 4-AP (right panel) across Control condition and various time points (5, 15, 30 60, 120, 180, and 240 min) after 4-AP administration. The actual control values of heart rate and respiratory rate of animals receiving 1 mg/kg were 276.5±14 and 110±3.3 respectively. The actual control values of heart rate and respiratory rate of animals given 2 mg/kg 4-AP were 287.5±10 and 110.5±6.4 respectively. Error bar = Standard Error of the Mean (SEM).

The first indication of STX intoxication was a state of progressive bradypnea which emerged 5–7 min after the toxin administration. In concomitance with a developing bradypneic profile were: i) a time-dependent decrease in ECoG amplitudes; ii) an increasing degree of diaphragm neuromuscular blockade; iii) a state of combined hypercapnia, hypoxia and uncompensated acidemia; iv) a declining blood pressure; and v) an incrementally degenerative and dissociative central respiratory activity profile. This pathophysiological profile ultimately culminated in apnea attributable to a complete blockade of diaphragmatic neurotransmisison (»10 min post-STX). The extent of myocardial perturbation during the peri-apnea period was also striking. Prior to the emergence of an idioventricular rhythm (50-80 bpm- a prelude to myocardial failure), it was not uncommon to observe signs of aberrant myocardial performance such as right/left bundle block, T-wave inversion, J-point elevation, atrioventricular dissociation, etc. Despite the pervasiveness of STX toxicity, our data from this and other investigations have consistently shown that it was the incremental blockade of the diaphragmatic function that constituted the most life-threatening aspect of STX poisoning (Chang *et al.* 1993). It was with this critical piece of information that 4-AP, a pharmacological agent known to possess cardiorespiratory stimulant effects, was chosen as a potential antidote against STX poisoning.

Figure 4 is an electrophysiographic depiction of the lethal effects of STX and recovery following 4-AP treatment (Chang *et al*. 1996). In this case,

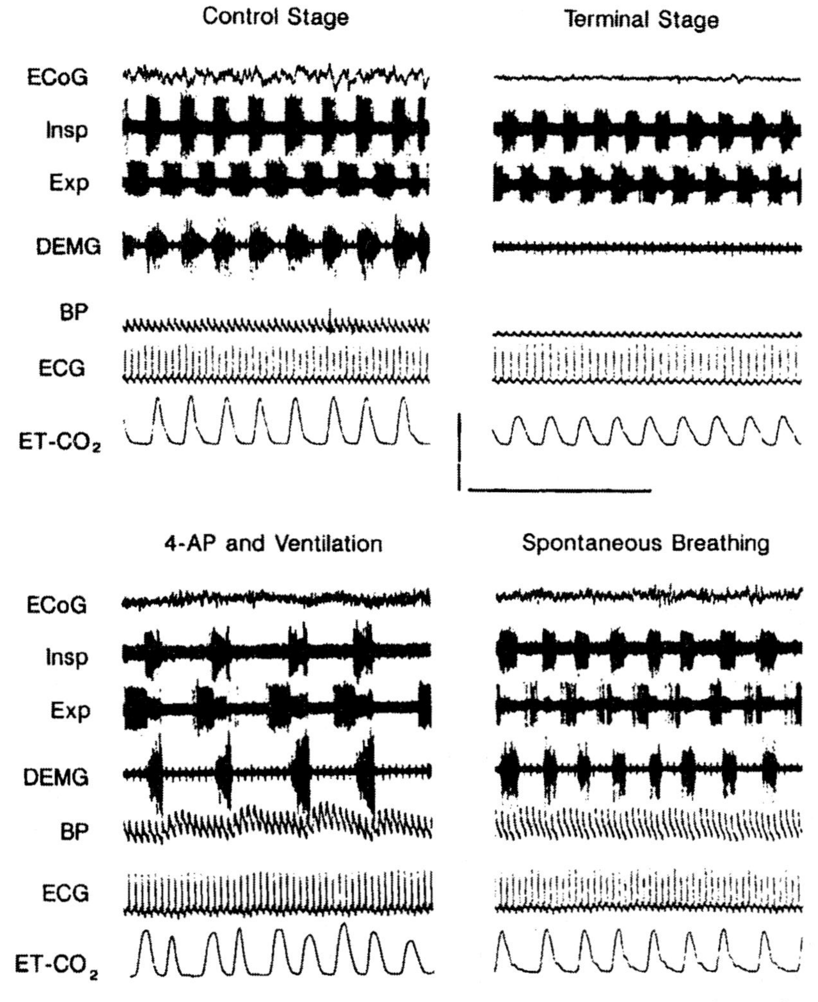

Figure 4. Electrophysiographic depiction of reversal of STX-induced cardiorespiratory dysfunctions with 4-AP (see text for description). The electrophysiograms were derived from a urethane-anesthetized guinea pig acutely instrumented for concurrent recording of a variety of CNS and cardiorespiratory activities. Signal trace description: ECoG, electrocorticogram; Insp, inspiratory unit activity recorded from Nucleus para-Ambiguus; Exp, expiratory unit activity recorded from the Bötzinger complex; DEMG, diaphragmatic electromyogram; BP, carotid blood pressure; ECG, Lead II electrocardiogram; ET-CO_2, end-tidal CO_2. Voltage calibrations: ECoG, 100 µV; Insp and Exp, 280 µV; DEMG, 1.44 mV.

STX was infused intravenously at a dose rate of 0.3 mg/kg/min to produce a state of progressive cardiorespiratory depression. The animals were artificially ventilated when the magnitude of integrated diaphragm activities was reduced to 50% of control while STX infusion continued. Parenthetically, ventilation support was instituted to permit a systematic evaluation of 4-AP's effectiveness in restoring diaphragmatic functions independent of other cardiorespiratory dysfunctions such as bradycardia, hypoxia, hypercapnia and blood *p*H fluctuations (acid-base imbalance). Immediately after the disappearance of the diaphragm EMG activities (see "Terminal Stage," Fig. 4), the toxin infusion was terminated, and 4-AP (2 mg/kg, iv) was administered. Particularly noteworthy at the "Terminal Stage" were a complete failure of the diaphragm (see DEMG trace), hypotension (see BP trace), and a markedly depressed CNS excitability (see ECoG trace). Because the animals were artificially ventilated, the extent of perturbation to the myocardial rhythm (see ECG trace) was not appreciable. Without artificial ventilation, cardiorespiratory collapse and death would ensue within minutes. The therapeutic effect of 4-AP was remarkable. As shown in Figure 4 (electrophysiograms labeled "4-AP and Ventilation" and "Spontaneous Breathing"), STX-induced blockade of diaphragmatic neurotransmission, vascular hypotension, myocardial anomalies and aberrant discharge patterns of medullary respiratory-related neurons were all restored to a level comparable to that of control condition and the animals were typically able to breathe spontaneously within minutes after 4-AP.

A survey of clinical case reports indicates that many accidentally poisoned victims do not require emergency respiratory support despite manifestation of severe symptoms of intoxication (see Sims and Ostman 1986). To gain a further understanding of 4-AP's therapeutic potential and to systematically evaluate 4-AP's effectiveness against the sub-lethal effects of STX, an unanesthetized, freely behaving animal model system was employed in our investigations (Benton *et al.* 1997, Chang *et al.* 1997). The use of this model system not only eliminated the confounding effects of anesthesia, but also removed concerns that the pharmacological attributes of the 4-AP may be modified by anesthetic agents.

With this model system, we used sub-lethal doses of STX (2 or 3 μg/kg, im) to simulate pathophysiological conditions resulting from accidental exposure to paralytic shellfish poison (STX). At either STX dose, we were able to produce a state of profound cardiorespiratory depression in our model system. Without treatment intervention, these symptoms could last in excess of 12 hours before signs of recovery would begin to slowly emerge (Chang *et al.* 1997). Figure 5 is an electrophysiographic depiction of cardiorespiratory and ECoG responses to STX (3 μg/kg, im) and 4-AP treatment (2 mg/kg, im) over a period of 6 hours. The most notable sub-lethal effects were seen during the "Depression" stage (» 30 min post-STX).

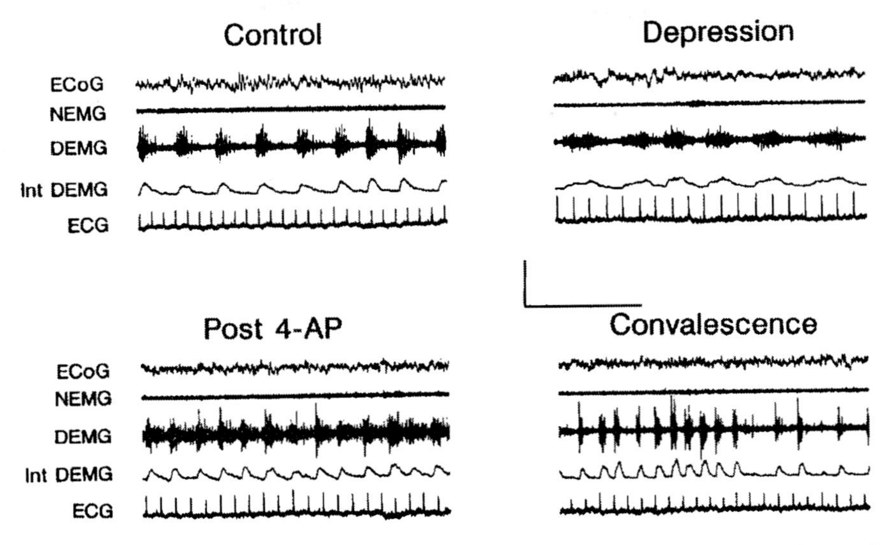

Figure 5. Electrophysiograms showing the cardiorespiratory activity profiles during i) control condition, ii) "Depression" stage; 30 min post-STX, iii) "Post 4-AP" stage (10 min after 4-AP), and iv) "Convalescence" stage (recovery stage, 2 hours after 4-AP). The effectiveness of 4-AP (2 mg/kg) against the sub-lethal effects of STX (3 μg/kg) is unequivocal. Without 4-AP treatment, the toxin-induced cardiorespiratory depression could last more than 12 hours. Signal trace description: ECoG, electrocorticogram; NEMG, neck (skeletal) muscle electromyogram; DEMG (see Chang and Harper 1989 for technical details), diaphragmatic electromyogram; Int DEMG, integrated (t = 50 msec) diaphragmatic electromyogram; ECG, Lead II electrocardiogram. Voltage calibrations: ECoG, 225 μV; DEMG, 2.14 mV. Time calibration: 5 sec.

These were i) bradypnea, ii) bradycardia, iii) a marked reduction in the amplitudes of diaphragmatic activity, and iv) visual indications of cyanosis, ataxia and paralysis. Minutes after 4-AP (1 or 2 mg/kg), STX-induced cardiorespiratory dysfunctions such as bradycardia, bradypnea and diaphragmatic functional compromise were all restored to a level comparable to that of control (see Fig. 5; "Post 4-AP" stage, recorded at 10 min after 2 mg/kg 4-AP). Moreover, close scrutiny of continuously recorded physiological data did not reveal any indication of aberrant muscle (from NEMG signals) or CNS (from ECoG signals) activities following 4-AP treatment. Physiological activity profiles documented at 2 hours after recovery (see Fig. 5, "Convalescence" stage) were virtually indistinguishable to those of control conditions.

The extent of STX-induced infirmity and the effectiveness of 4-AP in restoring the cardiorespiratory activity profile are summarized in Figure 6. The effectiveness of 2 μg/kg 4-AP was unequivocal. As illustrated in Figure

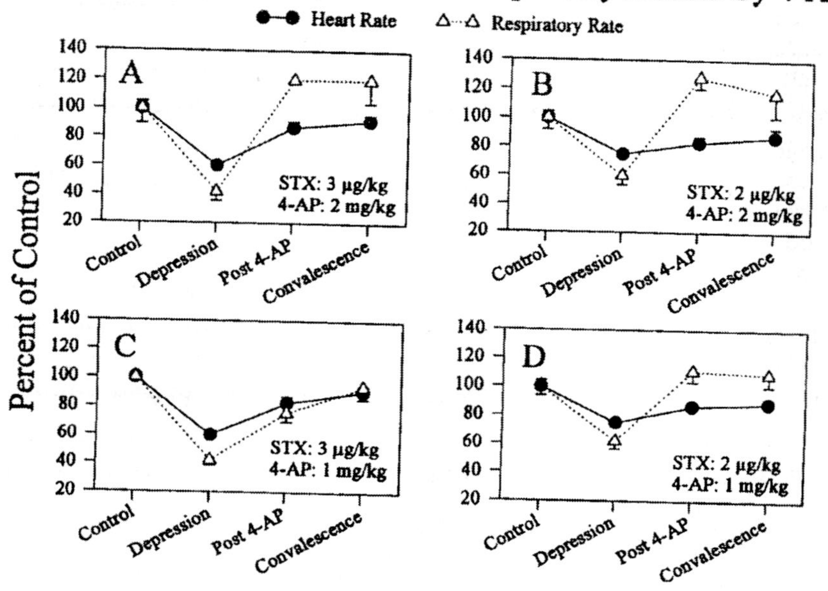

Figure 6. (A-D). Changes in heart rate (filled circle/solid line) and respiratory rate (open triangle/dashed line) in response to STX (2 or 3 μg/kg) and 4-AP treatment (1 or 2 mg/kg 4-AP) across i) control condition, ii) "Depression" stage (30 min post-STX), iii) "Post 4-AP" stage (10 min after 4-AP), and iv) "Convalescence" stage (2 hours after 4-AP). STX and 4-AP dose levels are indicated at the lower right-hand corner of each panel. Changes in cardiorespiratory activities are indicated as percent of control value. The actual control values of (heart rate)/(respiratory rate) are: Panel A, 290.5±7.7/108±8.6; Panel B, 280±8.3/111.8±7.8; Panel C, 288.5±6.7/106±6.6; Panel D, 296.5±8.2/111±6.3. Error bar = Standard Error of the Mean (SEM).

6, STX (2 or 3 μg/kg) induced cardiorespiratory depression could be restored in less than ten (10) minutes following 4-AP. And the post 4-AP cardiorespiratory activity profile was either comparable to, or more robust than, that of control condition (see Fig. 5, "Post 4-AP" stage). The effectiveness of 1 mg/kg 4-AP was less pronounced. That is, while 1 mg/kg 4-AP appeared adequate for managing the toxic effects resulting from low level intoxication (2 μg/kg), it was questionable if this dose would be adequate for treating animals intoxicated with 3 μg/kg STX. As Figure 6 indicates, in animals intoxicated with 3 μg/kg STX and subsequently treated with 1 mg/kg 4-AP (panel C), the respiratory frequency and heart rate were only restored to about 80% of control during the "Post 4-AP" stage (10 min post 4-AP) and with small subsequent increases to 80–90% of control during the "Convalescence" stage (2 hr post 4-AP). Since side effects resulting from 2 mg/kg 4-AP were minor (see Figs. 2 and 3); and, in view of the fact that it would be difficult, if not impossible, to accurately assess the severity of sub-lethal symptomology in actual cases of poisoning in humans, a maximally effective and minimally debilitating 4-AP dose (human equivalent of 2 mg/kg) would probably offer a more optimistic therapeutic outcome.

The final testimony of 4-AP's treatment efficacy would be its ability in reversing STX-induced cardiorespiratory collapse and death. However, because of a number of extreme physiological changes at the time of apnea, restoration of life-defining cardiorespiratory functions proved to be a formidable technical challenge. The activity profile of "Terminal" stage in Figure 7 is an electrophysiological portrayal of a failing cardiorespiratory

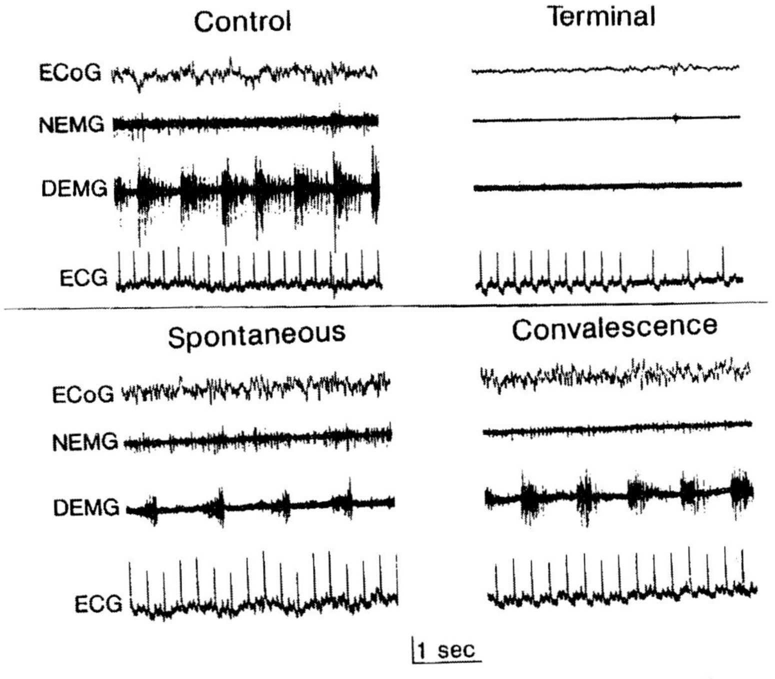

Figure 7. An electrophysiological depiction of STX-induced lethal effects and the treatment effects of 4-AP. Note the virtually iso-electric CNS (ECoG), skeletal (NEMG) and diaphragmatic (DEMG) neuromuscular activities and an emerging idioventricular ECG rhythm indicating a failing heart during the "Terminal" stage. Within a minute after the onset of apnea, the animals would have received ventilation support, pressor agent, 4-AP (2 mg/kg) and a blood *p*H buffer (see text). "Spontaneous" stage denotes the point where the animal was able to breathe without ventilation support. The "Convalescence" stage was the point at which the animal had been breathing spontaneously for at least one hour. Signal trace description: **ECoG**, electrocorticogram; **NEMG**, electromyographic recording from the neck muscle (dorsal portion of cervical trapezius); **DEMG**, diaphragmatic electromyogram; **Int DEMG**, integrated diaphragmatic EMG ($\tau = 50$ msec), and **ECG**, Lead II electrocardiogram.

status following a lethal dose of STX (4 μg/kg, im). Particularly noteworthy were: i) a complete loss of diaphragmatic (see DEMG trace) and skeletal (see NEMG trace) neuromuscular functions; ii) an almost complete absence of CNS excitability as indicated by an iso-electric ECoG activity pattern; and iii) the transition of a normocardiac profile to that of an idioventricular rhythm indicating an impending myocardial failure (see ECG trace). Another significant pathophysiological change was the development of hypoxia and hypercapnia which ultimately gave rise to a state of uncompensated acidemia (Chang *et al.* 1996). Blood samples taken from these animals at the time of apnea revealed an acidic shift in the blood *p*H from 7.38 to 6.9 which indicated not only a severe state of acid-base imbalance, but also a completely depleted blood pH buffering capacity.

Uncompensated acidemia to the extent described above could profoundly alter the therapeutic effectiveness of 4-AP. At normal physiological *p*H (7.38), 4-AP exists in both cationic and uncharged species ($pK_a \approx 9.17$; Howe and Ritchie 1991, Plant and Standen 1982, Stephens *et al.* 1994). The site of action of 4-AP is known to be on the cytosolic side of the excitable membrane (e.g., Glover 1982, Kirsch *et al.* 1993, Kirsch and Narahashi 1983, Plant and Standen 1982, Stephens *et al.* 1994). The uncharged species of 4-AP can readily cross the membrane to exert their pharmacological action on the potassium channels. During the development of acidemia, the acidic shift in the blood/extracellular *p*H can result in a *p*H-dependent increase in the positively charged 4-AP species (Howe and Ritchie 1991). In the cationic state, 4-AP will be much less likely to diffuse across the membrane to reach its target site on the potassium channel. As such, the animals could be refractory to treatment by 4-AP as a result of the acidic shift in the blood pH.

With a better understanding of the nature and extent of cardiorespiratory perturbation at the time of apnea, a regimen consisting of the following treatment components was conceived. First, a novel intra-tracheal intubation technique was developed and used to institute ventilation support to the animals within one minute following apnea (Benton *et al.* 1997). Second, during intra-tracheal intubation, a pressor agent (epinephrine; 0.1 mg/kg, im) was injected to reverse STX-induced hypotension, and to help reinstate the myocardial activity. Third, 4-AP (2 mg/kg, im) was administered to restore diaphragmatic function. Finally, 6-9 ml of sodium bicarbonate (7.5% or 892 mEq/l; ip) was given to reverse the acidemic condition. The therapeutic outcome following this regimen is depicted in Figure 7 (see "Spontaneous" and "Convalescence" stages). The most notable event following the treatment regimen was the return of diaphragmatic function to the point of being able to support spontaneous breathing in about an hour. Up to 6 hour post-treatment ("Convalescence" stage), the cardiorespiratory and CNS activity profiles remained essentially indistinguishable from those of control conditions.

In possession of an effective therapeutic strategy, the extent of 4-AP's therapeutic window was evaluated in subsequent investigations (Benton *et al.* 1997). Findings indicated that the temporal scope of the treatment window along which 4-AP remained effective appeared to be quite broad. In conjunction with ventilation and sodium bicarbonate, 4-AP (2 mg/kg) could bring about a reversal to STX-induced lethality when administered at times ranging from 10 minutes prior to STX, to the point of complete cardiorespiratory collapse (apnea). In consideration that many accidental STX poisoning occur in remote areas where medical attention may not be immediately available, this flexibility in administration times could make a significant difference in the treatment outcome of accidental cases of STX poisoning.

Figure 8 is a power spectrographic analysis of diaphragmatic activities throughout the course of STX intoxication (lethal dose; 4 µg/kg, im) and 4-AP (2 mg/kg, im) therapy. As shown in "Terminal" stage, the power spectrographic correlates of diaphragmatic functions disappeared entirely. Approximately an hour after the treatment regimen (ventilation, pressor

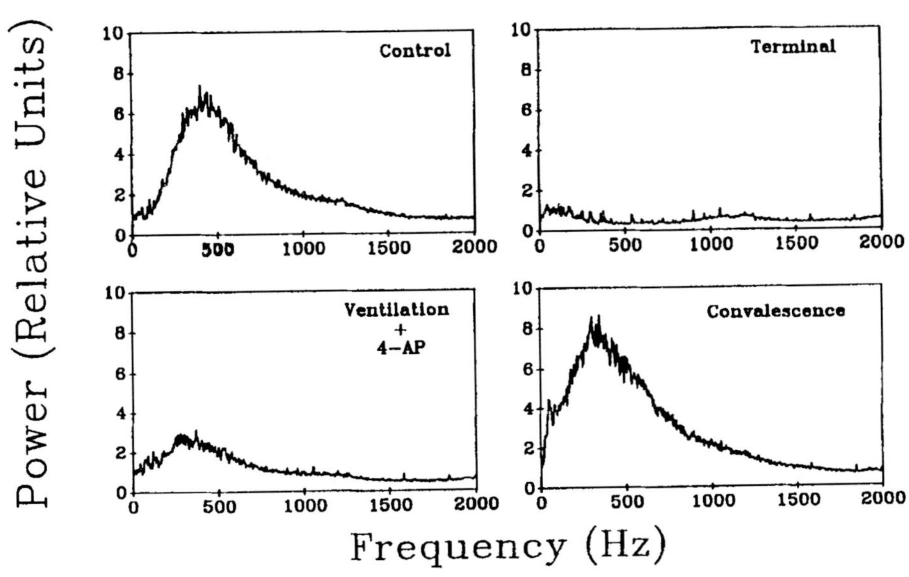

Figure 8. Changes in diaphragmatic (DEMG) spectral powers across i) control condition, ii) "Terminal" stage; note the complete disappearance of diaphragmatic activity, iii) "Ventilation and 4-AP" stage where the diaphragmatic neurotransmission was being restored, and iv) "Convalescence" stage—a point where the diaphragm has recovered sufficiently to support spontaneous breathing.

agent, 4-AP and sodium bicarbonate), the functional integrity of the diaphragm was restored to a level that could support spontaneous breathing.

The seizurogenic potential of 4-AP has been described earlier (Chesnut and Swann 1990, Galvan *et al.* 1982, Rutecki *et al.* 1987, Szente and Baranyi 1989). At the dose level of 2 mg/kg used in this study, however, we did not see any electrocorticographic correlates of seizure (or convulsion) activity (see Figs. 2, 4, 5, 7 and 9). Changes in cortical excitability throughout the course of STX intoxication and 4-AP treatment are shown in Figure 9. The only manifestation of 4-AP's cortical excitant effect was an increase in the number of low-amplitude, high-frequency (>10 Hz) asynchronous spectral events which typically indicate a heightened state of arousal. Results from other investigations in this laboratory indicated that 4-AP would become seizurogenic if the dose level is increased to 4 mg/kg or higher (unpublished observations).

Because of its remarkable effect in enhancing neuromuscular transmission, we were also concerned initially about the aberrant neuromuscular effects 4-AP may cause. To our surprise, untoward neuromuscular effects resulting from 2 mg/kg 4-AP were quite modest. In most animals, 4-AP stimulated

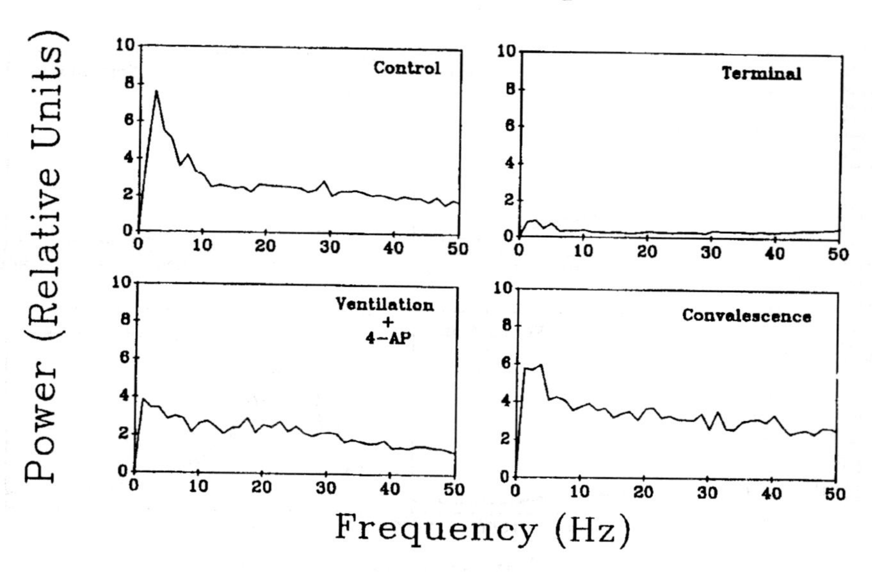

Figure 9. Changes in spectral power of electrocorticographic (ECoG) activities across i) control condition, ii) "Terminal" stage where the ECoG became virtually iso-electric, iii) "Ventilation and 4-AP" stage where the CNS excitability was being re-established, and iv) "Convalescence" stage showing an overall increase in high-frequency (>10 Hz), low-amplitude spectral varieties indicative of a heightened state of arousal. Otherwise, the cortical excitability appeared to have returned to a state comparable to that of control.

neuromuscular transmission could only be seen as an overall increase in the neck electromyogram background activity (see NEMG trace in Figs. 2, 5 and 7). A more noteworthy neuromuscular effect not portrayed in these electrophysiolograms was the sporadic "twitch movement" of the muscles of orofacial area, neck and extremities observed in about 20–30% of the animals. The "fascicular twitches" generally lasted between 5–20 minutes post 4-AP treatment and eventually became indistinguishable from background electromyographic activity. Notwithstanding, the therapeutic benefit of 4-AP, in our opinion, still greatly outweighs this short-lasting, but seemingly untoward, side effect. The possibility of incorporating a muscle relaxant in 4-AP therapy to reduce the severity of aberrant neuromuscular effects could perhaps be considered.

Therapeutic Potentials of 4-Aminopyridine (4-AP)

At first glance, the site and mechanism-specific antibodies or toxin-binding proteins seem to be more ideally suited for treating STX-induced toxicity. However, in actual cases of poisoning, the immunopharmacological agents could not possibly recognize all the STX analogs, and other toxic components that are chemically unrelated to STX within the poison-tainted organisms. Other considerations such as availability, shelf-life and cost of these products, could also significantly discourage their use, particularly in economically disadvantaged locations. The therapeutic potential of 4-AP appears to be more promising by comparison. As our research showed, in addition to its effects against STX toxicity, 4-AP can also antagonize the toxic and lethal effects of tetrodotoxin (Chang *et al.* 1996, 1997). Moreover, we must be mindful of the fact that the life-threatening aspects of STX poisoning could demand aggressive therapeutic measures even at the expense of serious untoward side effects such as seizures and convulsions. However, 4-AP induced side effects such as muscle fasciculation and deleterious CNS effects (such as seizures and convulsions) are a dose-dependent phenomenon. As long as the dose of 4-AP is administered with prudence, the severity of untoward side effects could be greatly minimized. Finally, in order to further define the therapeutic role of 4-AP, the inclusion of adjuncts that will either optimize its effectiveness or reduce the severity of its side effects should also be taken into consideration.

In conclusion, we have shown that STX-induced cardiorespiratory dysfunctions and lethality can be reversed by a potassium channel blocker, 4-aminopyridine (4-AP). The dose level used to achieve optimal therapeutic responses was 2 mg/kg (iv or im). At this dose level, 4-AP does not appear to be seizurogenic. Although side effects such as electrocorticographic indications of a heightened state of arousal and transient periods of fascicular twitches can be observed, these events, however disquieting, are trivial *vis-à-vis* the remarkable symptomolytic effects of 4-AP.

Acknowledgments

The opinions and assertions contained in this report are the views of the authors and should not be construed as official or as reflecting the views of the U.S. Army or the Department of Defense.

This research was conducted in compliance with the Animal Welfare Act and other Federal statutes and regulations relating to animals and experiments involving animals and adheres to principles stated in the *Guide for the Care and Use of Laboratory Animals,* National Research Council, 1996. The facility where this research was conducted is fully accredited by the Association for Assessment and Accreditation of Laboratory Animal Care International.

References

Benton, B.J., V.R. Rivera, J.F. Hewetson, and F.-C.T. Chang. 1994. Reversal of saxitoxin-induced cardio-respiratory failure by a burro-raised a-STX antibody and oxygen therapy. *Toxicol. Appl. Pharmacol.* 124: 39-51.

Benton, B.J., S.A. Keller, D.L. Spriggs, B.R. Capacio, and F.-C.T Chang. 1997. Recovery from the lethal effects of saxitoxin: A therapeutic window for 4-aminopyridine (4-AP). *Toxicon* 36: 571-588.

Borison, H.L. and L.E. McCarthy. 1977. Respiratory and circulatory effects of saxitoxin in the cerebrospinal fluid. *Br. J. Pharmacol.* 61: 679-689.

Borison, H.L., W.J. Culp, S.F. Gonsalves, and L.E. McCarthy. 1980. Central respiratory and circulatory depression caused by intravascular saxitoxin. *Br. J. Pharmacol.* 68: 301-309.

Bowman, W.C. and A.O. Savage. 1981. Pharmacology of aminopyridines and related compounds. *Rev. Pure and Appl. Pharmacol. Sci.* 2: 317-371.

Chang, F.-C.T., T.R. Scott, and R.M. Harper. 1988. Methods of single unit recording from medullary neural substrates in awake, behaving animals. *Brain Res. Bull.* 21: 749-756.

Chang, F.-C.T. and R.M. Harper. 1989. A procedure for chronic recording of diaphragmatic electromyographic activity. *Brain Res. Bull.* 22: 561-563.

Chang, F.-C.T., B.J. Benton, J.L. Salyer, R.E. Foster and D.R. Franz. 1990. Respiratory and cardiovascular effects of tetrodotoxin (TTX) in urethane-anesthetized guinea pigs. *Brain Res.* 528: 259-268.

Chang, F.-C.T., B.J. Benton, R.A. Lenz, and B.R. Capacio. 1993. Central and peripheral cardio-respiratory effects of saxitoxin (STX) in urethane-anesthetized guinea pigs. *Toxicon* 31: 645-664.

Chang, F.-C.T., R.M. Bauer, B.J. Benton, S.A. Keller, and B.R. Capacio. 1996. 4-Aminopyridine antagonizes saxitoxin (STX) and tetrodotoxin (TTX) induced cardiorespiratory depression. *Toxicon* 34: 671-690.

Chang, F.-C.T., B.J. Benton, D.L. Spriggs, and B.R. Capacio. 1997. 4-Aminopyridine reverses saxitoxin (STX) and tetrodotoxin (TTX) induced cardiorespiratory depression in chronically instrumented guinea pigs. *Fundam. Appl. Toxicol.* 38: 75-88.

Chesnut, T.J. and J. W. Swann. 1990. Suppression of 4-aminopyridine-induced epileptogenesis by the GABAA agonist muscimol. *Epilepsy Res.* 5: 8-17.

Davio, S.R. 1985. Neutralization of saxitoxin by anti-saxitoxin rabbit serum. *Toxicon* 23: 669-675.

Evans, M.H. 1972. Tetrodotoxin, saxitoxin, and related substances: Their application in neurobiology. *Int. Rev. Neurobiol.* 15: 83-165.

Folgering, H., J. Rutten, and S. Agoston. 1979. Stimulation of phrenic nerve activity by an acetylcholine releasing drug: 4-aminopyridine. *Pflügers Arch.* 379: 181-185.

Fukiya, S. and K. Matsumura. 1992. Active and passive immunization for tetrodotoxin in mice. *Toxicon* 30: 1631-1636.

Galvan, M., P. Grafe, and G. Ten Bruggencate. 1982. Convulsant actions of 4-aminopyridine on the guinea pig olfactory cortex slice. *Brain Res.* 241: 75-86.

Glover, W.E. 1981. Cholinergic effect of 4-aminopyridine and adrenergic effect of 4-methyl-2-aminopyridine in cardiac muscles. *Eur. J. Pharmacol.* 71: 21-31.

Glover, W.E. 1982. The aminopyridines. *Gen. Pharmacol.* 13: 259-285.

Howe, J.R. and J.M. Ritchie. 1991. On the active form of 4-aminopyridine: Block of K^+ currents in rabbit schwann cells. *J. Physiol.* 433: 183-205.

Hughes, J.M. 1979. Epidemiology of shellfish poisoning in the United States, 1971–1977. Pages 23-28 in *Toxic Dinoflagellate Blooms*, D.L. Taylor and H.H. Seliger, eds., Elsevier, New York/North-Holland, Holland.

Ikawa, M.K., K. Wegener, T.L. Foxall, and J.J. Sasner, Jr. 1982. Comparison of the toxins of the blue-green alga *Aphanizomenon flos-aquae* with the *Gonyauylax* toxins. *Toxicon* 20: 747-752.

Jaggard, P.J. and M.H. Evans. 1975. Administration of tetrodotoxin and saxitoxin into the lateral cerebral ventricle of the rabbit. *Neuropharmacology* 14: 345-349.

Kao, C.Y., T. Suzuki, A.L. Kleinhaus, and M.J. Siegman. 1967. Vasomotor and respiratory depressant actions of tetrodotoxin and saxitoxin. *Arch. Int. Pharmacodyn.* 165: 438-450.

Kao, C.Y. 1972. Pharmacology of tetrodotoxin and saxitoxin. *Fed. Proc.* 31: 1117-1123.

Kaufman, B., D.C. Wright, W.P. Ballou, and D. Monheit. 1991. Protection against tetrodotoxin and saxitoxin by a cross-protective rabbit antitetrodotoxin antiserum. *Toxicon* 29: 581-587.

Kirsch, G.E. and T. Narahashi. 1983. Site of action and active form of aminopyridines in squid axon membranes. *J. Pharmacol. Exp. Ther.* 226: 174-179.

Kirsch, G.E., C.-C. Shieh, J.A. Drewe, D.F. Verner, and A.M. Brown. 1993. Segmental exchanges define 4-aminopyridine binding and the inner mouth of K^+ pores. *Neuron* 11: 503-512.

Lundh, H. 1978. Effects of 4-aminopyridine on neuromuscular transmission. *Brain Res.* 153: 307-318.

Lundh, H., O. Nilsson, and I. Rosen. 1977. 4-Aminopyridine—a new drug tested in the treatment of Eaton-Lambert Syndrome. *J. Neurol. Neurosurg. Psychiat.* 40: 1109-1112.

Lundh, H., O. Nilsson, and I. Rosen. 1979. Effects of 4-aminopyridine in myasthenia gravis. *J. Neurol. Neurosurg. Psychiat.* 42: 171-175.

Miller, R.D., L.H. Booij, S. Agoston, and J.F. Crul. 1979. 4-Aminopyridine potentiates neostigmine and pyridostigmine in man. *Anesthesiol.* 50: 416-420.

Molgó, J., M. Lemeignan, and P. Lechat. 1977. Effects of 4-aminopyridine on frog neuromuscular junction. *J. Pharmacol. Exp. Ther.* 203: 653-663.

Murtha, E.F. 1960. Pharmacological study of poisons from shellfish and puffer fish. *Ann. N.Y. Acad. Sci.* 90: 820-836.

Murray, N.M.F. and J. Newsom-Davis. 1981. Treatment with oral 4-aminopyridine in disorders of neuromuscular transmission. *Neurology* 31: 264-277.

Narahashi, T. 1972. Mechanism of action of tetrodotoxin and saxitoxin on excitable membranes. *Fed. Proc.* 31: 1124-1132.

Pelhate, M. and Y. Pichon. 1974. Selective inhibition of potassium current in the giant axon of the cockroach. *J. Physiol.* 242: 90-91.

Plant, T.D. and N.B. Standen. 1982. The action of 4-aminopyridine (4-AP) on the early outward current (I_A) in *Helix aspersa* neurones. *J. Physiol.* 332: 18P.

Rivera, V.R., M.A. Poli, and G.S. Bignami. 1995. Prophylaxis and treatment with a monoclonal antibody of tetrodotoxin poisoning in mice. *Toxicon* 33: 1231-1237.

Rodrigue, D.C., R.A. Etzel, S. Hall, E. De Porras, O.H. Velasquez, R.T. Tauxe, E. M. Kilbourne, and P.A. Blake. 1990. Lethal paralytic shellfish poisoning in Guatemala. *Am. J. Trop. Med. Hyg.* 42: 267-271.

Rutecki, P.A., F.J. Lebeda, and D. Johnston. 1987. 4-Aminopyridine produces epileptiform activity in hippocampus and enhances synaptic excitation and inhibition. *J. Neurophysiol.* 57: 1911-1924.

Sims, J.K. and D.C. Ostman. 1986. Puffer fish poisoning: Emergency diagnosis and management of mild human tetrodotoxication. *Ann. Emer. Med.* 15: 1094-1098.

Smith, D.S., D.D. Kitts, and P. Townsley. 1989. Serological cross-reactions between crab saxitoxin-induced protein and paralytic shellfish poison-contaminated shellfish. *Toxicon* 27: 601-606.

Sobek, V. 1970. On the pharmacology of 4-aminopyridine compared with adrenaline. *Physiol. Bohemoslovaca.* 19: 417-419.

Stephens, G.J., J.C. Garratt, B. Robertson, and D.G. Owen. 1994. On the mechanism of 4-aminopyridine action on the cloned mouse brain potassium channel mKv1.1. *J. Physiol.*, 477: 187-196.

Sutherland, S.K. 1994. *Australian Animal Toxins*, Oxford University Press, Melbourne.

Szente, M.B. and A. Baranyi. 1989. Properties of depolarizing plateau potentials in aminopyridine-induced ictal seizure foci of cat motor cortex. *Brain Res.* 495: 261-270.

Taylor, D.L. and H.H. Seliger. 1979. *Toxic Dinoflagellate Blooms.* Elsevier, New York/North-Holland, Holland.

Valenti, M., P. Pasquini, and G. Andreucci. 1979. Saxitoxin and tetrodotoxin intoxication—Report of 16 cases. *Vet. Human Toxicol.* 21: 107-110.

Van Diemen, H.A.M., C.H. Polman, J.C. Koetsier, A.C. Van Loenen, J.J.P. Nauta, and F.W. Bertelsmann. 1993. 4-Aminopyridine in patients with multiple sclerosis: dosage and serum level related to efficacy and safety. *Clin. Neuropharmacol.* 16: 195-204.

Yanagisawa, T. and N. Taira. 1979. Positive inotropic effect of 4-aminopyridine on dog ventricular muscle. *Naunynschmiedeberg's Arch. Pharmacol.* 307: 207-212.

Yeh, J.Z., G.S. Oxford, C.H. Wu, and T. Narahashi. 1976. Interactions of aminopyridines with potassium channels of giant squid axon membranes. *Biophys. J.* 16: 77-81.

Development of Commercial Immunoassays for Seafood Poisonings

Paulo Vale

Ecotoxicology Unit, Instituto de Investigação das Pescas e do Mar, Av. Brasília, 1449-006 Lisboa, Portugal

Introduction

Marine biotoxins affect human consumers through different routes. The main and most widespread one is through consumption of bivalve molluscs. These filter-feeding animals ingest, among other food sources, phytoplankton and in certain epochs may ingest toxin-producing microalgae. Among these, paralytic shellfish poisoning (PSP), diarrhetic shellfish poisoning (DSP), neurotoxic shellfish poisoning (NSP) and amnesic shellfish poisoning (ASP) are the most common human shellfish poisonings. The responsible microalgae are planktonic dinoflagellates, except for ASP that is produced by diatoms. These toxins have also been found in higher-order consumers that prey on bivalves such as crustaceans and gastropods. Fishes are also known to contain PSP, ASP and NSP and are transfer vectors that have been held responsible for some massive deaths of marine mammals and birds. Besides shellfish vectors, only crustaceans present a serious risk to human consumers due to PSP accumulation. In tropical and sub-tropical areas (mainly discrete regions of the Pacific Ocean, western Indian Ocean and Caribbean Sea), benthic dinoflagellates are responsible for contaminating with ciguatera fish poisoning (CFP) herbivorous fish that feed on coral reef, which in turn contaminate carnivorous fish, that are the most dangerous for human consumption.

PSP and DSP are amongst the most common non-bacterial food poisoning in temperature zones. PSP is widely disseminated world-wide; in strongly endemic areas (generally of high latitude) it is responsible annually for a few casualties. DSP is most common in Europe and Japan, but does not present any fatal cases nor long lasting illness. Ciguatera is the most common non-bacterial food poisoning not only in circumtropical areas, but also in populations from temperate zones due to tourism in tropical islands and importation of fish from contaminated areas. It is estimated to affect several thousands of people annually, presents long lasting symptoms, but seldom is fatal.

The toxins involved range from amino acid analogues (domoic acid, responsible for ASP), alkaloids (saxitoxins, responsible for PSP) and polyethers (okadaic acid and dinophysistoxins, responsible for DSP; brevetoxins, responsible for NSP; ciguatoxins and maitotoxin, responsible for CFP).

Traditionally they have been monitored by mouse bioassays (MBA), with specific extraction and clean-up protocols for each of these toxins. Mouse bioassays are easy and ready to implement, not requiring highly trained personnel nor expensive equipment. Due to involvement of live animals they are highly criticised by animal rights supporters. They also present certain disadvantages: are not selective for the target toxin and thus can present false positive results; present large variability; are not sensitive to advance warning of contamination; some have long observation times and are not quantitative; are dependent on a large supply of mice in standardised conditions (strain, sex, age, weight), which may be critical in crisis situations.

For example, the mouse bioassay for PSP has been an AOAC Official Method of Analysis for several decades (McFarren 1959). Despite being a quantitative and fast assay, it can suffer interference from zinc in particular in oysters from some regions (McCulloch *et al.* 1989, Cacho 1993, Aune *et al.* 1999, Vale and Sampayo 2001); it underestimates toxicity in highly salty (mainly Na^+) matrices (Schantz *et al.* 1958); it only detects toxicity at a limit only half the regulatory threshold used by most countries in the world (AOAC 1990).

Despite the drawbacks, most countries still rely on them. One major advantage of a mammal bioassay is that it provides a measure of total toxicity based on the animal response to those toxins, although this can also be controversial due to the administration route in MBA being intraperitoneal. For many countries MBA can also be much cheaper than importation of expensive test kits, or acquisition of expensive equipment.

One of the first approaches to replace MBA, that can give simultaneously a knowledge of toxin profile, is high-performance liquid chromatography (HPLC) coupled to fluorometric or ultraviolet detection, and more recently to mass spectrometry detection, the last option being extremely expensive. Sample preparation is often more complex than for MBA, equipment operation requires centralised laboratories, and is time-limited to sequential injection.

Among other promising techniques that can process simultaneously a large number of samples are immunoassays, enzyme assays and cytotoxicity tests. The first ones are the most robust (adequate for fieldwork), fast and user-friendly tests. They can be used as a preliminary screening tool for toxins on shipboard, dockside, aquaculture facilities, canning industry and local monitoring laboratories, as only rudimentary extraction and test performance are needed. In some immunoassays a simple yes/no answer can be readily obtained without any specialised equipment by simple eye reading. Immunoassays can avoid waste of raw materials and save money, by preventing further harvesting or industrial processing of contaminated seafood before tests are performed at official laboratories. Although the cost of mice is relatively low, testing laboratories have to charge according to the overall costs of an analytical laboratory.

Therefore, we will focus here our main attention on colourimetric enzyme immunoassays (EIA), avoiding radio-immunoassays (RIA) that need more expensive equipment and are impractical for field use due to utilisation of radioisotopes. Also, EIA have sensitivities that can be higher than RIA because enzyme conjugates amplify the detection. A common enzyme used in EIA is horseradish peroxidase. EIA usually have a shelf life around one year, needing only a common house refrigerator for storing.

A complication for biotoxin monitoring is when several shellfish poisonings occur simultaneously (as an example, in New Zealand bivalves all four shellfish poisonings described above have been reported). For this situation, a battery of immunoassay tests is one of the best alternatives to survey for different classes of toxic compounds.

Several difficulties arise when developing immunoassays for marine biotoxins. There are numerous toxins known but they are not readily available in a purified form for developing antibodies. Specific toxicities of each toxin can vary a great deal, as well as cross-reactivities of antibodies towards them. The most controversial issue regarding alternative assays to MBA comes when toxicity levels are close to the regulatory limits widely accepted by the scientific/governmental communities. This is why it is mandatory that in many cases at least samples close to the safety threshold be analysed in a central laboratory. When centralised well-equipped laboratories are distant, immunoassays can provide at least a yes/no answer. As most of the samples tested usually are negative, they can reduce significantly animal sacrifice.

It is advisable that whoever tries to use immunoassay tests be aware of the toxin profiles in seafood from his region by sending samples to analyse in a highly equipped specialised laboratory or from bibliography on the subject already available from research groups in that region. We will highlight below some of the known drawbacks of these tests that can generate either false negative or false positive results.

Tests for Diarrhetic Shellfish Poisoning

Diarrhetic shellfish poisoning toxins found in shellfish comprise mainly three parent compounds: okadaic acid (OA) and two closely related analogues, dinophysistoxin-1 (DTX1) and dinophysistoxin-2 (DTX2). These three compounds possess roughly the same diarrhetic potency. Acyl derivatives of these toxins, designated as 'DTX3', are also found (Figure 1). They possess similar diarrhetic potency, and can be hydrolysed to their respective parent toxins. These toxins are unquestionably diarrhoeic, while other 'DSP' toxins, namely pectenotoxins and yessotoxins, are not diarrhoeic and their effects on man are still poorly studied. Today it is largely questionable if they should continue to be included in biotoxin monitorization programmes or not. Usually shellfish under 0.8-2.0 µg OA/g digestive gland[1] is considered safe for human consumption according to scientific studies, although there is no international agreement on a chemical threshold. Most countries simple state in their legislation that samples tested shall not give a positive result by MBA.

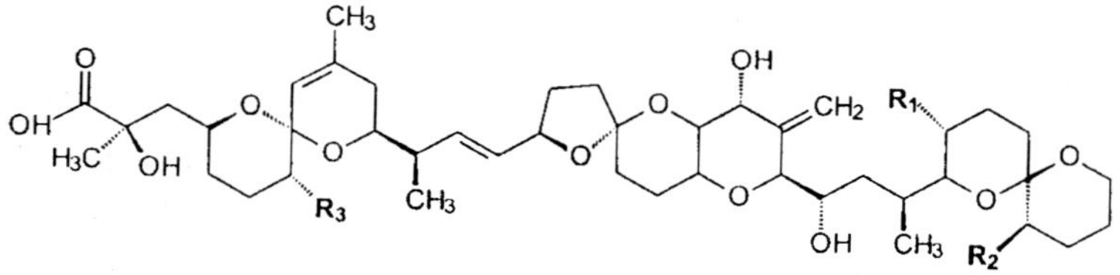

Figure 1. Structure of okadaic acid and its known derivatives most commonly found in shellfish. OA: R_1 = Me, R_2 = H; DTX1: R_1 = R_2 = Me; DTX2: R_1 = H, R_2 = Me; 'DTX3': R_3 = acyl and in the remaining R_3 = OH.

The first widely used commercial enzyme-linked immunosorbent assay (ELISA) for DSP was developed using mouse monoclonal antibodies against OA prepared according to Usagawa *et al.* (1989). This is an indirect competitive assay that uses an okadaic acid-bovine serum albumin coated microtiter wells to which standards/samples solutions are added plus enzyme-conjugated anti-OA. Bound OA and free OA in solution (from standards or samples) compete for coupling with the anti-OA antibody in solution. After incubation and washing of the micro-wells, one drop of enzyme substrate solution is added and the colourless chromogen is converted into a blue product. Addition of concentrated acid stops the reaction and changes colour to yellow. Measurement is made photometrically with an automated microplate reader at 450 nm. If present,

[1]alternatively, 20 µg OA/100 g edible parts

free toxin will lessen the amount of enzyme-conjugated anti-okadaic acid binding to its antigen coated in the bottom of the micro-wells, and the corresponding coloured reaction, in a dose-dependent manner that can be quantified from the standard curve (Figure 2).

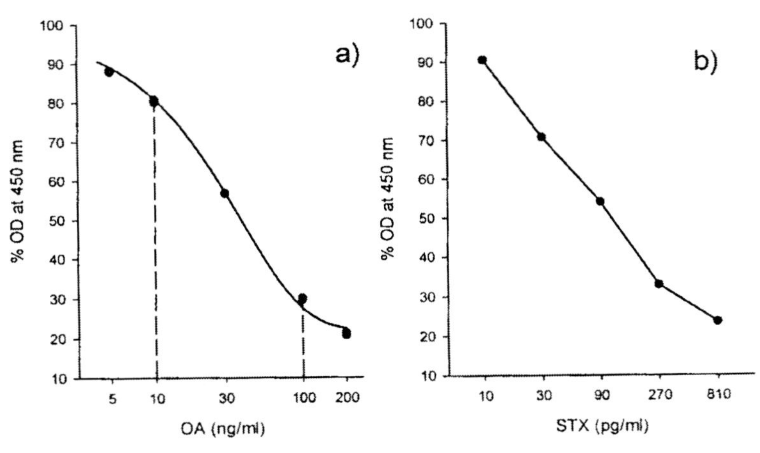

Figure 2. Typical calibration curves obtained with commercial ELISA kits: a) DSP-Check kit (Vale and Sampayo, 1999b), and b) RIDASCREEN® PSP test kit (Vale, data not published). In a) linearity is 1 order of magnitude from 10 to 100 ppb, while in b) it is 2 orders of magnitude from 10 to 810 ppt.

This test employs a simple aqueous-methanolic extraction, similar to the one used for liquid chromatography (Lee *et al.* 1987) and can detect OA quantitatively between 0.1-1.0 µg/g digestive glands. Ube Industries Ltd. (Tokyo, Japan), produced initially this test kit and designated it "DSP-Check". It is presently produced by Sceti Co., Ltd.[2] and distributed by Panapharm Co., Ltd.[3] in the format of six individual strips with 8 micro-wells each. A 10 and a 100 ppb okadaic acid standard are supplied with the kit.

Due to the use of OA for coating the micro-wells, the previous assay is expensive. Another assay was developed by Shestowsky *et al.* (1992), which uses a bound monoclonal anti-idiotypic antibody bearing an internal image of okadaic acid epitope to capture an anti-okadaic acid monoclonal antibody in the presence of free okadaic acid. Bound anti-okadaic acid antibody is detected with enzyme-conjugated anti-mouse immunoglobulin anti-serum. It can detect OA quantitatively between 0.09-0.81 µg/g digestive glands. This "Okadaic Acid ELISA Kit", was commercialised by Rougier Bio-Tech Ltd.

---

[2]DF Bldg. 2-2-8, Minami-Aoyama, Minato-Ku, Tokyo 107-0062, Japan.
[3]Kurisaki 1285, Uto, Kumamoto 869-04, Japan.

(Montreal, Canada). Alternatively, anti-idiotypic okadaic acid and anti-okadaic acid antibodies could be purchased separately from Calbiochem, a protocol for the preparation of the assay was reviewed by Cembella *et al.* (1995). Presently, Rougier Bio-Tech company is no longer operating, and the separate antibodies are no longer listed in the 2000/2001 Calbiochem Catalogue.

These two commercial ELISA kits have been used worldwide and extensively evaluated. "DSP-Check" kit was reported to have a cross-reactivity towards dinophysistoxin-1 of 73% (Usagawa *et al.* 1989), DTX2 of 40 ± 5% (Carmody *et al.* 1995), and DTX3 is not detectable (Usagawa *et al.* 1989). The "Okadaic Acid ELISA Kit" was reported to have cross-reactivity towards DTX1 20-fold lower, for DTX2 10-fold lower, while DTX3 was not detectable (Chin *et al.* 1995). However, this last kit has been shown to detect with equal sensitivity other kinds of OA derivatives such as: the methyl ester, diol ester (Chin *et al.* 1995), DTX4 and DTX5 (Lawrence *et al.* 1998), despite the remarkable variation in structure and chemistry of these derivatives. It was suggested that the epitopic region of the molecule for this antibody is the portion which is furthest from the carboxyl group (Chin *et al.* 1995, Lawrence *et al.* 1998). Both kits were found to produce quantitative agreement with HPLC determination of okadaic acid in mussels from France and Denmark (Fremy *et al.* 1994).

Despite OA being the most widespread toxin, in Europe for example, relatively high amounts of DTX2 can be found in mussels from Ireland (values as high as 95% have been reported by Carmody *et al.* 1995), Spain (Gago-Martinez *et al.* 1996) and Portugal (Vale and Sampayo 1999a). Also DTX1 is detected in Norway (Lee *et al.* 1988), Italy (Draisci *et al.* 1998) and United Kingdom (Nunez and Scoging 1997). Therefore, when compared with HPLC, the Rougier Bio-Tech kit has been reported to underestimate toxin levels in shellfish from certain regions by some authors (Carmody *et al.* 1995, Nunez and Scoging 1997). Vale and Sampayo (2000) reported for example that in Portugal, DTX2 appears only in some years, and is mainly restricted to late summer, although appearing in late spring in extremely toxic years. So, underestimation may not be a permanent drawback, however a monitorization relying solely in ELISA can not advert for the presence of DTX2 and DTX1, due to the unpredictability of the occurrence of their producer microalgae. "DSP-Check" has been found more adequate when compared with HPLC by many authors (Carmody *et al.* 1995, Vale and Sampayo 1999b). The Rougier Bio-Tech kit was well suited to study DSP in cultivated strains of the benthic *Prorocentrum* species that produce mainly OA and its precursors (Morton and Tindall 1996, Lawrence *et al.* 1998).

Another problem with the use of immunoassays for DSP is that acyl esters can be the main form of the toxins found in shellfish. These acylated forms have never been found in marine microalgae and recently have been proven to originate quickly in scallops when exposed to toxic dinoflagellates (Suzuki *et al.* 1999). They are the main diarrhetic toxins found in scallops in Japan (Suzuki *et al.* 1999, Suzuki and Mitsuya 2001), and have been held responsible for a human DSP outbreak in Portugal involving Donax clams (Vale and Sampayo 1999a). In Portugal, for instance, in most shellfish except mussels, acyl derivatives may be the main form of DSP toxins (Vale and Sampayo 1999a,b, unpublished data). A modified procedure that includes their alkaline hydrolysis is highly recommended for routine preparation of shellfish samples (Vale and Sampayo 1999b). It has the advantage of introducing a concentration step, which is useful for performing analysis in whole edible parts instead of removing laboriously digestive glands where toxins are more concentrated.

For extraction, one gram homogenated shellfish midgut is mixed with 5 times 90% aqueous methanol and blended in a centrifuge tube. An aliquot of the extract is used for the tests after filtration or centrifugation.

Tests for Paralytic Shellfish Poisoning

PSP has been studied scientifically since the beginning of the twentieth century. Initially it was thought to be composed of a single major poison, named saxitoxin after the butter clam *Saxidomus giganteus*. A quantitative bioassay calibrated with a purified solution of the *S. giganteus* poison was established, and adopted has an AOAC Official Method of Analysis (McFarren 1959). Later on it was discovered that *S. giganteus* presented a somewhat peculiarly simple toxin profile (due to its specific accumulation of saxitoxin in the siphon) in the emergent array of PSP compounds found in shellfish. Today at least 21 different toxins are known world-wide (Figure 3). This is a complicated picture for HPLC analysis, because all these toxins are not commercially available for calibration purposes. From the toxicological point of view, they are classified into three groups: the carbamate toxins (STX, NeoSTX, GTX2/3, GTX1/4)[4] are the most toxic; decarbamoyl toxins (dcSTX, dcGTX2/3) are moderately toxic, and N-sulfocarbamoyl toxins (C1-C4, B1, B2) are the least toxic (for a review of structures and potencies, see Oshima 1995). However, the last ones can be converted to the more toxic carbamate correspondents under extremely acidic conditions, such as stomach pH. On the other hand, from one analytical point of view, they can be grouped into two groups: non-N-1-hydroxyl containing such as STX, GTX2/3, dcSTX, dcGTX2/3, C1/2, B1; and N-1-hydroxyl containing such as NeoSTX, GTX1/4, C3/4, B2.

[4]STX = saxitoxin; GTX = gonyautoxin = decarbamoyl

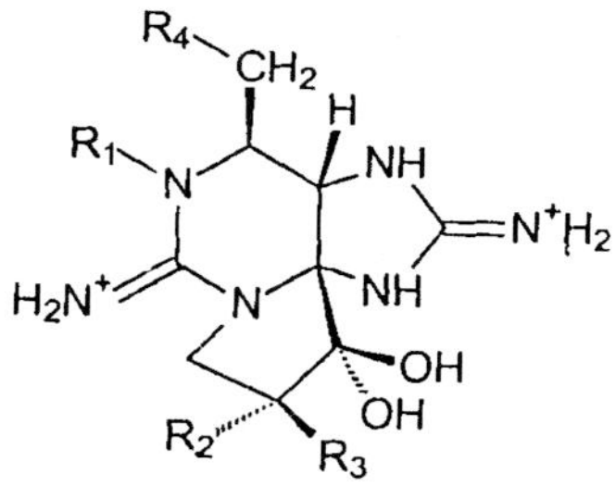

Figure 3. Structure of known PSP toxins. R_1 = H or OH; R_2 and R_3 = H or OSO_3^-; R_4 = $NH_2CO_2^-$ in carbamate toxins, $SO_3NHCO_2^-$ in N-sulfocarbamoyl toxins, HO^- in decarbamoyl toxins, H^- in deoxydecarbamoyl toxins. This totals 21 toxins.

An alternative method should classify a sample either below or above the MBA threshold of 80 µg STX equivalents/100 g tissue. This has been a strict observance world-wide due to the lethality of this poisoning. For a single antibody-based detection system the large number of toxins, their variable toxicity degrees and different cross-reactivities with each one complicate the task of mimicking MBA results.

One of the first commercial ELISA used immobilised saxitoxin fixed to polystyrene batons to competitively bind free STX-antibody from the toxic sample-antibody incubation mixture. In a second step, the batons are incubated in horseradish-peroxidase-conjugate, followed by a final development of a coloured reaction product in a cuvette containing the substrate solution and measurement of the optical density in a spectrophotometer. This was based on the policlonal antibody reported in Cembela *et al.* (1990). It cross-reacted with at least GTX2, GTX3 and NeoSTX, but there is no cross-reactivity to the low potency N-sulfocarbamoyl toxins. It was commercialised by Institut Armand Frapier (Canada), under the name SAXITOXIN TESTâ . Detailed collaborative studies were not completed and the kit is no longer produced commercially for shellfish toxin assays. This antibody was shown to be useful for phytoplankton samples from *Alexandrium* spp. (Cembella and Lamoreux 1993).

A policlonal rabbit antiserum against STX (Usleber *et al.* 1991) was used to develop another commercial ELISA: RIDASCREEN® PSP-test. It is produced and distributed by R-Biopharm GmbH[5] in the format of six strips with 8 wells each, accompanied with 5 saxitoxin standards: 10, 30, 90, 270 and 810 ppt. It detects STX in the lowest range of 0.2-16 µg/g, or higher ranges using greater sample dilutions. The wells in the microtiter strips are coated with anti-STX. By adding standards or sample solutions and enzyme

[5]Dolivostrasse, 10, D-64293 Darmstadt, Germany.

labelled STX (enzyme conjugate), free and enzyme labelled toxin compete directly for the antibody binding sites. Any unbound enzyme conjugate is washed away; enzyme substrate and chromogen are added and incubated. Bound enzyme conjugate converts the colourless chromogen into a blue product, and acid addition converts it into yellow. The optical density measured in a reader is inversely proportional to the STX concentration (Figure 2).

RIDASCREEN® PSP-test was subjected to a collaborative testing for saxitoxin along with several HPLC methods by the European Commission's Measurement and Testing Programme (BCR) (Van Egmond *et al.* 1994). It overestimated STX content due to its cross-reactivity towards other analogues present in multi-contaminated samples. It was found to be useful as a pre-screen in several studies (Usleber *et al.* 1997, O'Neill *et al.* 1998). This ELISA has a cross-reaction of 28% to dcSTX, 12% to GTX2/3, 18% to B1, 1% to C1/2 and to N-1-hydroxyl toxins of less than or equal to 3%. To overcome the lower detection of N-1-hydroxyl group (which must be reminded includes toxins equally potent) several authors have tried a combination of two immunoassays: one with antibodies against STX and another with antibodies toward NeoSTX (Chu *et al.* 1996), although the last ones are not commercial to the best of our knowledge.

A much faster approach, under the format of lateral flow immunochromatography (LFI) test strips, was recently launched on the market by Jellett Biotek[6], under the designation of "MIST Alert™ for PSP". It provides a qualitative (positive/negative) indication of the presence of PSP, similar to home pregnancy test kits. It is designed as a screening method for regulatory laboratories to eliminate negative samples, thereby leaving a smaller number of positive samples to be tested with more sophisticated and time-consuming quantitative methods. The toxin analogues STX, GTX2/3 and C1/2 epimeric mixtures, B1 and dcSTX were detectable at concentrations of approximately 200 nM. The test was somewhat less sensitive to the N1-hydroxy derivatives NeoSTX and GTX1/4, requiring concentrations of 400 and 600 nM, respectively, to give a substantially positive test (Laycock *et al.* 2000). To possess such a broad spectrum of selectivity it uses a combination of 8 antibodies against different PSP toxins (Joanne Jellett, personal communication).

Each test strip is mounted in a cassette and one drop of extract diluted with a buffer solution is added to the sample well (Figure 4). After a few minutes, persistence of a strong T line indicates that any PSP toxins in the sample are at concentrations less than the regulatory limit. Toxin concentrations at or above the equivalent of 80 μg STX per 100 g tissue (1075 nM for STX) in an extract eliminate the toxin T-line and further testing

---

[6]101 Research Dr., PO Box 790 Dartmouth, NS B2Y 3Z7, Canada.

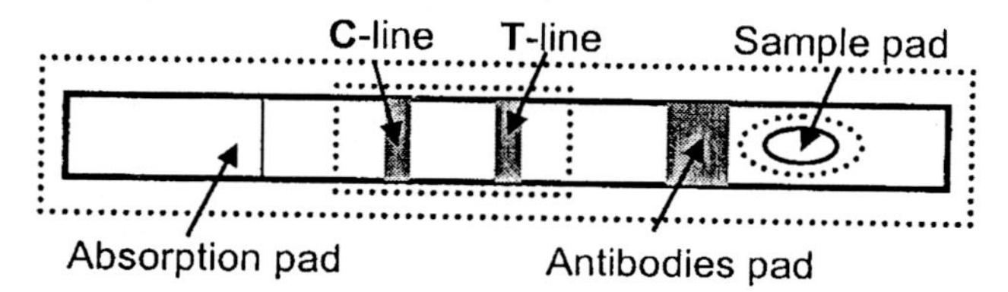

Figure 4. Diagram of MIST Alert™ test strip inside plastic cassette. The visible T-line indicates absence of toxins. The C-line visible indicates that the sample fluid has sufficiently resuspended and mobilised the antibody colour complex.

should be done on those samples. Because of the differing affinities of the antibody mixture for the individual analogues it is not possible to provide a definitive detection limit for the mixed toxin profiles found in shellfish tissues. Depending on the average toxin profile in any geographic area, the detection limit of the test units produced with the current antibody mixture will be in the range of 100-800 nM which corresponds to 7-60 µg STXeq per 100 g tissue (Laycock *et al.* 2000).

Due to its intermediate sensitivities to N-sulfocarbamoyl toxins it is prone to generate 'false positive' results in samples poorly toxic by MBA that contain a significant proportion of these toxins, as can be the case with shellfish from areas contaminated by *Gymnodinium catenatum*. But with such lethal poisoning, care must always be taken, and to avoid economical losses to shellfish industry confirmation should be performed in positive samples. Jellett Biotek can produce tests adapted for a given toxin profile, so we can expect in the future the distribution of tests adequate for each of the major producer microalgae. It must be reminded that confirmation of profile and identification of causative microalgae should be done regularly by aquaculturists due to the risk of introduction of new toxic species or changes in the ecosystem that favour the development of other toxic algae.

For all the immunoassays above, a universal chloridric acid boiling extraction in use for the MBA (AOAC 1990) can be employed. Alternatively, Jellett Biotek proposes a faster field extraction for use with the MIST™ Cell Bioassay and the MIST Alert™. Briefly, 10 ml of blended tissue are poured onto a graduated centrifuge tube, mixed with 10 ml HCl, capped and cooked in boiling water. An aliquot of the extract is used for the tests after decanting or centrifuging.

Tests for Amnesic Shellfish Poisoning

ASP is one of the most recently discovered shellfish poisonings and is also the one that presents fewer technical problems for liquid chromatography analysis. Although several isomers of the toxin domoic acid (DA) (Figure 5)

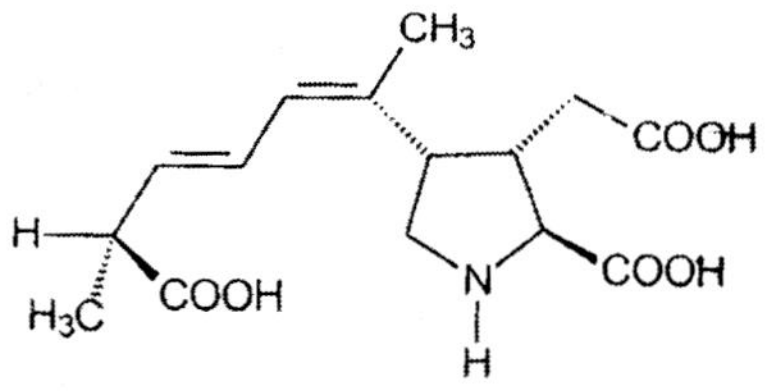

Figure 5. Structure of main ASP toxin: domoic acid.

have been reported, they are always in low concentration, and can be neglected for regulation purposes. On the other side, rodents commonly used for detection of all other marine toxins are not sensitive enough to the level found safe for humans after the studies on the first known mass human intoxication case that took place in late 1987 in Canada. The mandatory use of HPLC (there exists one AOAC method: AOAC 1991) to determine if shellfish meat is under 20 µg DA/g edible tissue poses a problem for countries relying in inexpensive MBA for regulatory purposes. Although no human intoxications have been reported since the 1987 incident, domoic acid has been found worldwide and included in monitoring programmes of many countries, and so it is a surveillance that is becoming mandatory for international trade of shellfish.

A fast yes/no test, similar to the one described above for PSP, has been developed by Jellett Biotek, under the designation "MIST Alert^{TM} for ASP", and is in the process of validation. The test can be performed with two different extractions. One is the boiling chloridric acid extraction, traditionally used for extracting the watersoluble PSP toxins (AOAC 1990, 1991; see section above for simpler extraction proposed by Jellett Biotek)[7]. Another is a methanolic extraction used for HPLC (Quilliam 1995), allowing analysis to be performed safely with toxins kept in methanol in the refrigerator or freezed. Different running buffers are needed depending on the extract used. The detection limit with the acid extract is about 1 µg DA/g edible tissue, and with the methanol extract is 2 µg DA/g (Joanne Jellett, personal communication).

An indirect competitive ELISA assay has been developed at AgResearch Ruakura[8]. It uses ovine antibodies and ester-derived conjugates of domoic acid as the microplate coater. The working range is 0.15-15 ng DA/ml, being able to detect 0.04 µg DA/g at a 250-fold dilution (Garthwaite *et al.* 1998). It is adequate to test for the presence of DA both in shellfish and phytoplankton samples, thus being able to provide early warning of developing algal

[7] It must be reminded that with this extraction, analysis of domoic acid must be carried out immediately, as the extract should not be kept refrigerated to avoid toxin losses. Only saxitoxin analogues are stable in this acid extract, not domoic acid!

[8] Private bag 3123, Hamilton, New Zealand.

blooms. After formal validation this test might be available from AgResearch in the near future.

Tests for Ciguatera Fish Poisoning

Unlike shellfish that possess negligible mobility and thus can be easily surveyed for microalgae contamination over a relatively large area, fish are not so easily traceable to their feeding places. Surveying a few individuals does not give us an idea of the contamination of all tropical reef fishes caught in one particular zone, especially because large carnivorous fish may travel significantly. The geographic distribution of toxic fish is very patchy. It is not uncommon for fish on one side of an island to be poisonous while the same species on the other side of that island is safe to eat.

For CFP, in particular when dealing with carnivorous animals, ideally every single individual should be tested for contamination before consumption. For precaution, in some endemic areas of CFP sale of large sized specimens of carnivorous fishes is avoided or ingestion of only small portions is recommended. This is the seafood poisoning case were dissemination of a rapid 'yes/no' answer field test is the best solution for local populations being able to benefit safely from all the wealth of protein supplied by fish. But ciguatoxins are numerous, in part due to metabolisation along the food chain. Their oxidation increases toxicity contributing to potenciate the effective toxic load accumulated by fish. Among them, P-CTX1 (Figure 6) appears to be the most potent, and typically contributes to 70–90% of the toxicity of carnivorous fish from the Pacific. Another family of toxins, closely related to ciguatoxins, has recently been characterised from ciguateric fish of the Caribbean Sea. For a review of structures and potencies, see Lewis (2001).

Since late 1970's the group of Hokama in Hawaii has been developing several immunoassays for detection of ciguatoxin purified from moray eel,

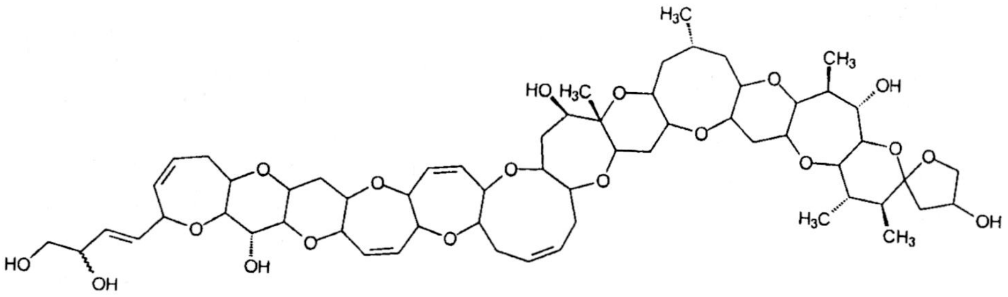

H_3C CH_3 HO H_3C OH O O O O O O O O O O O O O O H_3C OH CH_3 HO OH OH

Figure 6. Structure of CTX-1, one member of the Pacific ciguatoxins group.

starting with radio-immunoassays, followed by a first EIA also as time-consuming as the RIA (revised in Hokama *et al.* 1998). To simplify, a stick enzyme immunoassay (S-EIA) started with policlonal antibodies and later with monoclonal antibodies (MAb-CTX). The S-EIA used skewered bamboo coated with a white correction fluid (for adhesion of ciguatoxin) and MAb-CTX conjugated with an enzyme and respective substrate (Hokama *et al.* 1998a). No full collaborative study was performed with this test because of lack of a chemically identifiable standard (Park 1995).

The assay format was later modified to a solid-phase immunobead assay (S-PIA). The S-PIA used a bamboo paddle coated with a correction fluid and MAb-CTX bonded to coloured latex beads to detect ciguatoxin (Hokama *et al.* 1998a). The presence or absence of toxins was determined by binding the toxins to the correction fluid, and exposing the toxin-ladened paddle to the antibody-coloured-bead complex that has high specificity for the toxins. Hawaii Chemtect commercialised this test under the designation Ciguatectô , but presently it is no longer available.

Recently a membrane immunobead assay (MIA) was developed in which the correction fluid was replaced by a hydrophobic synthetic membrane laminated onto a solid plastic stick. The presence or absence of toxins is determined by binding the toxins to the membrane, and exposing the toxin-ladened membrane to an antibody-coloured-bead complex that has high specificity for the toxins (Hokama *et al.* 1998a). Presently, this test is produced by Oceanit Test Systems, Inc.[9] under the designation Cigua-Check.ô It is the only commercially available test capable of identifying the presence of ciguatoxin in fish flesh.

For field analysis simply a grain-sized fish muscle is removed with tweezers and put soaking in a vial with methanol together with a Test Stick for 20 minutes (Figure 7). The Test Stick is air dried and then placed in another vial with the reactives. After reacting for 10 minutes, it is washed, and colour compared with the colour strip provided. A purple colour on the Test Stick indicates the meat contains ciguatera poison and should not be eaten. The darker the colour the more poison the fish meat contains. Fish flesh measuring above 1 ppb on the toxin colour scale should not be eaten. A positive and a negative control test are provided to check the sensitivity of the kit upon storing. Currently it has a six-month shelf life.

All immunoassays developed by Hokama's group showed good detection of clinically implicated ciguateric fish. The sensitivities towards toxic samples of older procedures (RIA, S-EIA, S-PIA) ranged from 95.6 to 97.2% and the new MIA presented 92.3%. On the other hand, the specificities for reef fish of unknown toxicity of S-EIA and S-PIA are close to 50%, while that of MIA is 85.7%. This low specificity is due to non-specific binding of

---

[9]1100 Alakea Street, 31st Floor Honolulu, Hawaii 96813-2833, USA.

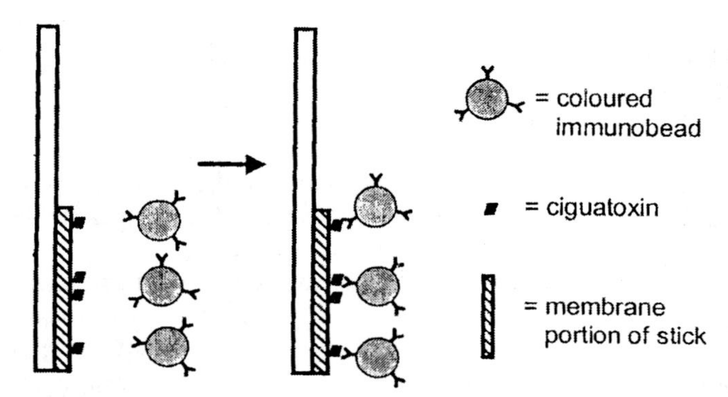

Figure 7. Concept of the membrane immunobead assay (MIA) procedure: monoclonal antibody against ciguatoxin attached to coloured latex particles, bonds to ciguatoxin attached to a synthetic membrane (coated on a plastic stick). Positive samples will show a visible colour, while negative ones will not.

the antibody conjugate and antibody bead to the correction fluid used. The MIA blanks showed much less nonspecific binding (Hokama *et al.* 1998b). Antibodies against CTX1 have affinity for the low potency analogues and also other compounds normally present in fish flesh. The MAb-CTX can also detect okadaic acid, and Ciguatectô was the subject of a collaborative assay using fish flesh spiked with okadaic acid (Park 1995). The SPIA has been useful for recreational fisherman in Hawaii (Hokama *et al.* 1998b).

Tests for Neurotoxic Shellfish Poisoning

NSP toxicity is mainly restricted to Gulf of Mexico and New Zealand shellfish. The polyether brevetoxins are grouped into two groups: brevetoxin PbTx-1 analogues contain a backbone of 10 fused polyether rings and brevetoxin PbTx-2 analogues contain 11 fused rings (Figure 8). Other new brevetoxins have been discovered in New Zealand shellfish.

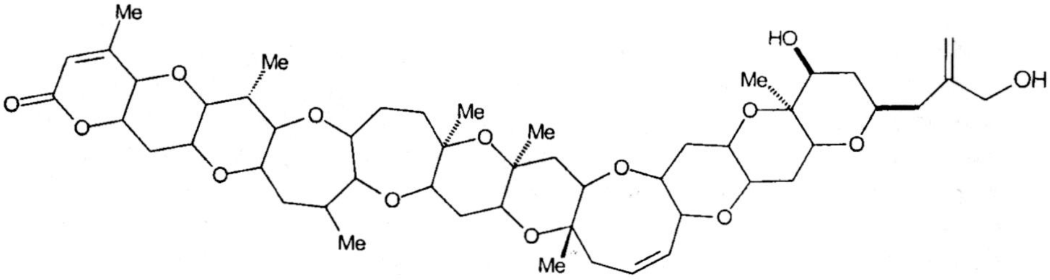

Figure 8. Structure of PbTx-3, from the brevetoxins PbTx-2 group.

Chiral Corp (Miami, USA) commercialised the first antibodies and brevetoxin standards for performing radioimmunoassay and ELISA. Presently, this company is no longer operating but the same service is provided by the HABLAB (Harmful Algal Bloom Laboratories for Analytical Biotechnology) at the University of North Carolina at Wilmington[10]. A brevetoxin-ELISA kit as been developed and is available from the same laboratory. It contains a PbTx-2 standard and sufficient reagents to perform fifteen 96-well ELISA plates. This test kit is more laboratory-oriented than most of the previous ones discussed above, but provides a totally quantitative result.

Briefly, an ELISA plate is first sensitized with brevetoxins. Then, shellfish homogenate dilutions are incubated with anti-PbTx antibodies. In case of toxic homogenate, the antibodies react with the toxins and do not bind to sensitised ELISA plate. Subsequently, the antibodies that are not attached to the plate are washed-out during the rinses. In case of non toxic homogenate, the antibodies react with brevetoxin on the sensitised ELISA plate and are immobilised on the plate. Their presence is revealed with a biotinilated secondary antibody. In this competitive assay, colour is indirectly proportional to toxin concentration. The limit of detection is around 0.3 ng/ml.

With this assay, the authors were able to measure brevetoxin in seawater, mammalian body fluids (serum, urine), shellfish extracts and shellfish homogenates without extraction (Naar *et al.* 2000, Naar *et al.*, submitted). The homogenate protocol reduces test time by 12 hours, compared with the mouse bioassay, and increases sensitivity and specificity. The antibodies used were produced by the protocols of Trainer and Baden (1992). They are specific for the last 4 rings of the PbTx type-2 backbone excluding the side-chain and have the identical affinities within the PbTx type-2 (PbTx-2; -3; -9 and -6) class.

Summary

Shellfish and tropical fish from certain areas are responsible for several human poisonings of non-bacterial origin, such as paralytic shellfish poisoning (PSP), diarrhetic shellfish poisoning (DSP), neurotoxic shellfish poisoning (NSP), amnesic shellfish poisoning (ASP) and ciguatera fish poisoning (CFP). These toxins are naturally produced by microalgae. Traditionally these contaminations have been monitored through mouse bioassay, which presents several disadvantages. The development of

---

[10]HABLAB, Center for Marine Science, One Marvin K. Moss Lane, Wilmington, North Carolina 28429, USA.

commercial immunoassays offers user-friendly tests with robustness adequate for fieldwork, high speed of analysis and high sensitivity and selectivity. They can be used as a preliminary screening tool for toxins on shipboard, dockside, aquaculture facilities, canning industry and local monitoring laboratories, as only rudimentary extraction and test performance are needed. In some immunoassays a simple yes/no answer can be readily obtained without any specialised equipment by simple eye reading. Immunoassays can avoid waste of raw materials and save money, by preventing further harvesting or industrial processing of contaminated seafood before tests are performed at official laboratories. The difficulties arrived in development and applications of each of these tests to monitoring situations are discussed in detail.

Acknowledgements

We are grateful for the help provided by several colleagues that supplied us updated information on the subject: Joanne Jellett, Jerome Naar, Yoshitsugi Hokama, Richard Lewis, Neale Towers.

References

AOAC. 1990. Paralytic shellfish poison, biological method, final action. Method nº 959.08 in *Official Methods of Analysis.* AOAC, ed. 15th Ed., Arlington, VA.

AOAC. 1991. Domoic acid in mussels, liquid chromatographic method, first action. Method nº 991.26 in *Official Methods of Analysis.* AOAC, ed. 15th Ed., 2nd supplement, AOAC, Arlington, VA.

Aune, T., H. Ramstad, B. Heidenreich, T. Landsverk, T. Waaler, E. Egaas, and K. Julshamn. 1998. Zinc accumulation in oysters giving mouse deaths in paralytic shellfish poisoning bioassay. *J. Shellfish Res.* 17: 1243-1246.

Cacho., E. 1993. Interferencias de ciertos metales pesados en el bioensayo PSP. Pages 79-82 in *III Reunion Ibérica Sobre Fitoplânton Tóxico y Biotoxinas.* J. Mariño and J. Manero, eds. Consejería de Agricultura y Pesca, Xunta de Galicia.

Carmody, E.P., K.J. James, and S.S. Kelly. 1995. Diarrhetic shellfish poisoning: evaluation of ELISA methods for determination of dinophysistoxin-2. *J. AOAC Internat.* 78: 1403-1408.

Cembella, A.D., Y. Parent, D. Jones, and G. Lamoureux. 1990. Specificity and cross-reactivity of an absorption inhibition enzyme-linked immunoassay for the detection of paralytic shellfish toxins. Pages 339-344 in *Toxic Marine Phytoplankton.* E. Graneli, B. Sundstrom, L. Edler and D.M. Anderson, eds. Elsevier, New York.

Cembella, A.D., and G. Lamoureux. 1993. A competitive inhibition enzyme-linked immunoassay for the detection of paralytic shellfish toxins in marine phytoplankton. Pages 857-862 in *Toxic Phytoplankton Blooms in the Sea.* T. Smayda and Y. Shimizu, eds. Elsevier, New York.

Cembella, A.D., L. Milenkovic, G. Doucette, and M.L. Fernandez. 1995. *In vitro* biochemical methods and mammalian bioassays for phycotoxins. Pages 177-211 in *Manual on Harmful Marine Microalgae.* G.M. Hallegraeff, D.M. Anderson and A.D. Cembella, eds. *IOC Manuals and Guides nº 33* — UNESCO, Paris, France.

Chin, J.D., M.A. Quilliam, J.M. Fremy, S.K. Mohapatra, and H.M. Sikorska. 1995. Screening for okadaic acid by immunoassay. *J. AOAC Internat.* 78: 508-513.

Chu, F.S., K.H. Hsu, X. Huang, R. Barrett, and C. Allison. 1996. Screening of paralytic shellfish poisoning toxins in naturally occurring samples with three different direct competitive enzyme-linked immunosorbent assays. *J. Agric. Food Chem.* 44: 4043-4047.

Draisci, R., L. Lucentini, L. Giannetti, P. Boria, K.J. James, A. Furey, M. Gillman, and S.S. Kelly. 1998. Determination of diarrhetic toxins in mussels by microliquid chromatography-tandem mass spectrometry. *J. AOAC Internat.* 81: 441-447.

Fremy, J.M., D.L., Park, E., Gleizes, S.K., Mohapatra, C.H., Goldsmith, H.M. Sikorska. 1994. Application of immunochemical methods for the detection of okadaic acid in mussels. *J. Nat. Toxins.* 3: 95-105.

Gago-Martinez, A., J.A. Rodriguez-Vazquez, P. Thibault, and M.A. Quilliam. 1996. Simultaneous occurrence of diarrhetic and paralytic shellfish poisoning toxins in Spanish mussels in 1993. *Nat. Toxins.* 4: 72-79.

Garthwaite, I., K.M. Ross, C.O. Miles, R.P. Hansen, D. Foster, A.L. Wilkins, and N.R. Towers. 1998. Polyclonal antibodies to domoic acid, and their use in immunoassays for domoic acid in sea water and shellfish. *Nat. Toxins* 6: 93-104.

Hokama, Y., K. Nishimura, W. Takenaka, and J.S.M. Ebesu. 1998a. Simplified solid-phase membrane immunobead assay (MIA) with monoclonal anti-ciguatoxin antibody (MAb-CTX) for detection of ciguatoxin and related polyether toxins. *J. Nat. Toxins* 7: 1-21.

Hokama, Y., W.E. Takenaka, K.L. Nishimura, J.S.M. Ebesu, R. Bourke, and P.K. Sullivan. 1998b. A simple membrane immunobead assay for detecting ciguatoxin and related polyethers from human ciguatera intoxication and natural reef fishes. *J. AOAC Internat.* 81: 727-735.

Lawrence, J.E., A.D. Cembella, N.W. Ross, and J.L.C. Wright. 1998. Cross reactivity of an anti-okadaic acid antibody to dinophysistoxin-4 (DTX-4), dinophysistoxin-5 (DTX-5), and an okadaic acid diol ester. *Toxicon* 36: 1193-1196.

Laycock, M., J.F. Jellett, E.B. Belland, P.C. Bishop, B.L. Thériault, A.L. Russell-Tattrie, M.A. Quilliam, A.D. Cembella, and R.C. Richards. 2000. MIST alert™ a rapid assay for paralytic shellfish poisoning toxins. In: Proceedings IX International Conference on Harmful Algal Blooms, Hobart, Australia, 7–11/February/2000.

Lee, J.S., K. Tangen, E. Dahl, P. Hovgaard, and T. Yasumoto. 1988. Diarrhetic shellfish toxins in Norwegian mussels. *Nippon Suisan Gakkaishi* 54: 1953-1957.

Lee, J.S., T. Yanagi, R. Kenma, and T. Yasumoto. 1987. Fluorometric determination of diarrhetic shellfish toxins by high performance liquid chromatography. *Agric. Biol. Chem.* 51: 877-881.

Lewis, R.J. 2001. The changing face of ciguatera. *Toxicon* 39: 97-106.

McCulloch, A.W., R.K. Boyd, AS.W. de Freitas, R.A. Foxall, W.D. Jamieson, M.V. Laycock, M.A. Quilliam, J.L.C. Wright, V.J. Boyko, J.W. McLaren, M.R. Miedema, R. Pocklington, E. Arsenault, and D.J.A. Richard. 1989. Zinc from oyster tissue as causative factor in mouse deaths in Official Bioassay for Paralytic Shellfish Poison. *J. Assoc. Offic. Anal. Chem.* 72 : 384-386.

McFarren, E.F. 1959. Report on collaborative studies of the bioassay for paralytic shellfish poison. *J. Assoc. Offic. Anal. Chem.* 42: 263-271.

Morton, S. and D. Tindall. 1996. Determination of okadaic acid content of dinoflagellate cells: a comparison of the HPLC-fluorescent method and two monoclonal antibody ELISA test kits. *Toxicon* 34: 947-954.

Naar, J., A. Bourdelais, C. Tomas, J. Lancaster, and D.G. Baden. A competitive ELISA to detect brevetoxin from *Gymnodinium breve* in seawater, shellfish, and mammalian body fluid. Submitted to *Environmental Health Perspective.*

Naar, J., A. Bourdelais, C. Tomas, J. Lancaster, and D.G. Baden. 2000. Brevetoxin analysis in seawater, mammalian body-fluid and shellfish homogenate using the brevetoxin-ELISA test Kit. Paper presented at the 114th meeting of AOAC International, Philadelphia, PN, 12 September 2000.

Neill, S., S. Gallacher, and I. Riddoch. 1998. Assessment of a saxitoxin ELISA as a pre-screen for PSP toxins in *Mytilus edulis* and *Pecten maximus*, from UK waters. Pages 551-553 in *Harmful Algae.* B. Reguera, J. Blanco, M.L. Fernández, and T. Wyatt, eds. Xunta de Galicia and IOC of UNESCO, Spain.

Núñez, P.E. and A.C. Scoging. 1997. Comparison of a protein phosphatase inhibition assay, HPLC assay and enzyme-linked immunosorbent assay with the mouse bioassay for the detection of diarrhetic shellfish poisoning toxins in European shellfish. *Int. J. Food Microb.* 36: 39-48.

Oshima, Y. 1995. Postcolumn derivatization liquid chromatographic method for paralytic shellfish toxins. *J. AOAC Internat.* 78: 528-532.

Park, D.L. 1995. Detection of ciguatera and diarrhetic shellfish toxins in finfish and shellfish with Ciguatect kit. *J. AOAC Internat.* 78: 533-537.

Quilliam, M.A., M. Xie, and W.R. Hardstaff. 1995. Rapid extraction and cleanup for liquid chromatographic determination of domoic acid in unsalted seafood. *J. AOAC Internat.* 78: 543-554.

Schantz, E.J., E.F., McFarren, M.L. Schaffer, and K.H., Lewis. 1958. Purified shellfish poison for bioassay standardization. *J. Assoc. Offic. Anal. Chem.* 41: 160-168.

Shestowsky, W.S., M.A. Quilliam, and H.M. Sikorska. 1992. An idiotypic-anti-idiotypic competitive immunoassay for quantitation of okadaic acid. *Toxicon* 30: 1441-1448.

Suzuki, T., and T. Mitsuya. 2001. Comparison of dinophysistoxin-1 and esterified dinophysistoxin-1 (dinophysistoxin-3) contents in the scallop *Patinopecten yessoensis* and the mussel *Mytilus galloprovincialis*. *Toxicon* 39: 905-908.

Suzuki, T., H. Ota, and M. Yamasaki, 1999. Direct evidence of transformation of dinophysistoxin-1 to 7-O-acyl-dinophysistoxin-1 (dinophysistoxin-3) in the scallop *Pactinopecten yessoensis*. *Toxicon* 37: 187-198.

Trainer, V.L. and D.G. Baden. 1991. An enzyme immunoassay for the detection of Florida red tide brevetoxins. *Toxicon* 29: 1387-1394.

Usagawa, T., M. Nishimura, Y. Itoh, T. Uda, and T. Yasumoto. 1989. Preparation of monoclonal antibodies against okadaic acid prepared from the sponge *Halichondria okadai*. *Toxicon* 27: 1323-1330.

Usleber, E., E. Schneider, and G. Terplan. 1991. Direct enzyme immunoassay in microtitration plate and test strip format for the detection of saxitoxin in shellfish. *Lett. Appl. Microbiol.* 13: 275-277.

Usleber, E., D. Donald, M. Straka, and E. Martlbauer. 1997. Comparison of enzyme immunoassay and mouse bioassay for determining paralytic shellfish poisoning toxins in shellfish. *Food Addit. Contam.* 14: 193-198.

Vale, P. and M.A.M. Sampayo. 1999a. Esters of okadaic acid and dinophysistoxin-2 in Portuguese bivalves related to human poisonings. *Toxicon* 37: 1109-1121.

Vale, P. and M.A.M. Sampayo. 1999b. Comparison between HPLC and a commercial immunoassay kit for detection of okadaic acid and esters in Portuguese bivalves. *Toxicon* 37: 1569-1581.

Vale, P. and M.A.M. Sampayo. 2000. Dinophysistoxin-2: a rare diarrhetic toxin associated with *Dinophysis acuta*. *Toxicon* 38: 1599-1606.

Vale, P. and M.A.M. Sampayo. 2001. Determination of paralytic shellfish toxins in Portuguese shellfish by automated pre-column oxidation. *Toxicon* 39: 623-633.

van Egmond, H.P., H.J. van den Top, W.E. Paulsch, X. Goenaga, and M.R. Vieytes. 1994. Paralytic shellfish poison reference materials: an intercomparison of methods for the determination of saxitoxin. *Food Addit. Contam.* 11: 39-56.

Highly Sensitive Bioassays for Paralytic Shellfish Toxins

Franca Guerrini[1], *Clementina Bianchi*[2], *Lorenzo Beani*[2], *Laurita Boni*[1] *and Rossella Pistocchi*[1]

[1]Centro Interdipartimentale di Ricerca per le Scienze Ambientali, Università di Bologna, Via Tombesi dall'Ova 55, 48100 Ravenna, Italy
[2]Dipartimento di Medicina Clinica Sperimentale: Sezione di Farmacologia, Università di Ferrara, Via Fossato di Mortara 17-19, 44100 Ferrara, Italy

Introduction

Paralytic shellfish poisoning (PSP) is the most widespread shellfish poisoning in the world. It is caused by toxins produced by a number of genera of gonyaulacoid or gymnodinioid dinoflagellates, including *Alexandrium, Gymnodinium* and *Pyrodinium*, or by several freshwater cyanobacteria. The toxins responsible for this syndrome consist of a group of heterocyclic guanidines called saxitoxins of which more than twenty analogues have been reported to occur naturally. In 1975 Schantz *et al.* described the crystal structure of the parent compound called saxitoxin (STX); the various analogue compounds, successively identified, differed by the substitution of the hydroxyl and sulfate groups present in four sites of the molecule (R1-R4) (Figure 1). Based on the substitution in R4, the saxitoxins can be subdivided into four groups: the carbamate toxins, sulfo-carbamoyl toxins, decarbamoyl toxins and deoxydecarbamoyl toxins, which display substantial changes in relative toxicity. The toxins belonging to the carbamate group are STX, neo-saxitoxin (neo-STX) and gonyautoxins 1,2,3,4 (GTX1-4) which are those displaying the highest level of toxicity, measured as mouse units (MU) by the mouse bioassay (AOAC 1990) as described below. The sulfo-carbamoyl toxins are weakly toxic while, among the decarbamoyl toxins, there are compounds with intermediate levels of toxicity.

The biochemical action of these toxins depends on a high affinity binding to the voltage-gated sodium channel present on the plasma membrane of many excitable cells, resulting in the inhibition of channel opening. The

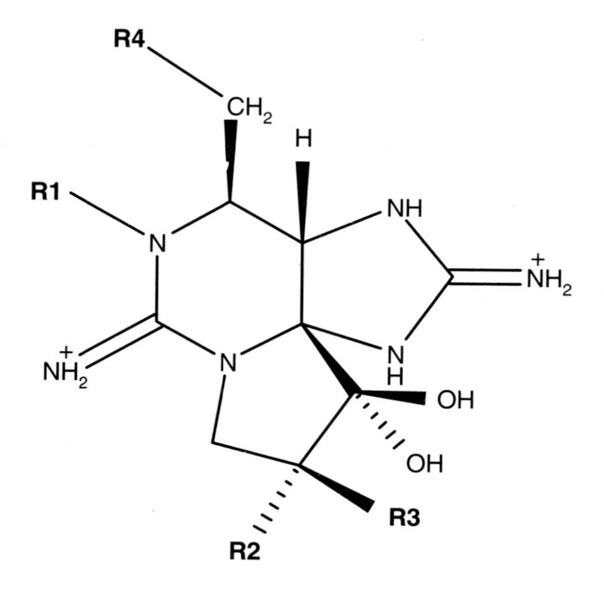

		R1	R2	R3	R4
Carbamate	**STX**	H	H	H	$OCONH_2$
	NeoSTX	OH	H	H	$OCONH_2$
	GTX1	OH	OSO_3^-	H	$OCONH_2$
	GTX2	H	OSO_3^-	H	$OCONH_2$
	GTX3	H	H	OSO_3^-	$OCONH_2$
	GTX4	OH	H	OSO_3^-	$OCONH_2$
Sulfocarbamoyl	**GTX5**	H	H	H	$OCONHSO_3^-$
	GTX6	OH	H	H	$OCONHSO_3^-$
	C1	H	OSO_3^-	H	$OCONHSO_3^-$
	C2	H	H	OSO_3^-	$OCONHSO_3^-$
	C3	OH	OSO_3^-	H	$OCONHSO_3^-$
	C4	OH	H	OSO_3^-	$OCONHSO_3^-$
Decarbamoyl	**dcSTX**	H	H	H	OH
	dcNeoSTX	OH	H	H	OH
	dcGTX1	OH	OSO_3^-	H	OH
	dcGTX2	H	OSO_3^-	H	OH
	dcGTX3	H	H	OSO_3^-	OH
	dcGTX4	OH	H	OSO_3^-	OH
Deoxydecarbamoyl	**doSTX**	H	H	H	H
	doGTX2	H	H	OSO_3^-	H
	doGTX3	H	OSO_3^-	H	H

Figure 1. Structures of paralytic shellfish toxins.

affinity of the different saxitoxin analogues to the binding site varies proportionally with their toxicity in mice (Doucette *et al.* 1997).

The official method adopted by various countries to detect the presence and to assess the levels of PSP toxins in shellfish extracts consists of the mouse bioassay (MBA), first applied by Sommer and Meyer (1937) and subsequently standardized and validated through the collaboration of the Association of Official Analytical Chemists (AOAC 1990). This assay consists of the intraperitoneal administration of aqueous shellfish extracts to standardized mice, after which the time of death is recorded and the toxicity is expressed in MU from Sommer's table. One mouse unit is represented by the amount of toxin required to kill a mouse of 20 g in 15 min and it can be converted in μg STX equivalents by calculations obtained through standardization of the method in which known amounts of saxitoxin are employed. The tolerance level of PSP toxins in seafood varies among countries but it usually has a value of 40 or 80 μg STX equivalents per 100 g shellfish tissue.

Although largely employed this method poses serious problems in addition to the ethical implications of the killing of animals. The main problems consist of the lack of specificity with respect to the various toxins, in the low sensitivity and in the high variability of the responses due to different animal strains along with the occurrence of false positives. Alternative methods for PSP are therefore needed and various approaches have already been developed by many researchers, although until now none has been adopted as a regulatory test.

Very reliable methods with high sensitivity are instrumental methods, such as HPLC (Lawrence *et al.* 1995, Oshima 1995), liquid chromatography-mass spectrometry (Quilliam 1998) and capillary electrophoresis (Thibault *et al.* 1991). These methods have the ability to identify already known toxins involved in each biointoxication, but the former needs several and not easy to find standards, while the latter requires expensive and complex instrumentation. The various analytical methods will not be considered in this review which will focus mainly on methods based on the immunochemical properties or on the functional effects of the toxins.

Immunochemical Methods

During the last years several immunochemical techniques have been developed for PSP toxins; some were adapted to be used for rapid screening as commercially available kits, but none has yet been employed as a regulatory test.

The sensitivity of immunodiagnostic tests is typically orders of magnitude greater than that of the mouse bioassay but there are many problems using these methods, often derived from lack of purified toxins for conjugation and for the need of producing stable immunogens from relatively low molecular weight toxins, such as STX and analogues. Cross-reactivity is also very important in the development of immunological methods because toxigenicity is due to several chemically related substances. Immunoassays have been prepared from both monoclonal and polyclonal antibodies; in general, polyclonal antibodies have higher affinity for multiple epitopic sites and have greater cross-reactivity with STX analogues than monoclonal antibodies. Detection systems for immunoassays commonly make use of a radio-label assay (RIA), a coupled enzyme reaction (EIA), a fluorescent marker (FIA) or an enzyme-linked immunosorbent assay (ELISA).

A first ELISA test for PSP detection (Cembella *et al.* 1990) configured as a rapid diagnostic immunoassay was the SAXITOXIN TEST (Institute Armand-Frapper); it was based on a polyclonal anti-STX antibody, obtained by covalent linkage of the STX molecule to the synthetic carrier polypeptide polyalanine-lysine. STX molecules were immobilized to polystyrene supports and these could bind free antibodies from an incubation mixture which contained the toxic sample and the antibodies. The toxin concentration was indirectly detected by a coupled enzyme reaction and the development of a colored reaction product. This test is no longer produced commercially for PSP assay because detailed collaborative studies were not completed.

Another competitive direct enzyme immunoassay was developed by Usleber *et al.* (1991), employing polyclonal antibodies against STX and a STX-horseradish-peroxidase conjugate. This test could be configured as a microplate titration and a strip assay and was converted into a test kit (RIDASCREEN, R-Biopharm) after an intercomparison with different methods made by the European Commission's Measurements and Testing Programme (van Egmond *et al.* 1994).

Further research studies were done in order to make the method simpler and rapid; modifications were the use of a membrane filter (Usleber *et al.* 1995) and the replacement of the hapten-STX conjugate with free STX, used to cover the microtitre plates (Kralovec *et al.* 1996).

The commercial ELISA test kit was found to be more than 150 times more sensitive than the mouse bioassay with a detection limit for STX of 2 µg per kg shellfish tissue. This assay was able to detect almost all the PSP toxins known at levels lower than that of the mouse bioassay; however molecules, such as neosaxitoxin and GTX 1,4, which had a poor cross-reactivity with anti-STX antibodies, gave discrepancies in comparison with the mouse bioassay for mussel extracts containing a mixture of derivatives (Usleber *et al.* 1997). Another study (O'Neall *et al.* 1998) compared the toxicity detected in mussels from the UK by the two methods and, although similar problems of cross-reactivity were found, no underestimation by the ELISA test was obtained for samples with a toxicity value above the regulatory limit of 80 µg STX equivalent/100 g shellfish tissue. The ELISA gave a good agreement with the mouse test especially with samples <40 µg STX equivalent/100 g; it was therefore suggested that it could be used as a pre-screen in order to detect negative samples resulting in a 65% reduction of animals used in the monitoring program in the UK.

Functional Methods

The methods included in this group are based on the highly specific interaction of the toxin with its receptor or on the mechanism of action on target cells. Two of these methods were already developed as test kits and are commercially available; in addition, we describe here a number of different assays which were developed more recently and that, after some refinements, could become useful in toxin control.

Binding Assays

The receptor binding assay was originally developed to characterize the ligand-receptor interaction (Krueger *et al.* 1979), and was later applied to the detection of PSP and other phycotoxins.

The receptor of STX and its congeners has been identified as the site 1 of the a-subunit of the voltage-dependent sodium channel (Catterall 1986). Binding of the various saxitoxins to the receptor is based on their functional activity and the response of this assay provides an estimate of the integrated toxic potency, since each toxin derivative present in the sample is bound by the receptor with affinity proportional to its intrinsic toxic potency. This method is based on the use of rat brain synaptosome membrane fractions containing receptor sites. The assay consists of a competition between [^{3}H]STX and the unlabeled STX standard or the toxic sample for a given number of available receptor sites in the synaptosome preparation; then the unbound toxin is removed by vacuum filtration and washing. This method was first adopted by Davio and Fontelo (1984) to detect saxitoxins in human plasma and it was then applied to mussel extracts by Vieytes *et al.* (1993) adapting the assay to a solid-phase format employing microtiter plates. The STX radio-receptor binding assay was further developed, validated (Van Dolah *et al.* 1994,1997, Doucette *et al.* 1997) and formatted for its use with the MultiScreen™Assay System (Millipore, Bedford, MA, USA). This method and the mouse bioassay showed good agreement for toxin contents ranging between 40 and 8000 µg STX equivalents in 100 g shellfish tissue but the limit of detection of the receptor binding assay was about 5 ng STX ml^{-1} in a sample extract (Doucette *et al.* 1997), well below that of the mouse assay, estimated as 0.5 mM (Sullivan *et al.* 1985). A further comparison between these two methods, made during a monitoring program on geoduck clams collected in Washington State, found an overall good agreement between the results; however, the receptor binding assay gave some overestimation of the results obtained by MBA at levels of toxicity near the regulatory value (Curtis *et al.* 2000). Recent research (Doucette *et al.* 2000) showed that the receptor binding assay could also be performed by substituting [^{3}H]STX with tritiated tetrodotoxin which has similar structural characteristics and a high affinity with the same receptor. This modification would allow the use of this assay despite the restriction on saxitoxin distribution imposed recently by the International Chemical Weapon Convention which included STX in Schedule 1 chemicals.

Recently a molecule called saxiphilin was discovered in the circulatory fluids of arthropods, amphibians, reptiles and fishes (Llewellyn *et al.* 1997) Saxiphilin has a high affinity towards STX and its derivatives and has been used to develop a radioligand binding assay in a microtiter plate format in which saxiphilin is used as a receptor to bind [^{3}H]STX and unlabeled STX and where the complex is separated through a simple filtration method (Llewellyn *et al.* 1998). This assay was found to be quite tolerant to interference by shellfish extracts, salts and various pH values and has a limit of detection for STX of 1.3 μg STX equivalents in 100 g tissue (Llewellyn and Doyle 2001).

An improvement of this method could be based on a saxiphilin molecule modified with the addition of specific linkers, such as the Flag epitope and hexa-histidine residue (Krishnan *et al.* 2001). The tagged version of saxiphilin would permit the development of a solid phase assay and applications such as purification, immunoprecipitation or others.

Cytotoxicity Assays

Another approach useful for PSP toxins is represented by a cytotoxicity bioassay that involves use of mouse neuroblastoma cell lines (Kogure *et al.* 1988, Jellet *et al.* 1992). This method is based on the pharmacological actions of veratridine and ouabain; these compounds directly activate sodium channels or indirectly enhance sodium ion influx; as a result the cells show pronounced changes becoming round and granular. Tetrodotoxin, STX and related toxins, which are sodium channel blockers, when added to the cells can prevent or reduce these effects in proportion to their concentration. Quantitative determination of the toxins was obtained by titrating standard amounts of ouabain and veratridine with STX (or its analogues) and making observations on the effects on the neuroblastoma cells.

The method was originally developed by Kogure *et al.* (1988, 1989) by employing the microscopic examination of cell morphology and time survival; after incubation of the cells with the reagents and the toxic samples, containing sodium channel blockers, the percent of swollen cells was determined by counting at least 200 cells per well in each 96-well culture dish. In order to reduce time-consuming operations and subjectivity in cell counts the method was later modified by Jellet *et al.* (1992, 1995). Their modification is based on the fact that when STX standards or extract containing unknown amount of PSP toxins were added first to the wells containing neuroblastoma cells then followed by ouabain and veratridine addition the cell swelling and lysis that occurred were associated with the loss of adherence of the cells to the bottom of the plastic wells in the tissue culture dishes. After 24 h incubation at 37°C the wells were rinsed, fixed with 10% aqueous formalin, and stained with Gram's crystal violet; during rinsing, fixing and staining those cells that had lost their adhesion were rinsed away, leaving only cells which were protected by STX or analogues. Once dried, the fixed and stained cells could be preserved for several days; following digestion of the cells with 33% aqueous acetic acid, a colorimetric detection of microtiter plates at 595 nm could be performed with a scanning spectrophotometer. The purple colour obtained after staining and rinsing was directly related to the amount of STX present and was linear in the range between 0 and 600 pg/10 µl. This method is now commercialized as a kit named MIST (Maritime *In Vitro* Shellfish Test) by Jellet Biotek (Jellet *et al.* 1998) in several versions: the fully quantitative method, the semi-quantitative version and the qualitative version. The MIST technology has been used to measure total PSP toxicity in acid extracts of phytoplankton, lobster hepatopancreas, mussels, scallops, clams, cyanobacteria and processed food. Over 200 samples have been tested with the cell bioassay in parallel with the mouse bioassay and the various trials performed showed that there was a good correlation between the two methods. The cell bioassay is 20x more sensitive than the mouse bioassay with a limit of detection of 2 µg/100 g tissue (Jellet *et al.* 1998). Recently the test was developed by Jellet Biotek also in a rapid form named MIST Alert for PSP, this test provide a qualitative indication of toxicity in less than 20 minutes and it could be useful in a regulatory laboratory to screen out negative samples and to reduce the number of positive samples to be tested with other analytical methods (Jellet *et al.* 2000).

A cytotoxicity bioassay based on similar assumptions was developed by García-Rodríguez *et al.* (1998) employing a different kind of cells, consisting in primary cultures of rat cerebellar neurons, and it was developed with the aim of quantifying other kinds of marine neurotoxins, in addition to PSP toxins. This was possible because of the presence in these cells of both voltage sensitive sodium channels (which can be blocked by saxitoxin) and glutamic acid receptors (which can be activated by domoic acid and blocked by glutamate antagonists), in addition they were sensitive to okadaic acid. More specifically, the effect of STX was that of increasing the survival of cerebellar neurons, which were otherwise subjected to apoptosis, after a 24 h pretreatment with veratridine. The number of viable cells was assessed through vital staining with 3,[4,5-dimethyltiazol-2-yl]-2,5-diphenyltetrazolium (MTT), a compound enzymatically reduced to formazane blue in living cells. The detection limit for PSP toxin measurement by this method was 0.04 µg STX equivalent per gram of mussel tissue.

Fluorimetric Assays

Primary cultures of rat cortical neurones which are functionally similar to the cells used in the previous method, were employed in a highly sensitive toxicity test based on the fluorimetric determination of intracellular calcium concentration ($[Ca^{2+}]_i$) (Beani *et al.* 1994, 2000). The effects of different drugs and toxins on neuronal voltage-gated Na^+-channels and on glutamate receptors, as well as on cell viability, can be quickly and directly tested by monitoring $[Ca^{2+}]_i$ changes in a representative neuronal population preloaded with Fura 2 (Tomasini *et al.* 1998). In this assay the cells were plated in 35 mm Nunc dishes provided with a glass coverslip, then the coverslip with the cells attached was inserted into a cuvette equipped with two electrodes so as to apply an electrical field stimulation. $[Ca^{2+}]_i$ were obtained by measuring the fluorescence intensity in a luminescence spectrometer. While under normal conditions the basal $[Ca^{2+}]_i$ level was about 50-100 nM, its concentration nearly doubled during the peaks induced by trains of electrical pulses at 10 Hz for 10 sec. Saxitoxin and tetrodotoxin dose-dependently prevented the peak height (IC50 = 3.5 and 24 nM, respectively) without affecting the basal calcium levels. Conversely, domoic acid increased the basal $[Ca^{2+}]_i$ up to +150% (EC50= 7 µM) and parallelely reduced the peaks. The validity and limits of this method were tested with acetic extracts of the non-toxic algae *Scrippsiella trochoidea* and *Gymnodinium* sp. and of a toxic strain (PSP causative agent) of *Alexandrium lusitanicum*. The extracts of the toxic algae, reduced the peaks as expected from their content in gonyautoxins 1–4, determined with HPLC method. However, these extracts slowly increased the basal $[Ca^{2+}]_i$ through mechanisms not involving glutamate receptors. Such an increase, due to unknown components of the acetic extracts, could be abolished by simply ultra filtering the extracts through Millipore membranes (cut off 10000 NMWL). This step enabled the titration of gonyautoxins as STX or TTX equivalents (Beani *et al.*, in press)

In conclusion this method allows detection of a) Saxitoxin-like toxins which inhibit the $[Ca^{2+}]_i$ peaks associated to the electrically-evoked opening of neuronal sodium (and calcium) channels and b) domoic acid, which increases basal $[Ca^{2+}]_i$ through glutamate receptors. An improvement of this assay could be based on the use of commercially available cell lines, such as PC-12 or human neuroblastoma in order to avoid the time consuming cell culturing technique.

Louzao *et al.* (2001) developed a fluorimetric assay based on the detection of changes in the membrane potential induced by toxins in excitable cells. This method consists of the incubation of human neuroblastoma cells with the fluorescent dye bis-oxonol, whose distribution across the membrane is potential-dependent. The cells, placed in a fluorometer, were depolarized by adding veratridine, causing a change of fluorescence of bis-oxonol. The addition of PSP toxins caused a dose-dependent inhibition of the depolarization and hence of bis-oxonol fluorescence. The method was validated with mussel extracts and the results displayed a good correlation with MBA, solid-phase radioreceptor assay and HPLC analyses. Furthermore the detection limit was very low, around 1 ng STX equivalents/ml.

Electrophysiological Assays

Based on the observation that the bull frog (*Rana catesbeiana*) bladder membrane contains many sodium channels (Updike and Treichel 1979), a tissue biosensor to measure PSP toxins was constructed (Cheun *et al.* 1998). The sensor consisted of a sodium electrode covered with a frog bladder membrane having its internal site in contact with the electrode, while the response was recorded from a sensor output amplified with an electrometer. The method allowed the detection of PSP toxins within 5 min and all the different gonyautoxins could be detected. Toxin concentrations determined by the sensor method were in good agreement with those measured by the traditional mouse bioassay with the exception of GTX2, and for TTX the amount detected was well below the detection limit of the mouse bioassay.

Another approach was used by Vélez *et al.* (2001) who developed a method based on cultured cells (HEK 293) permanently expressing the STX-sensitive rat skeletal muscle Na channel. The cells were patch clamped in the whole-cell configuration and Na currents were recorded. Standard STX inhibited the Na currents in a concentration-dependent manner. The assay was found to be three orders of magnitude more sensitive than the mouse bioassay and the analysis of toxic mussel extracts showed that the results obtained by the two methods correlated very well.

Rat hippocampal slice preparations were also used to detect marine algal toxins (Kerr *et al.* 1999). In this assay extracellular field potentials were recorded through a couple of stimulating electrodes inserted into the pyramidal cell layer and in the dendritic field of the hippocampal region CA1. Different toxins, such as saxitoxin, brevetoxin and domoic acid, could be identified according to different electrophysiological features. Saxitoxin abolished all neurophysiological responses in the slices and gave a linear response at concentrations between 25 and 200 nM.

Conclusions

The need to replace the mouse bioassay with more specific, sensitive and reliable methods induced the scientific world to develop a large variety of assays for the detection and quantification of marine toxins. Most of the efforts were addressed towards PSP toxins, as they represent a world-wide cause of severe intoxications. As reported in this review, the various methods are based either on the immunological properties of the molecules or on the highly specific binding and mechanism of action of PSP toxins so that the result is the existence of a large number of assays which are much more sensitive than the mouse bioassay. As shown in Table 1, the alternative methods have a sensitivity between 10- and 1000-fold higher than that of the mouse bioassay, comparable to that of HPLC analysis.

Table 1: Detection limits of some of the assays reported in the text

Methods	Limit of detection µg STX eq/100 g shellfish tissue
Mouse bioassay (AOAC 1990)	35-40
ELISA Immunoassay (Usleber *et al.* 1997)	0.2
Receptor binding assay (Doucette *et al.* 1997)	1
Saxiphilin binding assay (Llewellyn and Doyle 2001)	1.3
Cytotoxicity of mouse neuroblastoma cells (Jellet *et al.* 1998)	2
Cytotoxicity of rat cerebellar neurons (García-Rodriguez *et al.* 1998)	4
Intracellular calcium measurement (Beani *et al.* 2000)	0.2
Changes in membrane potential (Louzao *et al.* 2001)	0.2
Recombinant sodium channels (Vélez *et al.* 2001)	0.04

A few problems must be overcome before any of these assays can be adopted as a regulatory test. Immunological methods have the disadvantage that not all the toxins belonging to the same class of intoxication can be recognized with the same specificity. Among the functional methods, some are simple or well established but they can detect only one group of toxins, while others are very sensitive and promising but they need strong improvement in order to become simpler or usable in a kit format. Some of the latter methods, such as those based on apoptosis (García-Rodríguez *et al.* 1998), on intracellular calcium measurement (Beani *et al.* 2000) or on rat hippocampal slices (Kerr *et al.* 1999) have the advantage that they can be used to recognize a wider range of marine toxins, in addition to PSP toxins, so that the regulatory laboratories would need to become trained in a limited number of techniques. Many people envisage the utility of employing some of these methods in monitoring activities, at least as pre-screening assays, and it would be desirable to see in the near future the complete substitution of the mouse bioassay with a more reliable and scientific technique.

References

AOAC. 1990. Paralytic shellfish poison. Biological method. Final action. Pages 881–882, sec. 959.08, in *Official Methods of Analysis,* 15th edition, K. Ellrich, ed., Association of Official Analytical Chemists, Arlington, Virginia, USA.

Beani, L., C. Tomasini, B.M. Govoni, and C. Bianchi. 1994. Fluorimetric determination of electrically evoked increase in intracellular calcium in cultured cerebellar granule cells. *J. Neurosci. Methods.* 51: 1-7.

Beani, L., C. Bianchi, F. Guerrini, L. Marani, R. Pistocchi, M.C. Tomasini, A. Ceredi, A. Milandri, R. Poletti, and L. Boni. 2000. High sensitivity bioassay of paralytic (PSP) and amnesic (ASP) algal toxins based on the fluorimetric detection of $[Ca^{2+}]_i$ in rat cortical primary cultures. *Toxicon* 38: 1283-1297.

Beani, L., C. Bianchi, F. Guerrini, L. Marani, R. Pistocchi, A. Ceredi, A. Milandri, R. Poletti, and L. Boni. 2001. A sensitive method to bioassay algal neurotoxins by measuring $[Ca^{2+}]_i$ in primary cultures of rat cortical neurones. *Biol. Mar. Medit.,* in press.

Catterall, W.A. 1986. Molecular properties of voltage-sensitive sodium channels. *Ann. Rev. Biochem.* 55: 953-985.

Cembella, A.D., Y. Parent, D. Jones, and G. Lamoreux. 1990. Specificity and cross-reactivity of an absorption inhibition enzyme-linked immunoassay for the detection of paralytic shellfish toxins. Pages 339-344 in *Toxic Marine Phytoplankton,* E. Graneli, B. Sundstrom, L. Edler and D.M. Anderson, eds., Elsevier, New York.

Cheun, B.S., M. Loughran, T. Hayashi, Y. Nagashima, and E. Watanabe 1998. Use of a channel biosensor for the assay of paralytic shellfish toxins. *Toxicon* 36: 1371-1381.

Curtis, K.M., V.L. Trainer, and S.E. Shumway. 2000. Paralytic shellfish toxins in geoduck clams (*Panope abrupta*): variability, anatomical distribution, and comparison of two toxin detection methods. *J. Shellfish Res.* 19: 313-319.

Davio, S.R. and P.A. Fontelo. 1984. A competitive displacement assay to detect saxitoxin and tetrodotoxin. *Anal. Biochem.* 141: 199-204.

Doucette, G.J., M.M. Logan, J.S. Ramsdell, and F.M. Van Dolah. 1997. Development and preliminary validation of a microtiter plate-based receptor binding assay for paralytic shellfish poisoning toxins. *Toxicon* 35: 625-636.

Doucette, G.J., C.L. Powell, E.U. Do, C.Y. Byon, F. Cleves, and S.G. McClain. 2000. Evaluation of 11-[^{3}H]-tetrodotoxin use in a heterologous receptor binding assay for PSP toxins. *Toxicon* 38: 1465-1474.

García-Rodríguez, A., M.T. Fernández-Sánchez, M.I. Reyero, J.M. Franco, K. Haya, J. Martin, V. Zitko, C. Salgado, F. Arévalo, M. Bermúdez, M.L. Fernández, A. Míguez, and A. Novelli. 1998. Detection of PSP, ASP and DSP toxins by neuronal bioassay, comparison with HPLC and mouse bioassay. Pages 554–557 in *Harmful algae,* B. Reguera, J. Blanco, M.L. Fernández and T. Wyatt, eds., Xunta de Galicia and Intergovernmental Oceanographic Commission of UNESCO.

Jellet, J.F., L.J. Marks, J.E. Stewart, M.L. Dorey, W. Watson-Wright, and J.F. Lawrence. 1992. Paralytic shellfish poison (saxitoxin family) bioassays: automated endpoint determination and standardisation of the *in vitro* tissue culture bioassay, and comparison with the standard mouse bioassay. *Toxicon* 30: 1143-1156.

Jellet, J.F., J.E. Stewart and M.V. Laycock. 1995. Toxicological evaluation of saxitoxin, neosaxitoxin, gonyautoxin II, gonyautoxin II plus III and decarbamoylsaxitoxin with the mouse neuroblastoma cell bioassay. *Toxic. in Vitro* 9: 57-65.

Jellet, J.F., C.R. Wood, E.R. Belland, and L.I. Doucette. 1998. The MISTÔ shippable cell bioassay kits for PSP: an alternative to the mouse bioassay. *VIII International Conference on Harmful Algae*, B. Reguera, J. Blanco, M.L. Fernandez and T. Wyatt, eds., Xunta de Galicia and IOC of UNESCO.

Jellet, J.F., M.V. Laycock, P. Bishop, M. Quilliam, R.E. Barret, C.G. Allison, C.R. Bentz, and S.L. Plummer. 2000. Comparison trial of a rapid test for paralytic shellfish poisoning (PSP) and the AOAC mouse bioassay. Proceedings of the International Conference on Harmful Algal Blooms, p.25, Hobart, Tasmania, 7-11 February.

Kerr, D.S., D.M. Briggs, and H.I. Saba. 1999. A neurophysiological method of rapid detection and analysis of marine algal toxins. *Toxicon* 37: 1803-1825.

Kogure, K., M.L. Tamplin, U. Simidu, and R.R. Collwell. 1988. A tissue culture assay for the tetrodotoxin, saxitoxin and related toxins. *Toxicon* 26: 191-197.

Kogure, K. M.L. Tamplin, U. Simidu, and R.R. Collwell. 1989. Tissue culture assay method for PSP and related toxins. Pages 587–589 in *Red Tides: Biology, Environmental Science, and Toxicology,* T. Okaichi, D.M. Anderson and T. Nemoto., eds., Elsevier, New York.

Kralovec, J.A., M.V. Laycock, R. Richards, and E. Usleber. 1996. Immobilization of small molecules on solid matrices: a novel approach to enzyme-linked immunosorbent assay screening for saxitoxin and evaluation of anti-saxitoxin antibodies. *Toxicon* 34: 1127-1140.

Krishnan, G., M.A. Morabito, and E. Moczydlowski. 2001. Expression and characterization of flag-epitope- and hexahistidine-tagged derivatives of saxiphilin for use in detection and assay of saxitoxin. *Toxicon* 39: 291-301.

Krueger, B.K., R.W. Ratzlaff, G.R. Strichartz, and M.P. Blaustein. 1979. Saxitoxin binding to synaptosomes, membranes, and solubilized binding sites from rat brain. *J. Membr. Biol.* 50: 287-310.

Lawrence, J.F., C. Ménard, and C. Cleroux. 1995. Evaluation of prechromatographic oxidation for liquid chromatographic determination of paralytic shellfish poisons in shellfish. *J. AOAC Internat.* 78: 514-520.

Llewellyn, L.E., P.M. Bell, and E.G. Moczydlowski. 1997. Phylogenetic survey of soluble saxitoxin-binding activity in pursuit of the function and molecular evolution of saxiphilin, a relative of transferrin. *Proc. R. Soc. Lond.* B 264: 891-892.

Llewellyn, L.E., J. Doyle, and A.P. Negri. 1998. A high-throughput, microtiter plate assay for paralytic shellfish poisons using the saxitoxins-specific receptor, saxiphilin. *Anal. Biochem.* 261: 51-56.

Llewellyn, L.E. and J. Doyle. 2001. Microtiter plate assay for paralytic shellfish toxins using saxiphilin: gauging the effects of shellfish extract matrices, salts and pH upon assay performance. *Toxicon* 39: 217-224.

Louzao, M.C., M.R. Vieytes, J.M.V. Baptista de Sousa, F. Leira, and L.M. Botana. 2001. A fluorimetric method based on changes in membrane potential for screening paralytic shellfish toxins in mussels. *Anal. Biochem.* 289: 246-250.

O'Neall, S., S. Gallacher, and I. Riddoch. 1998. Assessment of a saxitoxin ELISA as a pre-screen for PSP toxins in *Mytilus edulis* and *Pecten maximus,* from UK waters. Pages 554-557 in *VIII International Conference on Harmful Algae,* B. Reguera, J. Blanco, M.L. Fernandez and T. Wyatt, eds., Xunta de Galicia and IOC of UNESCO.

Oshima, Y. 1995. Postcolumn derivatization liquid chromatographic method for paralytic shellfish toxins. *J. AOAC Internat.* 78: 528-532.

Quilliam, M.A. 1998. Liquid chromatography-mass spectrometry: a universal method for analysis of toxins? Pages 509–514 in *VIII International Conference on Harmful Algae*, B. Reguera, J. Blanco, M.L. Fernandez and T. Wyatt, eds., Xunta de Galicia and IOC of UNESCO.

Schantz, E.J., V.E. Ghazarossian, H.K.M. Schnoes, F.M. Strong, J.P. Springer, J.O. Pezzanite, and J. Clardy. 1975. The structure of saxitoxin. *J. Am. Chem. Soc.* 97: 1238-1239.

Sommer, H. and K.F. Meyer. 1937. Paralytic shellfish poisoning. *Arch. Pathol.* 24: 560-598.

Sullivan, J.J., M.M. Wekell, and L.L. Kentala. 1985. Application of HPLC for the determination of PSP toxins in shellfish. *J. Fd. Sci.* 50: 26-29.

Thibault, P., S. Pleasance, and M.V. Laycock. 1991. Analysis of paralytic shellfish poisons by capillary electrophoresis. *J. Chromatogr.* 542: 483-501.

Tomasini, M.C., T. Antonelli, D.G. Trist, A. Reggiani, L. Beani, and C. Bianchi. 1998. Protective effect of GV 150526A on the glutamate-induced changes in basal and electrically-stimulated cytosolic Ca^{++} in primary cultured cerebral cortical cells. *Neurochem. Int.* 32: 345-351.

Updike, S. and I. Treichel. 1979. Antidiuretic hormone specific electrode. *Anal. Chem.* 51: 643.

Usleber, E., E. Schneider, and G. Terplan. 1991. Direct enzyme immunoassay in microtitration plate and test strip format for the detection of saxitoxin in shellfish. *Lett. Appl. Microbiol.* 13: 275-277.

Usleber, E., E. Schneider, G. Terplan, and M.V. Laycock. 1995. 2 formats of enzyme-immunoassay for the detection of saxitoxin and other paralytic shellfish poisoning toxins. *Food Addit. Contam.* 12: 405-413.

Usleber, E., M. Donald, M. Straka, and E. Märtlbauer. 1997. Comparison of enzyme immunoassay and mouse bioassay for determining paralytic shellfish poisoning toxins in shellfish. *Food Addit. Contam.* 14: 193-198.

Van Dolah, F.M., E.L. Finley, B.L. Haynes, G.J. Doucette, P.D. Moeller, and J.S. Ramsdell. 1994. Development of rapid and sensitive high throughput assays for marine phycotoxins. *Nat. Toxins* 2: 189-196.

Van Dolah, F.M., T.A. Leighfield, B.L. Haynes, D.R. Hampson, and J.S. Ramsdell. 1997. A microplate receptor assay for the amnesic shellfish poisoning toxin, domoic acid, utilizing a cloned glutamate receptor. *Anal. Biochem.* 245: 102-105.

Van Egmond, H.P, H.J. van den Top, W.E. Paulsch, X. Goenaga, and M.R. Vieytes. 1994. Paralytic shellfish poison reference materials: an intercomparison of methods for the determination of saxitoxin. *Food Addit. Contam.* 11: 39-56.

Vélez, P., J. Sierralta, C. Alcayaga, M. Fonseca, H. Loyola, D.C. Johns, G.F. Tomaselli, E. Marbán, and B.A. Suárez-Isla. 2001. A functional assay for paralytic shellfish toxins that uses recombinant sodium channels. *Toxicon* 39: 929-935.

Vieytes, M.R., A.G. Cabado, A. Alfonso, M.C. Louzao, A.M. Botana, and L.M. Botana. 1993. Solid-phase radioreceptor assay for paralytic shellfish toxins. *Anal. Biochem.* 211: 87-93.

Bioassays for Paralytic Shellfish Toxins in Crustacea

Jason Doyle and Lyndon Llewellyn

Australian Institute of Marine Science, PMB 3, Townsville MC, Queensland, Australia 4810

Paralytic Shellfish Toxins and their Origins

What are Paralytic Shellfish Toxins?

Paralytic shellfish toxins (PSTs), derived from the archetypal compound saxitoxin (STX, Figure 1), are responsible for the human disease known as Paralytic Shellfish Poisoning (PSP). It is named as such because it more commonly results from ingestion of molluscan shellfish, usually bivalves such as oysters, which may contain PSTs. It is well known however that the cause of this illness is not restricted to shellfish and as will become evident, crustacea number amongst the culprits of PSP. Persons intoxicated with PSTs can expect the onset of symptoms within 30 minutes, starting with a burning or tingling sensation on the lips and face, increasing to total numbness. This effect extends to upper body extremities and over time spreads all the way to fingers and toes. Gradually the body of the victim becomes paralysed with voluntary movements becoming increasingly difficult. All the while, the victim can remain completely conscious throughout the event. Other minor symptoms include dizziness, headache, salivation, intense thirst and perspiration, vomiting, diarrhoea and stomach cramps. A lethal dose can see you expire within 12 hours from respiratory failure. Victims surviving beyond 12 hours will usually make a full recovery. There is no known antidote for PSP with the only adequate treatment being artificial respiration (Morse 1977).

How do PSTs Work?

STX is one of the most potent neurotoxins known and gram for gram, is 1000 times more deadly than cyanide (RaLonde 1996) with only 0.5 to 1 mg

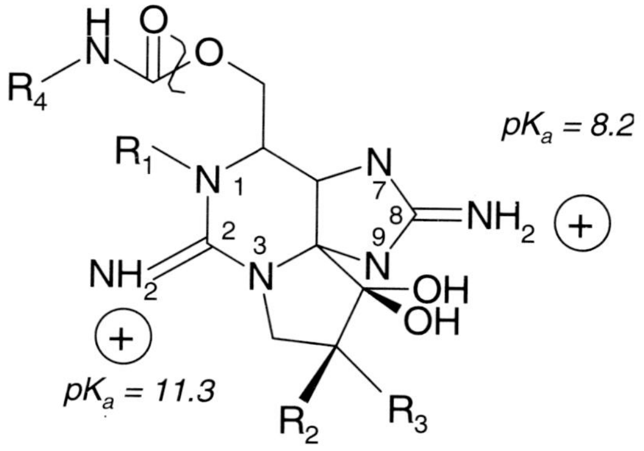

	R_1	R_2	R_3	R_4
STX	H	H	H	H
neoSTX	OH	H	H	H
B1	H	H	H	SO_3^-
Gonyautoxin 2	H	H	OSO_3^-	H
Gonyautoxin 3	H	OSO_3^-	H	H
decarbamoylSTX	*	H	H	H

* In this derivative, a proton replaces all of the structure beyond the wavy line, including R_4.

Figure 1. STX and some of its analogues.

of pure toxin necessary to kill an average adult human (Morse 1977). There are over 20 known analogues of STX, each differing slightly from the parent and involving one or a combination of substituted sulfate, carbamoyl or hydroxy groups (Figure 1). PSTs exert their toxicity in the same manner as tetrodotoxin (TTX) by binding to the voltage dependent sodium channel (Na channel) in nerve and excitable tissue (Strichartz *et al.* 1986). As an action potential travels along a tissue, Na channels respond by becoming permeable to Na ions allowing the continuation of the action potential along the tissue. STX and TTX block the influx of Na ions into a cell by physically occluding the Na channel stopping the progression of the action potential, rendering cells and tissue non-functional (Lipkind and Fozzard 1994).

Where do PSTs come From?

The first source of PSTs was identified in 1966 with the dinoflagellate, *Gonyaulax catenella*, shown to harbour PSTs (Schantz *et al.* 1966). These neurotoxins have since been found to originate in many microalgae and cyanobacteria (Table 1) from both marine and freshwater environments. When suitable environmental conditions prevail, the populations of these plants may bloom creating a health hazard not just to humans but also to wildlife and agricultural livestock. These so-called harmful algal blooms (HABs) are by no means a phenomenon of modern times. Though a recent increase in their frequency and intensity can be argued (Hallegraeff 1993), HABs have been recorded as far back as 208 BC with the Greeks naming the Red Sea for the formation of algal blooms which can have a red appearance (Morse 1977).

Table 1. Organisms identified as producers of PSTs

Organism	Reference
Cyanobacteria	
Anabaena circinalis	Negri *et al.* 1995, Runnegar *et al.* 1988
Lyngbya wollei	Carmichael *et al.* 1997
Aphanizomenon flos-aquae	Ferreira *et al.* 2001
Dinoflagellates	
Gymnodinium catenatum	Oshima *et al.* 1987, Hallegraeff *et al.* 1988
Alexandrium andersoni	Ciminiello *et al.* 2000
Alexandrium catenella	Hallegraeff *et al.* 1988
Alexandrium excavatum	Laycock *et al.* 1994
Alexandrium minutum	Chang *et al.* 1997, Hallegraeff *et al.* 1988
Alexandrium (=Gonyaulax) tamarensis	Buckley *et al.* 1976
Protogonyaulax cohorticula	Kodama *et al.* 1988a
Pyrodinium bahamense	Harada *et al.* 1982
Alga	
Jania sp.	Kotaki *et al.* 1983

One of the largest HABs recorded occurred in the Barwon-Darling River system, a major river system in Australia, with the bloom stretching for almost 1000 kms (Bowling and Baker 1996). This bloom involved the cyanobacteria *Anabaena circinalis*, a species with a well-recorded history of toxicity (Runnegar *et al.* 1988, Jones 1992, Negri *et al.* 1995, Beltran and Neilan 2000) and has also been implicated in livestock deaths resulting from the drinking of contaminated water (Negri *et al.* 1995). PSTs have only been detected in two other species of cyanobacteria (Table 1), namely *Lyngbya wollei* (Carmichael *et al.* 1997, Onodera *et al.* 1997) and *Aphanizomenon flos-aquae* (Mahmood and Carmichael 1986, Ferreira *et al.* 2001), both also freshwater species. In the ocean, PSTs are produced by several dinoflagellate species (Table 1). These microalgae have been studied primarily due to their impacts on shellfish farming and aquaculture because commercially grown molluscan shellfish are filter feeders that take up not just nutrition from the microalgae but also their toxins.

Microalgae are not the only plants found to produce PSTs, with these toxins also present in the red calcareous alga, *Jania* sp. (Kotaki *et al.* 1983). This taxonomic diversity in the origin of PSTs has lead to the speculation that PSTs have a bacterial origin (Silva 1982). Several attempts to isolate PST synthesizing bacteria from toxic dinoflagellates have resulted in mixed success (Nelinda *et al.* 1985, Kodama *et al.* 1988b). Evidence recently however has provided further support that bacteria isolated from dinoflagellates autonomously produce PSTs (Gallacher *et al.* 1997).

Why test for PSTs in crustacea?

The best way of illustrating the answer to the above question is the following case from the Philippines. "On August 9, 1975, two adult males, aged 35 and 37, both farmers and residents of the coastal town of Bindoy, Negros Oriental Province, died within 30–60 mins following ingestion of crab soup. Both victims complained of intense abdominal pain and dizziness, and vomited profusely. They were taken to the town's rural health office, about 5 km away, but were pronounced dead on arrival by the physician " (Gonzales and Alcala 1977). The culprit was identified as *Lophozozymus pictor*, a crab from the family Xanthidae previously shown to be toxic in the laboratory (Teh and Gardiner 1970, Teh and Gardiner 1974). *Lophozozymus pictor* has long had a reputation as being "completely harmful" (sic) in the Amboina Islands, where locals avoid eating crabs with "fingers black or brown" (sic), a description fitting of members of the xanthid family of crabs (Holthuis 1968). Human poisonings due to crustacea draw our attention to this phenomenon (Figure 2) but over time, laboratory studies have extended the number of crustacean species now known to carry PSTs to over 40 (Table 2). This table also highlights that PSTs have been recorded in lobsters (Lawrence *et al.* 1994), another important food item and copepods (Boyer *et al.* 1985), which are integral to oceanic food chains.

Figure 2. Distribution of crabs that have been reported to have caused deaths (●) or severe illness (▲) resulting from PST intoxication, or have been demonstrated to contain PSTs (■).

Table 2. Crustaceans known to contain PSTs

Family	Genus species	Reference
Crabs		
Xanthidae	*Actaeodes tomentosus*	Yasumoto *et al.* 1983
	Atergatopsis germaini	Tsai *et al.* (1996)
	Atergatus dilitatus	Garth and Alcala 1977
	Atergatus floridus	Raj *et al.* 1983
	Atergatus integerrimus	Yasumura *et al.* 1986
	Carpilius maculatus	Yasumura *et al.* 1986
	Demania alcala	Carumbana *et al.* 1976
	Demania reynaudii	Alcala *et al.* 1988
	Demania toxica	Garth 1971
	Eriphia scabricula	Yasumoto *et al.* 1983
	Eriphia sebana	Llewellyn and Endean 1989a
	Etisus rhynchophorous	Garth and Alcala 1977
	Etisus splendidus	Garth and Alcala 1977
	Euzanthus exsculptus	Negri and Llewellyn 1998
	Leptodius sanguineus	Llewellyn and Endean 1988
	Lophozozymus pictor	Llewellyn and Endean 1989b
	Lophozozymus octodentatus	Negri and Llewellyn 1998
	Neoxanthias impressus	Yasumoto *et al.* 1983
	Pilodius areolatus	Llewellyn and Endean 1988
	Phymodius ungulatus	Llewellyn and Endean 1988
	Platypodia granulosa	Yasumoto *et al.* 1983
	Platypodia pseudogranulosa	Negri and Llewellyn 1998
	Zosimus aeneus	Llewellyn and Endean 1989b
Portunidae	*Portunus pelagicus*	Negri and Llewellyn 1998
	Thalamita stimpsoni	Llewellyn and Endean 1988
	Thalamita wakensis	Llewellyn and Endean 1988
	Thalamita sp.	Yasumoto *et al.* 1983
Majidae	*Schizophrys aspera*	Yasumoto *et al.* 1983
	Chionoecetes bairdi	Huang *et al.* 1996
Grapsidae	*Grapsus albolineatus*	Llewellyn and Endean 1988
	Metopograpsus frontalis	Negri and Llewellyn 1998
	Percnon planissimum	Yasumoto *et al.* 1983
Calappidae	*Calappa calappa*	Yasumura *et al.* 1986
Dromidae	*Dromidiopsis* sp.	Yasumura *et al.* 1986
Cancridae	*Cancer magister*	Huang *et al.* 1996
Parthenopidae	*Daldorfia horrida*	Garth and Alcala 1977
Pilumnidae	*Pilumnus pulcher*	Negri and Llewellyn 1998
	Pilumnus vespertilio	Negri and Llewellyn 1998
Hippidae	*Emerita analoga*	Sommer 1932
Lobster		
Homaridae	*Homarus americanus*	Lawrence *et al.* 1994
Copepods		
Harpacticidae	*Tigriopus californicus*	Boyer *et al.* 1985

The first crustacean in which PSTs were detected was *Emerita analoga* (Sommer 1932), a filter feeding crab hypothesised to accumulate toxins from the dinoflagellate, *Gonyaulax catanella* (Sommer and Meyer 1937). Interestingly,

Table 2 is dominated by non-filter feeding crustacea like the xanthids, which are omnivores that feed on macroscopic food items. Testing of identifiable species from xanthid crab gut content led to the hypothesis that the source of PSTs was the red calcareous alga, *Jania* sp. (Kotaki *et al.* 1983, Yasumoto *et al.* 1983) found to contain less than 2 mouse units (MU) per gram of tissue. This is probably only part of the story with regards the origin of PSTs in these crabs because they may carry enormous amounts of PSTs, like an individual of *Zosimus aeneus* which had 745,000 MU in its 45 g body (Koyama *et al.* 1983).

Human intoxication not only has the direct effect of fatalities and illness, but also has economic and social effects where publicised poisonings can decrease consumer confidence in other seafood industries (Laxminarayan 2001). For this reason it is essential to have efficient monitoring programs set in place to detect PSTs prior to species entering the market place. Routine monitoring for PSTs does in fact occur for several commercially important crabs in Alaska such as the Dungeness crab (*Cancer magister*), Tanner crab (*Chionoecetes* sp.) and King crab (*Paralithodes* sp., and *Lithodes aequispina*). This monitoring resulted in a warning issued in November 1998 to halt the harvesting of Tanner crabs in a number of fisheries off the Alaskan coast (Soares 1998).

As described above, unlike molluscan shellfish, there is no relationship between HABs and toxified crustacea. Therefore, observational monitoring for toxic microalgae will not suffice in reducing the occurrence of intoxication from crustacea. The only satisfactory solution then is to directly test the crustacea using either biological assays or chemical analysis. The focus of this chapter is bioassays and one can refer to authors such as, Sullivan and Iwaoka (1983) Lawrence *et al.* (1995) and Oshima (1995) for analytical methods available.

Bioassays used for PST Monitoring in Crustacea

Numerous bioassays are available for the detection of PSTs with several utilised for toxin analysis in crustacea. Critical to any of the assays is the means of toxin extraction. Less potent PSTs can be converted to other members of this family by exposure to acid and heat. This is desired if the aim is to ensure detection of all PSTs using an assay with variable sensitivity for the different toxins. It is this principle which underlies the Association of Official Analytical Chemists (AOAC) endorsed extraction protocol one must follow if regulatory testing for PSTs is desired (Official Methods of Analysis 1990). Briefly, this protocol involves extracting 100 g of blended tissue with 100 ml of 0.1 M hydrochloric acid and boiling for 5 minutes. The resultant mixture is then clarified for bioassay. Not using such treatment will maintain the chemical diversity of the PSTs present in a crustacean extract. Any aqueous solvent then, along with standard homogenization procedures, can be used to extract PSTs with the proviso that the solvent should be somewhat acidified to prevent toxin degradation over time (Alfonso *et al.* 1994).

Mouse lethality

This is the assay official endorsed by the AOAC and the most commonly adopted method for detection of PSTs in crustacea. Mouse injection is intraperitoneal with the toxin taken up into the bloodstream through the blood vessels lining the peritoneal cavity. Compliance with this method requires a stock colony of healthy mice all within a certain weight range (19-21 g). Prior to conducting the assay, standardisation of the mouse colony in terms of their sensitivity to an STX standard must take place. This is achieved by repeated injections of a known amount of STX until a death time of between 5 and 7 minutes is regularly recorded ($n \geq 10$). The assay is then repeated a day or two later with the same solutions of STX and then the entire experiment is repeated again with a new batch of STX to ensure reproducibility (Official Methods of Analysis 1990). Similar to the standard STX solution, test samples are concentrated or diluted until the death of the mouse is within 5 to 7 minutes. PSTs are quantitated by comparing the death time, recorded as the last gasp, to a standard series of mouse units (Official Methods of Analysis 1990).

The mouse lethality bioassay has the advantages of many decades of use being derived from the studies of Sommer and Meyer (1937). This enormous amount of data has enabled continual validation of the assay. It is however an expensive assay requiring maintenance of mouse colonies and in these days of ethical concern about animal experimentation, this assay is becoming increasingly unpopular. This assay is also prone to human error, with its end point being the last gasp of the mouse, a subjective judgement by the person performing the experiment. Other sources of error arise form poor injection technique. For example, if a toxic extract is not injected exclusively into the peritoneal cavity and some ends up in subcutaneous compartments or in the digestive tract itself, then toxicity may be underestimated (Adams and Furfari 1984, Steward *et al.* 1968). Further error can arise from PST accumulation to the testicular cavity of male mice leading to sexual differences in mouse toxin sensitivity (Steward *et al.* 1968). In addition, discrepancies have been observed depending on the presence of metals or common salts found in marine samples (McCulloch *et al.* 1989, Park *et al.* 1986a).

Tissue culture detection

This procedure, originally developed by Kogure *et al.* (1988) is the basis of a tissue culture method (Jellett *et al.* 1992) currently under investigation for AOAC endorsement and has been utilised for the detection of PSTs in crab viscera (Manger *et al.* 1995). Laboratory grown mouse neuroblastoma type

2A cells containing Na channels swell in the presence of veratridine, a Na channel activator. This swelling is enhanced with the addition of ouabain, a Na^+/K^+ ATPase inhibitor which prevents removal of the excessive Na ions allowed in by veratridine. Addition of both drugs at an appropriate concentration can result in cell lysis or, at the very least, severe morphological changes that affect cell viability. If a Na channel blocker such as STX, TTX or an active analogue of either is added to the cell culture media prior to the addition of these drugs thus preventing the influx of Na ions, cell viability is maintained. Initially, cells were manually counted as alive or dead and prone to experimenter error. Jellett *et al.* (1992) modified the assay so that it relies on spectrophotometric measurement of the product of cellular metabolism of tetrazolium salts.

This method is inexpensive, requires little expertise and only basic laboratory equipment such as tissue culture facilities and a plate reader. It mimics whole organism activity and has a high correlation with the mouse bioassay making it an attractive candidate for an internationally standard technique (Jellett *et al.* 1992, Truman and Lake 1996). Sensitivity is governed by the duration of the assay with best results achieved after 24 hours. Non-specific agents in marine extracts which may cause cellular toxicity (Jellett *et al.* 1992) can confound this assay by counteracting the rescue effect of any toxins that may be present. In addition to detecting Na channel blocking agents, this assay has also been employed to detect Na channel enhancing compounds such as brevetoxins and ciguatoxins (Manger *et al.* 1995, Manger *et al.* 1993) and the presence of similar compounds in a crustacean extract can also confound interpretation of assay results.

Sodium channel ^{3}H-STX binding

As mentioned previously, the Na channel is the biological target for STX and it is this interaction which led to the development of this assay. The basic assay method has changed little from its beginnings (Weigele and Barchi 1978a) and relies on membrane preparations made from tissue known to contain Na channels, typically brain tissue, and their binding of tritiated STX (^{3}H-STX). Detection of PSTs can be achieved by incubation of receptor and radioligand in the presence of the test sample, salt and buffer. Remembering that this is a voltage-gated ion channel, the salt used must be unable to pass through the open Na channel to prevent the possible interference of altered membrane potential upon the conformation of the Na channel. It is for this reason that the usual salt incorporated to maintain osmotic balance is choline chloride. After an appropriate length of time the incubation mixture is filtered through glass fiber type C filter with the Na channel/^{3}H-STX complex being trapped on the filter. The filter is then washed with a cold buffer solution and the radioactivity bound on the filter quantitated by scintillation counting. Detection of PSTs in a sample is evident by a reduction in the amount of ^{3}H-STX bound to the synaptosomes and hence a reduction in the radioactive signal from scintillation counting. Alternatively, free ^{3}H-STX can be separated from bound radioligand by passage of the assay mixture through a cation exchange chromatography resin which retains the charged toxin and allows passage of the receptor bound ^{3}H-STX to be collected into a vessel for later scintillation counting. It is this latter protocol where Na channel binding was used to measure PSTs in crustacea in a study conducted off the Australian coastline (Negri and Llewellyn 1998).

Na channel affinity for STX decreases with increased temperature and in a technical sense, it is best if one can maintain such experiments as close to 0°C as possible. Toxin binding is diminished at pH's below 6 (Weigele and Barchi 1978a) due to an effect upon essential amino acids within the Na channel and experiments therefore must be well buffered to prevent test samples, especially those produced using the AOAC procedure where the solvent is 0.1N HCl, affecting the assay. Further, the binding of STX by the Na channel depends on both of the guanidino groups borne by STX (Figure 1) being in their charged state. With pKa's of 8.2 and 11.3 (Rogers and Rapaport 1980), the guanidino groups will deprotonate as the pH moves closer to these pKa's, making it unable to bind to the channel. This therefore constrains this assay to a narrow pH range. It has also been demonstrated that monovalent and divalent cations affect the binding of STX to the receptor (Weigele and Barchi 1978b), thus the potential of false negatives increases if the test samples contain significant quantities of salt.

Binding of PSTs to the Na channel determines mammalian toxicity and as such, toxicity observed by this method correlates well with the mouse lethality bioassay (Doucette *et al.* 1997). It would be unwise however to rely on this assay as a selective detection method for PSTs as it has equal affinity for TTX. This does indeed occur in crustacea with both these groups of toxins co-occurring in the same individuals (Negri and Llewellyn 1998, Tsai *et al.* 1995). The Na channel assay is amenable to microplate technology and has evolved into a rapid throughput format (Doucette *et al.* 1997, Llewellyn *et al.* 2001a, Llewellyn *et al.* 2001b). It is a cheaper assay compared to mouse lethality and tissue culture in terms of consumables and reagents but requires expensive equipment such as a scintillation counter and generates radioactive waste.

Saxiphilin binding of ^{3}H-STX

Saxiphilin is an STX receptor protein unrelated in structure to the Na channel but is a member of the transferrin family which binds and circulates iron throughout the body of many animals (Llewellyn *et al.* 1997, Morabito and Moczydlowski 1994). It also differs from the Na channel in that while it selectively binds STX with a sub-nanomolar affinity it is insensitive to TTX (Llewellyn *et al.* 1998). Saxiphilin is found in a wide range of vertebrates and invertebrates including species that have no demonstrated contact with potential sources of PSTs such as the tropical centipede, *Ethmostigmus rubripes*, or the crevice spiny lizard, *Sceloporus poinsetti*. As such, a biological function for saxiphilin is yet to be proven (Llewellyn *et al.* 1997). It is the isoform from this centipede, which has formed the basis of this approach. In essence, the saxiphilin assay is similar to the Na channel assay in that the protein is incubated with ^{3}H-STX, test sample and salt in a buffered solution (Llewellyn *et al.* 1998). Filtration of the reaction mixture through polyethylimine pre-treated glass fiber type B filters, a treatment that enables glass fibres to then bind soluble proteins, occurs after incubation and is followed by a series of washes. The presence of PSTs in a test sample reduces the radioactive signal seen from scintillation counting of the filter which bears the saxiphilin/^{3}H-STX complex. Like the Na channel, it can also be conducted using cation exchange chromatography resin for separation of bound and free ^{3}H-STX. It has been used in both assay styles for detection of PSTs in crustacea from North West Australia (Negri and Llewellyn 1998, Llewellyn *et al.* 1998).

Saxiphilin is able to detect several PST analogues with equal affinity with the exception of the sulfated C class toxins. Preparation of extracts via the AOAC method described above would overcome this problem as this brings about their conversion to the non-sulfated PSTs which are readily detected by saxiphilin. This assay has recently been used for the detection of PSTs in a variety of samples including an extensive study using AOAC prepared extracts and has correlated favourably with Na channel and mouse lethality bioassay as well as with HPLC analysis (Llewellyn *et al.* 2001a, Llewellyn *et al.* 2001b). The assay also stands up to the vagaries of complex matrices, high salt and gives reproducible results over a wide range of pH's (Llewellyn and Doyle 2001). It is however subject to the same problems as the Na channel assay in that it requires the use of radioactively labelled STX, suffering the inherent dilemmas associated with radioactive handling and waste management.

Direct competitive enzyme linked immunosorbent assay

A direct competitive enzyme linked immunosorbent assay (dc-ELISA) has been utilised for the detection of STX and neo STX in AOAC prepared extracts of dungeness and tanner crabs (Huang *et al.* 1996) using antibodies generated against toxin linked to a carrier protein to manufacture an antigenic epitope. This immunological approach is also the basis for a commercially available kit, namely Ridascreen ® Saxitoxin Test. Briefly, the dc-ELISA technique uses microtitre plate wells coated with antibody to which test sample and a conjugate of STX with the enzyme, horse radish peroxidase (STX-HRP), are then added simultaneously to then compete for the STX antibody. If the test sample contains STX in sufficient quantities, then it will outcompete STX-HRP for the binding sites. Alternatively, if no STX is present in the test sample, the STX-HRP will saturate the STX antibody. A HRP chromogenic substrate is then added to the wells of the plate after washing to remove unbound test sample and STX-HRP and a colour forms if HRP is present as the STX-HRP. No colour formation reveals the presence of STX in the test sample that prevented the STX-HRP binding to the antibody coated microtitre plate.

Correlation with the mouse bioassay is reasonable, being 61% (Usleber *et al.* 1997), which may relate to the fact that this system does not relate to mammalian toxicity *per se*, that is, it is purely an antigen/antibody reaction rather than a ligand binding to a biological target to manifest toxicity. This method also underestimated toxicity relative to mouse lethality by a factor of approximately two (Usleber *et al.* 1997). It is a simple assay to perform with the major equipment requirement being a photometric plate reader. Limitations to this method can be attributed to the specificity by the antibodies towards individual PSTs. For example (Huang *et al.* 1996) showed that the ELISA would underestimate total toxicity if the sample were to contain high levels of the PSTs, neoSTX, GTX 1 or GTX 4. This reflects in part, the fact that antibodies are raised towards one PST only, usually STX, and non-STX PSTs may not be detected at all (Chu and Fan 1985). In addition, raising polyclonal antibodies is a time consuming and expensive exercise with harvesting taking place anywhere from five to fifteen weeks after initial injection of the antigen into the host animal (Chu and Huang 1992).

Tissue biosensor

Recently, a tissue biosensor method was developed for the detection of PSTs (Cheun *et al.* 1998). It is a relatively simply approach with a Na ion electrode utilised to detect fluctuations in Na ion concentration across a frog bladder membrane, a tissue rich in Na channels. The presence of Na channel blocking agents such as STX or TTX alters the membrane permeability to Na ions, which is then detected as a change in potential by the electrode. An average measurement will take approximately five minutes, making it a comparatively fast assay. However, it will not distinguish between STX and TTX as they have the same functionality on the Na channel. It has very good correlation with mouse lethality bioassay and has been utilised to detect STX in extracts of the toxic crab, *Zosimus aeneus* (Cheun *et al.* 1998). Like other bioassays that rely on the biological action of PSTs, the tissue biosensor depends on intact Na channels present in the frog bladder membrane and as such has differing sensitivities for various PSTs. There has been little work done on the characterisation of the assay with regard to physicochemical properties such as pH variation, salts or other chemical factors that may affect the integrity of the method making it premature to draw comparisons with other assay techniques.

PST Bioassays not yet used for Monitoring of PSTs in Crustacea

Immunological approaches

Most other bioassays that have been adopted for detecting PSTs fall into the immunological category and include, STX induced protein (Smith and Kitts 1994: Smith *et al.* 1989) haemagglutination (Johnson and Mulberry 1966), enzyme-linked immunofiltration assay (Usleber *et al.* 1995), radioimmunoassays (Carlson *et al.* 1984, Chu and Huang 1992) and finally indirect ELISA (Chu and Fan 1985). The main dilemma associated with this style of assay is the inherent specificity for a particular PST. Both monoclonal and polyclonal antibodies possessed little crossreactivity with other PSTs apart from the one against which the antibody was raised, presenting difficulties for samples containing a variety of PSTs. Antibody production is a lengthy operation with the time from innoculation to harvesting being up to several months, with the success of antibody production being unpredictable. Multiple animals would be required raising the negative spectre of animal use in scientific research. Immunological approaches however, have the ability to be formatted into qualitative techniques that can be taken into the field (Usleber *et al.* 1995), a valuable capability not yet available to other assays described.

Whole organism bioassays

Whole organism bioassays have the advantage that they can record gross toxicity of a sample. Several organisms including chick embryos, brine shrimp, desert locust and even some bacteria have been examined for their suitability to detect PST's (Park *et al.* 1986b, McElhiney *et al.* 1998). All have the same basic principle of observing an effect on the organism, usually whether it is alive or dead, upon application of a PST or a PST containing sample. As an example, the house fly bioassay involves injection of samples into the common house fly, *Musca domestica* (Ross *et al.* 1985). Responses based on fly movement are scored as zero for no movement or one for movement. It is a rapid and cost effective method with a single fly able to be processed every minute requiring a small injection volume of only 1 to 1.5 µl. It is also claimed that the 'salt effect' that plagues the mouse bioassay is not observed with this system. It is however a tedious and technically challenging method.

Fluorescent probes

Toxin activities can be reflected in changes in cellular properties beyond the direct action by the toxin on the Na channel. Recent advances in fluorescent probes have provided sensitive methods for detecting such cellular alterations. One such measure looks at variations in intracellular calcium resulting from the action of PSTs on Na channels in rat cortical primary cultures (Beani *et al.* 2000). Blockade of the Na channel by PSTs results in a reduction of the peaks of intracellular calcium induced by electrical stimulus, as measured by a calcium sensitive fluorescent dye. More recently another fluorescent method of PST detection has been developed which relies on the membrane potential dependent distribution of a fluorescent dye bis (1,3-diethylthiobarbituric acid) trimethine oxonol (bis-oxonol). Neuroblastoma cells are equilibrated with the dye followed by the addition of the Na channel activator, veratridine, to depolarise the membrane (Louzao *et al.* 2001). STX contained in a test sample would reduce the effect of veratridine and subsequently change the distribution of bis-oxonol across the cell membrane which is then measured fluorometrically. Further evaluation of physicochemical effects induced by salts, extract matrices and pH is necessary and work is in progress to miniaturise the assay to gain the benefits of microplate handling technology (Louzao *et al.* 2001).

Synthesis

PSTs are traditionally a concern for the molluscan shellfish industry and consumers. Human poisoning resulting from the ingestion of crabs alerted the world to the fact that some crustacea can be quite toxic (Gonzales and Alcala 1977, Negri and Llewellyn 1998) and that these toxins were in fact PSTs. These poisonings have been sporadic, reflecting the fact that most of these toxic crabs are not mainstream dietary items. Increasing consumer demand for more 'exotic' seafood has lead to an increase in the harvesting of non-fisheries species of crustaceans (Carney and Kvitek 1991). Market demands, where the presentation of the food has the highest priority, may result in known toxic species being served to unwary customers. In addition, these non-fisheries species can enter the human food chain in other ways through recreational collecting and local gathering made by small communities to supplement existing diets. As this trend continues, PST monitoring of crustacea will become increasingly important to ensure the security of public health.

Bioassays that are presently available can be divided into two different, but not mutually exclusive, categories. The first division is those that rely on the biological activity of PSTs, namely Na channel blockade, and those that do not. The second being whole organism versus an *in vitro* approach. Whole organism bioassays, such as the mouse and insect bioassays, entrain the complexities of toxin pharmacokinetics like metabolism, uptake and elimination, as well as biological variation between individual test animals in their responses. It is utterly dependent on the action of PSTs on the Na channel. One can reduce these assays to systems whereby the Na channel action is isolated from the vagaries of whole animal systems such as the cell-based bioassays, Na channel radioligand binding and the biosensor. The remaining bioassays are not dependent on mammalian potency mediated by the Na channel but still rely upon a biological action of the toxin with a target receptor system, for example, saxiphilin or an antibody.

One of the defining properties of an assay is its detection limit and the reported detection limits for the assays described herein are listed in Table 3. The limits of an assay will impact upon the levels that one can apply for regulation. For example, one cannot set a limit below that which an assay cannot detect. If the purpose of an assay is for management, particularly of human safety, then the assay must reflect human toxicity. This raises two issues, namely, one cannot experiment directly on humans so all methods will always be a proxy for human toxicity and that each toxin has a different efficacy. This latter point is minimized somewhat by subjecting samples to the AOAC extraction protocol which converts many of the less toxic PSTs to their more potent cousins.

Table 3. Detection limits for bioassays that have been used to determine PST levels in crustacea

Bioassay	Detection limit (µg/100g tissue STX equivalents*)	Reference
Mouse lethality	36	Johnson and Mulberry 1966
Tissue culture	2	Jellett *et al.* 1992
Na channel radioligand binding	3	Negri and Llewellyn 1998
Saxiphilin radioligand binding	3	Negri and Llewellyn 1998
Ridascreenä dcELISA	0.2	Usleber *et al.* 1997
Biosensor	Not reported	Cheun *et al.* 1998

* STX equivalents is the standard way to represent the amount of PST within a sample. It assumes that all PSTs have been converted to STX or have been normalised to STX toxicity.

Filter feeding organisms are prone to bioaccumulating toxins from their diet. This includes the molluscan shellfish which is the predominant problem that one faces with PSTs. Most of the crabs that become toxic with PSTs are not filter feeders but omnivores. Therefore, one cannot use the presence of a HAB as a predictor of potential crustacean toxicity. Direct testing is the only possible means. The fact that we do not yet fully understand the source or the means by which PSTs accumulate in these omnivores begs the question as to how many other species of crustacea may become toxic of which we are not aware. Without a long history of testing, which one inadvertently achieves from human consumption of an established fisheries, then one cannot assume that crabs are always edible.

References

Adams, W. and S. Furfari. 1984. Evaluation of laboratory performance of the AOAC method for PSP toxin in shellfish. *J. Assoc. Off. Anal. Chem. Internat.* 67: 1147-1148.

Alcala, A.C., L.C. Alcala, J.S. Garth, D. Yasumura, and T. Yasumoto. 1988. Human fatality due to ingestion of the crab *Demania reynaudii* that contained a palytoxin-like toxin. *Toxicon* 26: 105-107.

Alfonso, A., M. Louzao, M. Vieytes, and L. Botana. 1994. Comparative study of the stability of saxitoxin and neosaxitoxin in acidic solutions and lyophilized samples. *Toxicon* 32: 1593-1598.

Beani, L., C. Bianchi, F. Guerrini, L. Marani, R. Pistocchi, M. C. Tomasini, A. Ceredi, A. Milandri, R. Poletti, and L. Boni. 2000. High sensitivity bioassay of paralytic (PSP) and amnesic (ASP) algal toxins based on the fluorometric detection of $[Ca^{2+}]_i$ in rat cortical primary cultures. *Toxicon* 38: 1283-1297.

Beltran, E. C. and B. A. Neilan. 2000. Geographical segregation of the neurotoxin-producing cyanobacterium *Anabaena circinalis*. *Appl. Environ. Microbiol.* 66: 4468-4474.

Bowling, L. and P. Baker. 1996. Major cyanobacterial bloom in the Barwon-Darling River, Australia, in 1991, and underlying limnological conditions. *Mar. Freshw. Res.* 47: 643-647.

Boyer, G.L., J.J. Sullivan, M.LeBlanc, and R.J. Andersen. 1985. The assimilation of PSP toxins by the copepod *Tigriopus californicus* from dietary *Protgonyaulax catenella*. Pages 407-412 in *Toxic Dinoflagellates*. D.M. Anderson, A.W. White, and D.G. Baden, eds., Elsevier Science Publishing Company Inc. New York, USA.

Buckley, L.J., M. Ikawa, and J.J. Jr. Sasner. 1976. Isolation of *Gonyaulax tamarensis* toxins from soft shell clams (*Mya arenaria*) and a thin-layer chromatographic-fluorometric method for their detection. *J. Agric. Food Chem.* 24: 107-111.

Carlson, R., M. Lever, B. Lee, and P. Guire. 1984. Development of immunoassays for the PSA: a radioimmunoassay for saxitoxin. *Seafood Toxins* 262: 181-192.

Carmichael, W.W., W.R. Evans, Q.Q. Yin, P. Bell, and E.G. Moczydlowski. 1997. Evidence for paralytic shellfish poisons in the freshwater cyanobacterium *Lyngbya wollei* (Farlow ex Gomont) comb. nov. *App. Environ. Microbiol.* 63: 3104-3110.

Carney, D. and R.G. Kvitek. 1991. Assessment of non-game marine invertebreate harvest in Washington. Washington Department of Wildlife Report. Department of Wildlife. Washington D.C.

Carumbana, E.E., A.C. Alcala, and E.P. Ortega. 1976. Toxic marine crabs in Southern Negros, Philippines. *Silliman J.* 23: 265-278.

Chang, F.H., D.M. Anderson, D.M. Kulis, and D.G. Till. 1997. Toxin production of *Alexandrium minutum* (Dinophyceae) from the Bay of Plenty, New Zealand. *Toxicon* 35: 393-409.

Cheun, B.S., M. Loughran, T. Hayashi, Y. Nagashima, and E. Watanabe. 1998. Use of a channel biosensor for the assay of paralytic shellfish toxins. *Toxicon* 36: 1371-1381.

Chu, F. and X. Huang. 1992. Production and characterisation of antibodies against neosaxitoxin. J. Assoc. Off. Anal. Chem. 75: 341-345.

Chu, F.S. and Fan, T.S. 1985. Indirect enzyme-linked immunosorbent assay for saxitoxin in shellfish. *J. Assoc. Off. Anal. Chem. Internat.* 68: 13-16, 1985.

Ciminiello, P., E. Fattorusso, M. Forino, and M. Montresor. 2000. Saxitoxin and neosaxitoxin as toxic principles of *Alexandrium andersoni* (Dinophyceae) from the Gulf of Naples, Italy. *Toxicon* 38:1871-1877.

Doucette, G.J., M.M. Logan, J.S. Ramsdell, and F.M. Van Dolah. 1997. Development and preliminary validation of a microtiter plate-based receptor binding assay for paralytic shellfish poisoning toxins. *Toxicon* 35: 625-636.

Ferreira, F.M., J.M. Soler, M.L. Fidalgo, and P. Fernandez-Vila. 2001.PSP toxins from *Aphanizomenon flos-aquae* (Cyanobacteria) collected in the Crestuma-Lever reservoir (Douro river, northern Portugal). *Toxicon* 39: 757-761.

Gallacher, S., K.J. Flynn, J.M. Franco, E.E. Brueggemann, and H.B. Hines. 1997. Evidence for production of paralytic shellfish toxins by bacteria associated with *Alexandrium* spp. (Dinophyta) in culture. *Appl. Environ. Microbiol.* 63: 239-245.

Garth, J. S. 1971. *Demania toxica*, a new species of poisonous crab from the Philippines. *Micronesica* 7: 179-183.

Garth, J.S. and A.C. Alcala. 1977. Fatalities from crab poisoning on Negros Island, Philippines. *Toxicon* 15: 169-170.

Hallegraeff, G.M., D.A. Steffensen, and R. Wetherbee. 1988. Three estuarine Australian dinoflagellates that can produce paralytic shellfish toxins. *J. Plankton Res.* 10: 533-541.

Hallegraeff, G. M. 1993. A review of harmful algal blooms and their apparent global increase. *Phycologia* 32: 79-99.

Harada, T., Y. Oshima, H. Kamiya, and T. Yasumoto. 1982. Confirmation of paralytic shellfish toxins in the dinoflagellate *Pyrodinium bahamense* var. *compressa* and bivalves in Palau. *Bull. Jpn. Soc. Sci. Fish.* 48: 821-825.

Holthuis, L. 1968. Are there poisonous crabs? *Crustaceana* 15: 215-222.

Huang, X., K. Hsu, and F. Chu. 1996. Direct competitive enzyme-linked immunosorbent assay for saxitoxin and neosaxitoxin. *J. Agric. Food Chem.* 44: 1029-1035.

Jellett, J.F., L.J. Marks, J.E. Stewart, M.L. Dorey, W. Watson-Wright, and J.F. Lawrence. 1992. Paralytic shellfish poison (saxitoxin family) bioassays: automated endpoint determination and standardization of the *in vitro* tissue culture bioassay, and comparison with the standard mouse bioassay. *Toxicon* 30: 1143-1156.

Johnson, H. and G. Mulberry. 1966. Paralytic shellfish poison: serological assay by passive haemagglutination and bentonite flocculations. *Nature* 211: 747-748.

Jones, G. 1992. Algal blooms in the Murrumbidgee River. *Farmers Newsl.* 139: 9-12.

Kodama, M., T. Ogata, Y. Fukuyo, T. Ishimaru, S. Wisessang, K. Saitanu, V. Panichyakarn, and T. Piyakarnchana. 1988a. *Protogonyaulax cohorticula*, a toxic dinoflagellate found in the Gulf of Thailand. *Toxicon* 26: 707-712.

Kodama, M., T. Ogata, and S. Sato. 1988b. Bacterial production of saxitoxin. *Agric. Biol. Chem.* 52: 1075-1077.

Kogure, K., M.L. Tamplin, U. Simidu, and R.R. Colwell. 1988. A tissue culture assay for tetrodotoxin, saxitoxin and related toxins. *Toxicon* 26: 191-197.

Kotaki, Y., M. Tajiri, Y. Oshima, and T. Yasumoto. 1983. Identification of a calcareous red alga as the primary source of paralytic shellfish toxins in coral reef crabs and gastropods. *Bull. Jpn. Soc. Sci. Fish.* 49: 283-286.

Koyama, K., T. Noguchi, A. Uzu, and K. Hashimoto. 1983. Resistibility of toxic and non-toxic crabs against paralytic shellfish poison and tetrodotoxin. *Nippon Suisan Gakkaishi* 49: 485-489.

Lawrence, J.F., M. Maher, and W. Watson-Wright. 1994. Effect of cooking on the concentration of toxins associated with paralytic shellfish poison in the lobster hepatopancreas. *Toxicon* 32: 57-64.

Lawrence, J.F., C. Menard, and C. Cleroux. 1995. Evaluation of prechromatographic oxidation for liquid chromatographic determination of paralytic shellfish poisons in shellfish. *J. Assoc. Off. Anal. Chem. Internat.* 78: 514-520.

Laxminarayan, R. 2001. Economic consequences of Red Tides. [Web Document]. Available: http://www.nwfsc.noaa.gov/hab/newsletter/economic_impacts.htm [2001, April 27].

Laycock, M.V., P. Thibault, S.W. Ayer, and J.A. Walter. 1994. Isolation and purification procedures for the preparation of paralytic shellfish poisoning toxin standards. *Natural Toxins* 2: 175-183.

Lipkind, G.M. and H.A. Fozzard. 1994. A structural model of the tetrodotoxin and saxitoxin binding site of the Na^+ channel. *Biophys. J.* 66 :1-13.

Llewellyn, L. E. and R. Endean. 1988. Toxic coral reef crabs from Australian waters. *Toxicon* 26: 1085-1088.

Llewellyn, L.E. and R. Endean. 1989a. Toxins extracted from Australian specimens of the crab, *Eriphia sebana* (Xanthidae). *Toxicon* 27: 579-586.

Llewellyn, L.E. and R. Endean. 1989b. Toxicity and paralytic shellfish toxin profiles of the Xanthid crabs, *Lophozozymus pictor* and *Zosimus aeneus*, collected from some Australian coral reefs. *Toxicon* 27: 596-600.

Llewellyn, L.E., P.M. Bell, and E.G. Moczydlowski. 1997. Phylogenetic survey of soluble saxitoxin-binding activity in pursuit of the function and molecular evolution of saxiphilin, a relative of transferrin. *Proc. Roy. Soc. Lond. Ser. B, Biol. Sci.* 264: 891-902.

Llewellyn, L.E., J. Doyle, and A.P. Negri. 1998. A high-throughput, microtiter plate assay for paralytic shellfish poisons using the saxitoxin-specific receptor, saxiphilin. *Anal. Biochem.* 261: 51-56.

Llewellyn, L.E. and J. Doyle. 2001. Microtitre plate assay for paralytic shellfish toxins using saxiphilin: gauging the effects of shellfish extract matrices, salts and pH upon assay performance. *Toxicon* 39: 217-224.

Llewellyn, L.E., J. Doyle, J. Jellett, R. Barrett, C. Alison, C. Bentz, and M. Quilliam. 2001a. Measurement of paralytic shellfish toxins in molluscan extracts: comparison of the microtitre plate saxiphilin and sodium channel radioreceptor assays with mouse bioassay, HPLC analysis and a commercially available cell culture assay. *Food Additives and Contaminants* 18: 970-980.

Llewellyn, L.E., A.P. Negri, J. Doyle, P.D. Baker, E.C. Beltran, and B.A. Neilan. 2001b. Radioreceptor assays for sensitive detection and quantitation of saxitoxin and its analogues from strains of the freshwater cyanobacterium, *Anabaena circinalis*. *Environ. Sci. Technol.* 35: 1445-1451.

Louzao, M.C., M.R. Vieytes, J.M.V. Baptista de Sousa, F. Leira, and L.M. Botana. 2001. A fluorimetric method based on changes in membrane potential for screening paralytic shellfish toxins in mussels. *Anal. Biochem.* 289: 869-878.

McCulloch, A., R. Boyd, A. de Freitas, R. Foxall, W. Jamieson, M. Laycock, M. Quilliam, J. Wright, V. Boyko, J. McLaren, M. Miedema, R. Pocklington, E. Arsenault, and D. Richard. 1989. Zinc from oyster tissue as causative factor in mouse deaths in official bioassay for paralytic shellfish poison. *J. Assoc. Off. Anal. Chem. Internat.* 72: 384-386.

McElhiney, J., L.A. Lawton, C. Edwards, and S. Gallacher. 1998. Development of a bioassay employing the desert locust (*Schistocerca gregaria*) for the detection of saxitoxin and related compounds in cyanobacteria and shellfish. *Toxicon* 36: 417-420.

Mahmood, N.A. and W.W. Carmichael. 1986. Paralytic shellfish poisons produced by the freshwater cyanobacterium *Aphanizomenon flos-aquae* NH-5. *Toxicon* 24: 175-186.

Manger, R.L., L.S. Leja, S.Y. Lee, J.M. Hungerford, and M.M. Wekell. 1993. Tetrazolium-based cell bioassay for neurotoxins active on voltage-sensitive sodium channels: semiautomated assay for saxitoxins, brevetoxins, and ciguatoxins. *Anal. Biochem.* 214: 190-194.

Manger, R., L. Leja, S. Lee, J. Hungerford, Y. Hokama, R. Dickey, H. Granade, R. Lewis, T. Yasumoto, and M. Wekell. 1995. Detection of sodium channel toxins: directed cytotoxicity assays of purified ciguatoxins, brevetoxins, saxitoxins, and seafood extracts. *J. Assoc. Off. Anal. Chem. Internat.* 78: 521-527.

Morabito, M.A. and E. Moczydlowski. 1994. Molecular cloning of bullfrog saxiphilin: a unique relative of the transferrin family that binds saxitoxin. *Proc. Natl. Acad. Sci. USA* 91: 2478-2482.

Morse, E. 1977. Paralytic shellfish poisoning: a review. *J. Am. Vet. Med. Associ.* 171: 1178-1180.

Negri, A., G. Jones, and M. Hindmarsh. 1995. Sheep mortality associated with paralytic shellfish poisons from the cyanobacterium *Anabaena circinalis*. *Toxicon* 33: 1321-1329.

Negri, A. and L. Llewellyn. 1998. Comparative analyses by HPLC and the sodium channel and saxiphilin ^{3}H-saxitoxin receptor assays for paralytic shellfish toxins in crustaceans and molluscs from tropical North West Australia. *Toxicon* 36: 283-298.

Nelinda, M.,V. Dimanlig, and F. J. R. Taylor. 1985. Extracellular bacteria and toxin production in *Protogonyaulax* species. Pages 103–108 in *Toxic Dinoflagellates*. D.M. Anderson, A.W. White, and D.G. Baden, eds., Elsevier Science Publishing Company Inc. New York, USA.

Official Methods of Analysis. 15th Ed. 1990. Association of Official Analytical Chemists, Arlington, VA, USA. Section 959.08.

Onodera, H., M. Satake, Y. Oshima, T. Yasumoto, and W.W. Carmichael. 1997. New saxitoxin analogues from the freshwater filamentous cyanobacterium *Lyngbya wollei*. *Natural Toxins* 5: 146-151.

Oshima, Y., M. Hasegawa, T. Yasumoto, G. Hallegraeff, and S. Blackburn. 1987. Dinoflagellate *Gymnodinium catenatum* as the source of paralytic shellfish toxins in Tasmanian shellfish. *Toxicon* 25: 1105-1111.

Oshima, Y. 1995. Postcolumn derivatisation liquid chromatographic method for paralytic shellfish toxins. *J. Assoc. Offi. Anal. Chem. Internat.* 78: 528-532.

Park, D.L., W.N. Adams, S.L. Graham, and R.C. Jackson. 1986a. Variability of mouse bioassay for determination of paralytic shellfish poisoning toxins. *J. Assoc. Offi. Anal. Chem. Internat.* 69: 547-550.

Park, D.L., I. Aguirre-Flores, W.F. Scott, and E. Alterman. 1986b. Evaluation of chick embryo, brine shrimp and bacterial bioassays for saxitoxin. *J. Toxicol. Environ. Health* 18. 589-594.

RaLonde, R. 1996. Paralytic shellfish poisoning: the Alaska problem. *Alaska's Mar. Resources* 8: 1-7.

Raj, U., H. Haq, Y. Oshima, and T. Yasumoto. 1983. The occurrence of paralytic shellfish toxins in two species of xanthid crab from Suva barrier reef, Fiji Islands. *Toxicon* 21: 547-551.

Rogers, R. and H. Rapaport. 1980. The pKa's of saxitoxin. *J. Am. Chem. Soc.* 102: 7335-7339.

Ross, M. R., A. Siger, and B. C. Abbott. 1985. The house fly: an acceptable subject for paralytic shellfish toxin bioassay. Pages 433–438 in *Toxic Dinoflagellates*. D.M. Anderson, A.W. White, and D.G. Baden, eds., Elsevier Science Publishing Company Inc. New York, USA.

Runnegar, M., A. Jackson, and I. Falconer. 1988. Toxicity to mice and sheep of a bloom of the cyanobacterium (blue-green alga) *Anabaena circinalis*. *Toxicon* 26: 599-602.

Schantz, E., J. Lynch, G. Vayvada, K. Matsumoto, and H. Rapoport. 1966. The purification and characterization of the poison produced by *Gonyaulax catenella* in axenic culture. *Biochemistry* 5: 1191-1195.

Silva, E.S. 1982. Relationship between dinoflagellates and intracellular bacteria. Proceedings of the IV IUPAC Symposium on Mycotoxins and Phycotoxins, p 8.

Smith, D. and D. Kitts. 1994. Development of a monoclonal-based enzyme-linked immunoassay for saxitoxin-induced protein. *Toxicon* 32: 317-323.

Smith, D., D. Kitts, and P. Townsley. 1989. Serological cross-reactions between crab saxitoxin-induced protein and paralytic shellfish poison-contaminated shellfish. *Toxicon* 27: 601-606.

Soares, M. 1998. Bairdi Tanner crab notification. [Web Document]. Available: http://www.state.ak.us/local/akpages/ENV.CONSERV/deh/seafood/psp/tannerSE.htm [2001, April 27]

Sommer, H. 1932. The occurrence of paralytic shellfish poison in the common sand crab. *Science* 76, 574-575.

Sommer, H. and K.F. Meyer. 1937. Paralytic shellfish poisoning. *Arch. Pathol.* 24: 560-598.

Steward, J., E. Ornellas, K. Beernink, and W. Northway. 1968. Errors in the technique of intraperitoneal injection of mice. *App. Microbiol.* 16: 1418-1419.

Strichartz, G., T. Rando, S. Hall, J. Gitschier, L. Hall, B. Magnani, and C.H. Bay. 1986. On the mechanism by which saxitoxin binds to and blocks sodium channels. *Ann. New York Acad. of Sci.* 479: 96-112.

Sullivan, J.J. and W.T. Iwaoka. 1983. High pressure liquid chromatographic determination of toxins associated with paralytic shellfish poisoning. *J. Assoc. Offi. Anal. Chem. Internat.* 66: 297-303.

Teh, Y. and J. Gardiner. 1970. Toxin from the coral reef crab *Lophozozymus pictor*. *Pharmacol. Res. Commun.* 2: 251-256.

Teh, Y.F. and J.E. Gardiner. 1974. Partial purification of *Lophozozymus pictor* toxin. *Toxicon* 12: 603-610.

Truman, P. and R.J. Lake. 1996. Comparison of mouse bioassay and sodium channel cytotoxicity assay for detecting paralytic shellfish poisoning toxins in shellfish extracts. *J. Assoc. Offi. Anal. Chem. Internat.* 79: 1130-1133.

Tsai, Y.H., D.F. Hwang, T.J. Chai, and S.S. Jeng. 1995. Occurrence of tetrodotoxin and paralytic shellfish poison in the Taiwanese crab *Lophozozymus pictor*. *Toxicon* 33: 1669-1673.

Tsai, Y.H., D.F. Hwang, T.J. Chai, and S.S. Jeng. 1996. Occurrence of paralytic toxin in Taiwanese crab *Atergatopsis germaini*. *Toxicon* 34: 467-474.

Usleber, E., M. Donald, M. Straka, and E. Martlbauer. 1997. Comparison of enzyme immunoassay and mouse bioassay for determining paralytic shellfish poisoning toxins in shellfish. *Food Additives Contaminants* 14: 193-198.

Usleber, E., E. Schneider, G. Terplan, and M.V. Laycock. 1995. Two formats of enzyme immunoassay for the detection of saxitoxin and other paralytic shellfish poisoning toxins. *Food Additives Contaminants* 12: 405-413.

Weigele, J.B. and R.L. Barchi. 1978a. Analysis of saxitoxin binding in isolated rat synaptosomes using a rapid filtration assay. *FEBS Letters* 91: 310-314.

Weigele, J.B. and R.L. Barchi. 1978b. Saxitoxin binding to the mammalian sodium channel. Competition by monovalent and divalent cations. *FEBS Letters*, 95: 49-53.

Yasumoto, T., Y. Oshima, M. Tajiri, and Y. Kotaki. 1983. Paralytic shellfish toxins in previously unrecorded species of coral reef crabs. *Nippon Suisan Gakkaishi* 49: 633-636.

Yasumura, D., Y. Oshima, T. Yasumoto, A.C. Alcala, and L.C. Alcala. 1986. Tetrodotoxin and paralytic shellfish toxins in Philippine crabs. *Agric. Biol. Chem.* 50: 593-598.

Antimicrobial Peptides for Fish Disease Control

Aleksander Patrzykat and Robert E.W. Hancock

Department of Microbiology and Immunology, University of British Columbia, Vancouver, B.C., Canada V6T 1Z3

Introduction

A recent FAO/NACA/WHO report identifies aquaculture as one of the fastest growing food production sectors in the world, and focuses in particular on food safety issues associated with products from aquaculture (Food Safety 1999, p.2). One of the biggest challenges of aquaculture is its cost-effectiveness, and fish loss from infectious disease is a significant problem. While vaccination is commonly employed in the aquaculture setting, its effectiveness is limited by seasonal, environmental and age/size constraints (Smith *et al.* 2000). Antibiotics are therefore extensively used. The FAO/NACA/WHO report identifies several hazards associated with the use of antibiotics in aquaculture, including the transmission of bacterial antibiotic resistance genes from marine environments to humans (Food Safety 1999, p. 26), and antimicrobial drug residues in edible tissues (Food Safety 1999, p. 27). In addition, the authors of the report point out that intensive aquaculture is a relatively new field, and pharmaceutical companies have tended to borrow antimicrobials from other areas of veterinary medicine instead of developing chemotherapeutants specifically for use in the aquatic milieu (Food Safety 1999, p. 22).

As we have advanced into the era of bacterial resistance to conventional antibiotics, we are looking to natural immune defenses to provide us with novel solutions. It is in this search that we now have an opportunity to develop new agents for infectious disease control, derived from fish immune defenses, and intended for aquaculture use. Among the most promising contenders are cationic antimicrobial peptides. These peptides have been isolated from virtually all groups of living organisms, and shown to play a role in their innate defenses. Given that fish at times must rely heavily on their innate immunity (Bly and Clem 1991, 1992), cationic peptides are of particular importance in this field. The increasing pace at which cationic antimicrobial peptides are being commercialized, as well as the potential for developing transgenic species overproducing these peptides, should ensure that peptide antimicrobials will eventually find their way to the realm of aquaculture medicine.

Cationic Antimicrobial Peptides — Overview

Antimicrobial cationic peptides are becoming increasingly well understood as important components of the innate defenses of all species of life (Hancock and Diamond 2000, Hancock and Scott 2000, Boman 1995, Ganz and Lehrer 1997). They are virtually ubiquitous, being a major form of antimicrobial defences in plants and insects, and an important component of local defenses in amphibians, crustaceans, fish, birds, food and domestic animals, and man. These endogenous antimicrobial peptides are typically cationic (i.e. contain excess lysine and arginine residues) molecules composed of 12 to 45 amino acid residues. They generally share little sequence homology but fall into four broad structural classes: b-sheet peptides comprising 2-4 b-strands interconnected by disulphide bridges; amphipathic a-helical peptides; extended peptides with a predominance of 1-2 amino acids such as tryptophan, proline or histidine; and loop peptides that are circularized with a single disulphide bridge. In the case of the a-helical and extended peptides, these structures form in the presence of microbial membranes which are important for peptide activity.

The peptides have two types of activities: an ability to kill microbes directly, and stimulation of other elements of innate immunity. Regarding the direct antimicrobial activities, different peptides have antibacterial, antifungal anti-viral and anti-parasite properties. The best antimicrobial peptides kill susceptible bacteria *in vitro* at concentrations ranging from 0.25 to 4 μg/ml. In addition, they have the ability to work in synergy with one another, and with other elements of the host immune response such as lysozyme. While more potent antibiotics exist, there are definite advantages to these peptides, including an ability to rapidly kill target cells, unusually broad spectra of activity, activity against some of the more serious antibiotic-resistant pathogens in clinics, synergy with conventional antibiotics particularly against resistant mutants, and the relative difficulty in selecting resistant mutants *in vitro*. Therefore they are being developed as a novel class of antimicrobial agents, with two peptides currently in clinical trials (Hancock 1998), as well as being used as the basis for making transgenic, disease-resistant plants and animals (Broglie *et al.* 1991, Osusky *et al.* 2000).

Natural cationic antimicrobial peptides may be produced constitutively or only after infection or injury. Their role in innate immunity extends beyond direct killing, and they can interact directly with host cells to modulate inflammatory process and innate defences (Hancock and Diamond 1998, Hancock and Scott 2000). Their reported activities in mammals include stimulation of mast cell degranulation leading to histamine release and consequent increase in blood vessel permeability (vasodilation), promotion of chemotaxis of neutrophils and T helper cells resulting in leukocyte recruitment to the infection site, promotion of non-opsonic phagocytosis, inhibition of fibrinolysis by tissue plasminogen activator, thus reducing the spreading of bacteria, tissue/wound repair through promotion of fibroblast chemotaxis and growth, and inhibition of tissue injury by inhibiting certain proteases such as furin and cathepsin. In addition they neutralize the stimulatory effects of bacterial molecules such as gram-negative bacterial lipopolysaccharide (LPS), gram-positive bacterial lipoteichoic acid (LTA) and bacterial CpG-motif DNA, both directly and indirectly by interaction with the host cells they stimulate. In the case of LPS, which is also called endotoxin, this results in neutralization of the ability of this molecule to cause harmful sepsis and endotoxic shock.

Natural Antimicrobial Peptides of Marine Organisms

Introduction

While cationic antimicrobial peptides have been isolated from most species, they are a particularly important defense mechanism in invertebrates and lower vertebrates, whose immune systems lack several of the features present in mammals. Indeed some of the first peptides ever isolated came from insects. Since this early work, the focus has tended to move to peptides of mammalian origin, as medical applications appear to drive much of today's research.

From this perspective, progress in researching and reporting cationic antimicrobial peptides of fish origin has been slow. This is unfortunate given that fish possess a limited range of antibody-mediated and cell-mediated secondary immune responses, and appear to depend greatly on non-specific defenses, thus being potentially good sources of cationic peptides. To date, several notable peptides have been isolated, and several synthetic derivatives of potential importance in controlling infectious disease in fish have been developed (summarized in Table 1).

Antimicrobial Peptides Isolated from Fish

The earliest and best-studied group of fish antimicrobial peptides were the natural pardaxins and scores of their synthetic derivatives. Pardaxins were

Table 1. Overview of antimicrobial cationic peptides of marine origin.

Peptides	Natural sources	References
Peptides originating from fish		
Pardaxins	Toxic secretions of sole fish	Shai 1994
Pleurocidin	Skin mucus secretions of winter flounder	Cole *et al.* 1997
Misgurin	Loach	Park *et al.* 1997
Histone peptides	Injured catfish	Park *et al.* 1998
	Coho salmon blood and mucus	Patrzykat *et al.* 2001
	Channel catfish	Robinette *et al.* 1998
Other antimicrobial peptides and proteins	Skin mucosa of carp	Lemaitre *et al.* 1996
	Mucus cocoon of queen parrotfish	Viedler *et al.* 1999
	Mucus of rainbow trout	Smith *et al.* 2000
Sample peptides originating from other marine organisms		
Tachyplesins	Japanese horseshoe crab	Nakamura *et al.* 1988
Polyphemusins	American horseshoe crab	Miyata *et al.* 1989
Penaeidins	Tropical shrimp	Destomieux *et al.* 1997
MGDs, mytilins, and myticins	Mussels	Mitta *et al.* 2000

isolated from the toxic secretions of at least two separate species of sole fish (Lazarovici *et al.* 1986, Shai 1994), and described as predator-repellant peptides (Shai 1994). The structures and activities of natural pardaxins have been studied extensively. Their potential for forming amphiphilic a-helices, and their ability to permeate phospholipid membranes in a manner comparable to that of dermaseptins and magainins, resulted in the inclusion of pardaxins among cationic antimicrobial peptides. This classification is however inaccurate as natural pardaxins have a net charge of only +1, compared to much higher charges carried by conventional cationic peptides. In addition, natural pardaxins are hemolytic and highly neurotoxic, and hence are not host-friendly antimicrobial peptides. However, several pardaxin-derived, highly cationic synthetic peptides have been shown to exhibit improved antimicrobial activities and reduced toxicities (Oren and Shai 1996), and are described below.

Perhaps the most conventional cationic antimicrobial peptide from fish described to date, is the 25-amino acid pleurocidin (Figure 1), isolated from skin mucous secretions of winter flounder *Pleuronectes americanus* (Cole *et al.* 1997). Pleurocidin is highly cationic and shares homology with the frog dermaseptin and medfly ceratotoxin. It is active against a wide spectrum of bacteria, including the fish pathogens *Aeromonas salmonicida* and *Vibrio anguillarum*, and it has recently been localized to the mucin granules of skin and intestinal goblet cells (Cole *et al.* 2000). The pleurocidin gene has been described (Cole *et al.* 2000) and its upstream region shows similarity to that found in mammalian host defense genes. Specifically, it includes

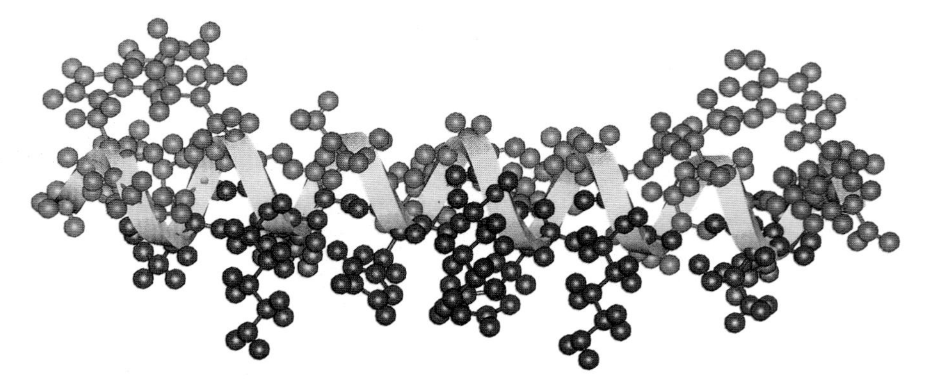

Figure 1. A rendition of the probable amphipatic a-helix of pleurocidin. Although the structure of pleurocidin has not yet been solved, when modelled as an a -helix the peptide exhibits a separation of hydrophilic residues, depicted in blue, from hydrophobic residues, depicted in red.

consensus binding sequences identical to the NF-IL6 and alpha and gamma interferon response elements. In addition, several pleurocidin-like peptides have recently been cloned from the winter flounder (Douglas *et al.* 2001). It is therefore not surprising that pleurocidin and its derivatives are now being extensively studied for their potential to control bacterial disease in fish.

Another cationic antimicrobial peptide isolated from fish is the mudfish *Misgurnus anguillicaudatus* peptide misgurin (Park *et al.* 1997). This highly charged (+7) peptide was originally reported to be active against a range of gram-negative and gram-positive bacteria, as well as fungi, but we were unable to reproduce the activity of synthetic misgurin in our laboratory (Jia *et al.* 2000). We are not aware of any further reports regarding the activity or expression of this peptide.

The discoverers of misgurin have also reported isolating a potent antimicrobial peptide from the mucosal secretions of injured catfish *Parasilurus asotus* (Park *et al.* 1998). This 19-amino acid peptide, parasin I, is highly basic, and 18 of its residues are identical to the N-terminal region of buforin I, an histone-H2A-derived peptide from toads isolated by the same group (Kim *et al.* 1996). While this discovery may seem unusual, histone proteins are well conserved across species and have since been implicated in antimicrobial activity in channel catfish *Ictalurus punctatus* (Robinette *et al.* 1998) and coho salmon *Oncorhynchus kisutch* (Patrzykat *et al.* 2001). In our hands, a histone-H1-derived peptide was co-induced by infection and LPS in the mucus and serum of salmon (Figure 2). By itself it did not appear to have any antibacterial activity, but it was able to strongly potentiate the activities of pleurocidin, mucous and serum extracts, and egg lysozyme, against *Vibrio anguillarum* and *Aeromonas salmonicida.*

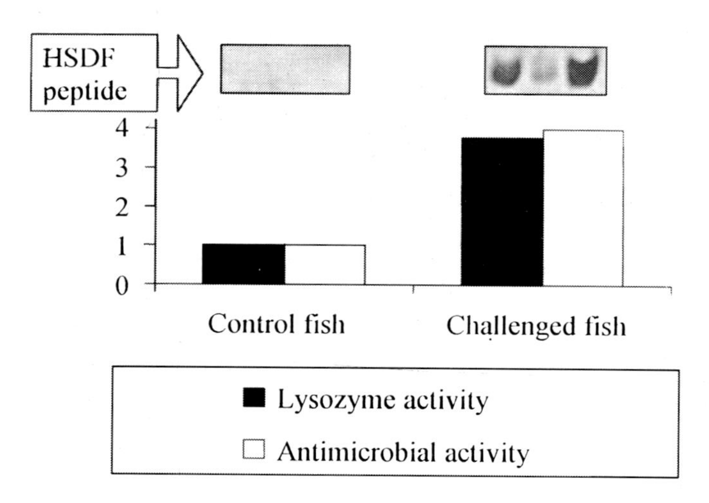

Figure 2. Induction of cationic peptide-associated antimicrobial activity in coho salmon serum. An increase in the lysozyme and antimicrobial activities was observed in disease-challenged coho compared to unchallenged control fish. Acid exctracts from the serum of challenged fish also revealed the presence of a cationic peptide HSDF-1 (Patrzykat *et al.* 2001).

Several other reports related to cationic antimicrobial peptides merit mentioning in this section. Two novel antimicrobial proteins (27 kDa and 31 kDa) were isolated from the skin mucosa of carp *Cyprinus carpio* (Lemaitere *et al.* 1996) and a 21 kDa antibacterial protein was isolated from the mucus cocoon of the queen parrotfish (Viedeler *et al.* 1999). While these are not cationic antimicrobial peptides, they reinforce the concept that a variety of innate antimicrobials are present at fish mucosal surfaces. This idea is further supported by a recent report of an unidentified cationic antimicrobial peptide being isolated alongside lysozyme and lysozyme-like proteins from the mucus of unstimulated rainbow trout *Oncorhynchus mykiss* (Smith *et al.* 2000) and other reports (Ebran *et al.* 1999).

Antimicrobial Peptides from other Marine Organisms

The cationic antimicrobial peptides of other co-habitants of the aquatic environment, should also be considered when searching for sources of cationic peptides with potential for controlling disease in fish. However, such a large number of antimicrobial peptides are known that a comprehensive review of all falls beyond the scope of this chapter. The following provides a few prominent examples of cationic peptides from the aquatic milleau.

Both the Japanese (*Tachypleus tridentatus*) and the American (*Limulus polyphemus*) species of horseshoe crab, the oldest marine arthropod, have provided researchers with potent cationic antimicrobial peptides: tachyplesins (Nakamura *et al.* 1988) and polyphemusins (Miyata *et al.* 1989), respectively. In addition, a large 6.5 kDa antibacterial peptide has been isolated from the hemocytes of the shore crab *Carcinus maenas* (Schnapp *et al.* 1996), and shown to have similarity to bactenecin-7. Tachyplesins and polyphemusins have been extensively studied since they were discovered.

Motivated by infectious disease problems associated with penaeid shrimp aquaculture, researchers isolated three new antimicrobial peptides from the tropical shrimp *Penaeus vannamei* (Destoumieux *et al.* 1997, 1999, 2000). These large peptides (5.5 to 6.6 kDa) are proline-rich in their N-termini and have three disulphide bonds near their C-termini. The penaeidins have been cloned and their C-termini were shown to be amidated. These peptides were further described by their discoverers as active against gram-positive bacteria and fungi, but not against gram-negative bacteria, and localized to shrimp granulocytes, where they are released upon microbial challenge (Destoumieux *et al.* 2000). While penaeidins are still rather obscure, they are a good example of the many individual efforts, currently under way, aimed at identifying new marine antimicrobials.

Another example of such efforts is the isolation and subsequent characterization of three types of cysteine-rich, cationic antimicrobial peptides from mussel hemocytes: *Mytilus galloprovincialis* defensins (MGDs), mytilins and myticins (Hubert *et al.* 1996, 1997, Mitta *et al.* 1999, 2000, Yang *et al.* 2000). Also, histidine-rich, amidated, a-helical antimicrobial peptides, clavanins, have been isolated from the solitary tunicate *Styela clava* (Lee *et al.* 1997), as tunicate hemocytes are thought to be good models for studying the evolution of leukocyte-mediated host defenses. In conclusion, it appears that virtually all groups of marine organisms are potential sources of antimicrobial peptides for use in aquaculture, and for other applications.

Role of Cationic Peptides in Fish Immunity

Secondary, pathogen-specific immune responses in teleosts are confined to immunologically permissive temperatures and limited by fish size or age restrictions (Bly and Clem 1991, 1992, Tatner 1996). In addition, the fish secondary immune response is relatively brief. This drives fish to rely more heavily on their non-specific innate defenses than mammals.

As reviewed earlier, cationic antimicrobial peptides have already been isolated from fish mucosal secretions (Cole *et al.* 1997), which constitute the interface between environmental insults and the organism. Also, the context of the gene encoding flounder pleurocidin further indicates that the expression of fish antimicrobial peptides can be induced in response to immunogenic stimuli. Based on experience with other species, fish antimicrobial peptides are likely to be expressed by mucosal lining cells, macrophages and granulocytes. In conjunction with known antimicrobials, such as the leukocyte-secreted lysozyme which is active against gram-positive bacteria, antimicrobial peptides offer protection against a wide spectrum of pathogens.

In addition, it is well known that fish can be exposed to relatively high doses of bacterial endotoxin (as much as 200 mg/kg of body weight), without exhibiting any clinical symptoms (Wedemeyer and Ross 1969, Dalmo *et al.* 1997). While it is possible that this lack of response may be due to the fact that fish cellular responses are less manifest than those observed in mammals, another explanation is that fish possess highly efficient LPS detoxification and neutralization mechanisms. Since cationic antimicrobial peptides can bind lipopolysaccharide and thus inhibit LPS-mediated effects, they may account in part for the low sensitivity of fish to endotoxin.

As host-directed and pathogen-directed activities of cationic peptides are better described, we can expect that the extent of peptide involvement in fish immunity will also be better understood.

Applications of Peptides in Aquaculture

Introduction

Cationic antimicrobial peptides are almost invariably active against wide spectra of bacteria. After accounting for factors such as media salinity and assay incubation temperatures, it can be therefore assumed that a peptide active against gram-negative non-fish pathogens will also be active against fish bacteria. The main thrust in research to date has however been to test and engineer peptides of various origins for use against mammalian pathogens. It is only thanks to the rapid development of transgenic technologies, that peptides are now being tested for use in agriculture and aquaculture, as potential candidates for transgenically expressed disease control agents (Broglie *et al.* 1991, Osusky *et al.* 2000).

We will here review several past and present attempts to optimize cationic peptides of marine and non-marine origin for use in aquaculture, as well as identify the logistic, environmental and financial issues involved in developing cationic peptide technologies.

Therapeutic Potential and Mode of Delivery

In 1990, Kelly *et al.* (1990) described the activities of two synthetic cecropin-derived peptides against eight fish bacterial pathogens, including *V. anguillarum, A. salmonicida, A. hydrophila, Edwardsiella ictaluri,* and *Yersinia ruckeri*. The peptides exhibited activity against all fish bacteria tested. One of these cecropin-derived peptides, LSB-37, was later shown to enhance the resistance of the channel catfish *Ictalurus punctatus* to enteric septicemia caused by *E. ictaluri* (Kelly *et al.* 1994). The peptide in this study was delivered to live fish, together with a bacterial challenge, through injection and through an implanted osmotic pump. The latter administration method, which is considered to approximate the constant production of peptide from an active gene, was successful.

A similar approach was used in our laboratory, where several derivatives of the flounder pleurocidin were constructed and, along with several non-fish-derived peptides were tested for activity against *V. anguillarum* and *A. salmonicida*. An amidated version of the flounder peptide pleurocidin, as well as a non-fish peptide, were delivered to *V. anguillarum*-infected coho salmon via an implanted osmotic pump, and significantly reduced fish mortality from the infection (Jia *et al.* 2000) (Figure 3). It therefore seems likely that a flounder-derived gene, when placed under the control of an appropriate promoter, could be expressed in coho salmon or other fish, and protect them from infection.

Siginificant sections of recent conferences on applied biotechnology (Pacific Rim Biotechnology Conference, Vancouver 2000) and on Molecular Aspects of Fish Genomes and Development (Singapore 2001) were devoted to transgenic technology in fish. Among the reports presented were the attempts, currently under way, to use cationic antimicrobial peptide genes to protect cultured hybrid striped bass (*Morone saxatilis ´ Morone chrysops*) from *Streptococcus iniaie* infections (Burns *et al.* 2001), as well as attempts to use cecropin B transgenes to protect fish from various bacterial infections (Chen 2001).

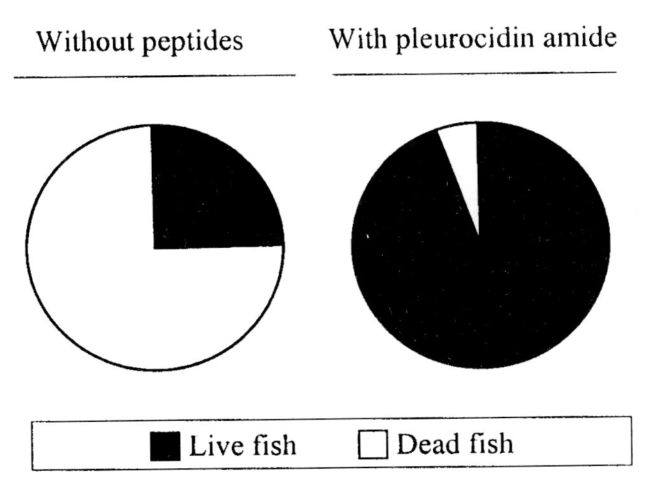

Figure 3. Protection of coho salmon from *Vibrio anguillarum* infections by the cationic peptide pleurocidin (amide). Coho salmon challenged with lethal doses of *Vibrio anguillarum* were protected by pleurocidin amide administered at 250 µg/day via an implanted osmotic pump (Jia *et al.* 2000). Thirty-day cumulative mortalities are shown for groups treated with pleurocidin amide and a saline control (without peptide).

While transgenic technology provides perhaps the best potential for developing cationic antimicrobial peptides into tools for aquaculture, it is not the only avenue. Provision of cationic peptide antibiotics in feed, or through baths or injections, should also be considered, although at this time the cost of producing the required quantities of cationic peptides is prohibitive. As recombinant methods of producing cationic peptides develop, however, the cost may be comparable or lower than the cost of using conventional antibiotics. While the ultimate applications and modes of delivery must certainly be considered, the bulk of fish cationic peptide research to date, as for all cationic peptides, is centered around understanding the rationale behind peptide activity on bacteria and the host, and developing rational approaches to designing new and better peptides.

Rational Engineering of Fish Peptides

As previously described, the sole peptide pardaxin was the first pore-forming peptide isolated from fish. Although structurally close to melittin, with a proline hinge separating two helices, pardaxin carries the net charge of +1 compared to that of +6 for melittin. In an attempt to derive cationic antimicrobial peptides with antimicrobial activities, but without toxicities, from pardaxin, several variant peptides were constructed (Thennarasu and Nagaraj 1995, 1996, 1997, Oren and Shai 1996). In one case, three 18-amino acid constructs based on the N-terminal sequence of pardaxin were made, exhibiting different antimicrobial and hemolytic activities based on whether the proline hinge or lysine were present (Thennarasu and Nagaraj 1996). In another case (Oren and Shai 1996) eight separate pardaxin analogues were constructed and tested. An N-terminal fragment with additional positive charges exhibited good antimicrobial activities and was not hemolytic. In addition, studies aimed at elucidating the mode of action of pardaxin analogues were conducted (Oren and Shai 1996) and the authors of these studies concluded that pardaxin inhibited bacterial growth by totally lysing the bacterial wall. Several other studies describing the activities of pardaxin have also been conducted, with particular focus on its interaction with lipid membranes (Rapaport *et al.* 1991, 1992, 1994, 1996). While these studies did not concentrate on testing of the peptides against fish pathogens, the results can be used as an indication that the engineering of natural fish peptides can produce better antimicrobials.

In our laboratory we constructed several analogues of the winter flounder pleurocidin and its two closest analogues, frog dermaseptin and insect ceratotoxin, and tested them for activity against many pathogens, including *Vibrio anguillarum* and *Aeromonas salmonicida* (Jia *et al.* 2000). In the same study we tested several non-fish-derived cationic peptides, and demonstrated their activities against *V. anguillarum* and *A. salmonicida*. Among the pleurocidin derivatives, the C-terminally amidated version of the native pleurocidin, as well as the C-terminally amidated hybrid of pleurocidin and dermaseptin, exhibited improved antimicrobial activities. The former was further demonstrated to protect coho salmon from *V. anguillarum* infections *in vivo* (Jia *et al.* 2000), while we used the latter to study the mode of action of pleurocidin derivatives (Patrzykat, submitted). In a separate study, several pleurocidin analogues were constructed with various amino acid substitutions, and the relationship between peptide structure and function was further elucidated (Yoshida *et al.* 2000). While we still do not know whether we have transformed native pleurocidin into an ideal antimicrobial for fish disease control, we now understand some of the principles governing the activity of pleurocidin.

Significantly more work has been done on engineering other marine antimicrobial peptides, such as polyphemusins and tachyplesins, and understanding their modes of action, as they are good candidates for disease control agents in mammals (Iwanaga *et al.* 1994, Zhang *et al.* 2000). We are confident that better understanding of the structure/function relationships and mode of action of cationic antimicrobial peptides in general will eventually allow us to construct peptides with better antimicrobial activities against fish pathogens.

Environmental Impact of Developing Peptide Antimicrobials

One of the environmental impacts of aquaculture is that significant quantities of wild stocks are used in feeding cultured fish (Naylor *et al.* 2000). This is why reducing mortalities from infectious diseases in cultured species would improve not only the financial balance sheet of a fish farmer, but would have a downstream positive impact on the wild stocks, in that less of them would be required to produce a given amount of marketable cultured fish. The need to control infectious disease in aquaculture should therefore be evident also from an environmentalist's point of view. The question remains whether cationic peptide technologies would be more or less environmentally sustainable than currently used antibiotics.

It is now known that pathogenic bacteria are becoming resistant to antibiotics. To delay this phenomenon, several measures have been put in place including the establishment of maximum residue levels in fish destined for consumption, and the requirement for elaborate approval procedures as well as guidelines for use of new antibiotics. This reflects the concern among the scientific and health communities that if we continue to overuse antibiotics we are facing a pre-antibiotic era all over again. This is why we need to develop new agents, with lower potential for producing resistant bacteria. Given that cationic antimicrobial peptides are derived from nature, have wide spectra of activity, and do not induce resistance as readily as classical antibiotics, they should be even less objectionable to the environment. There is also a substantial concern that the intensive use of antibiotics in aquaculture leads to antibiotic residues in fish that can potentially induce resistance in people that consume such fish. The use of natural peptides that can be processed in the body by proteases might be considered a substantial advantage over conventional antibiotics in this regard.

Our last comment concerns the use of peptide transgenes. Global scrutiny is evident, but will likely not stop the progress in this field. It will remain the responsibility of the scientist to ensure that transgenic species, with better survival characteristics, do not escape to natural habitats. This concern can and should be addressed at several levels: starting from confining transgenic animals to in-land systems, to producing triploid species and preventing transgenic animals from reproducing. Once the appropriate precautions are taken however, the transgenic technology carries an immense potential and lesser environmental impact than traditional antibiotics.

Acknowledgements

Our research on fish cationic antimicrobial peptides is supported by the Canadian Bacterial Diseases Network and National Science and Engineering Research Council. REWH is a recipient of the Medical Research Council Distinguished Scientist Award, as well as the Canada Research Chair Award.

References

Bly, J.E. and W.M. Clem. 1991. Temperature-mediated processes in teleost immunity: *in vitro* immunosuppression induced by *in vivo* low temperature in channel catfish. *Vet. Immunol. Immunopathol.* 28: 365-377.

Bly, J.E. and W.M. Clem. 1992. Temperature and teleost immune function. *Fish Shellfish Immunol.* 2: 159-171.

Boman H.G. 1995. Peptide antibiotics and their role in innate immunity. *Annu. Rev. Immunol.* 13: 61-92.

Burns, J.C., X. Lauth, V. Nizet, H. Shike, M. Westerman, P. Bulet, and J. VanOlst. 2001. Genetic engineering of disease resistance in fish. Presented at 3rd IUBS Symposium on Molecular Aspect of Fish Genomes and Development, February 18-21, 2001, Singapore.

Broglie, K., I. Chet, M. Holliday, R. Cressman, P. Biddle, C. Knowlton, C.J. Mauvai, and R. Broglie. 1991. Transgenic plants with enhanced resistance to the fungal pathogen *Rhizoctonia solani. Science* 254: 1194-1197.

Chen, T.T. 2001. Recent advances in transgenic fish research. Presented at 3rd IUBS Symposium on Molecular Aspect of Fish Genomes and Development, February 18-21, 2001, Singapore.

Cole, A.M., R.O. Darouiche, D. Legarda, N. Connell, and G. Diamond. 2000. Characterization of a fish antimicrobial peptide: gene expression, subcellular localization, and spectrum of activity. *Antimicrob Agents Chemother.* 44: 2039-2045.

Cole, A.M., P. Weis, and G. Diamond. 1997. Isolation and characterization of pleurocidin, an antimicrobial peptide in the skin secretions of winter flounder. *J. Biol. Chem.* 272: 12008-12013.

Dalmo, R.A., K. Ingebrigtsen, and J. Bogwald. 1997. Non-specific defence mechanisms in fish, with particular reference to the reticuloendothelial system (RES). *J. Fish Dis.* 20: 241-273, 1997.

Douglas, S. E., J. W. Gallant, Z. Gong, and C. Hew. 2001. Cloning and developmental expression of a family of pleurocidin-like antimicrobial peptides from winter flounder, *Pleuronectes americanus* (Walbaum). *Dev. Comp. Immunol.* 25: 137-147.

Destoumieux, D., M. Munoz, P. Bulet, and E. Bachere. 2000. Penaeidins, a family of antimicrobial peptides from penaeid shrimp (Crustacea, Decapoda). *Cell Mol. Life Sci.* 57: 1260-1271.

Destoumieux, D., M. Munoz, C. Cosseau, J. Rodriguez, P. Bulet, M. Comps, and E. Bachere. 2000. Penaeidins, antimicrobial peptides with chitin-binding activity, are produced and stored in shrimp granulocytes and released after microbial challenge. *J. Cell. Sci.* 113: 461-469.

Destoumieux, D., P. Bulet, J.M. Strub, A. Van Dorsselaer, and E. Bachere. 1999. Recombinant expression and range of activity of penaeidins, antimicrobial peptides from penaeid shrimp. *Eur. J. Biochem.* 266: 335-346.

Destoumieux, D., P. Bulet, D. Loew, A. Van Dorsselaer, J. Rodriguez, and E. Bachere. Penaeidins, a new family of antimicrobial peptides isolated from the shrimp *Penaeus vannamei* (Decapoda). *J. Biol. Chem.* 272: 28398-28406.

Ebran, N., S. Julien, N. Orange, P. Saglio, C. Lemaitre, and G. Molle. 1999. Pore-forming properties and antibacterial activity of proteins extracted from epidermal mucus of fish. *Comp. Biochem. & Physiol.* 122A: 181-189.

Food safety issues associated with products from aquaculture: a report of a joint FAO/NACA/WHO study group. 1999. World Health Organization Technical Report: 883. World Health Organization, Geneva.

Ganz, T. and R.I. Lehrer. 1997. Antimicrobial peptides of leukocytes. *Curr. Opin. Hematol.* 4: 53-58.

Hancock, R.E.W. and R.I. Lehrer. 1998. Cationic peptides: a new source of antibiotics. *Trends Biotech.* 16: 82-88.

Hancock, R.E.W. and G. Diamond. 2000. The role of cationic antimicrobial peptides in innate host defences. *Trends Microbiol.* 8: 402-410.

Hancock, R.E.W. 1998. The therapeutic potential of cationic peptides. *Expert Opinion Invest. Dis.* 7: 167-174.

Hancock, R.E.W. and M.G. Scott. 2000. The role of antimicrobial peptides in animal defences. *Proc. Natl. Acad. Sci. USA.* 97: 8856-8861.

Hubert, F., T. Noel, and P. Roch. 1996. A member of the arthropod defensin family from edible Mediterranean mussels (*Mytilus galloprovincialis*). *Eur. J. Biochem.* 240: 302-306.

Hubert, F., E.L. Cooper, and P. Roch. 1997. Structure and differential target sensitivity of the stimulable cytotoxic complex from hemolymph of the Meditteranean mussel *Mytilus galloprovincialis. Biochimi. et Biophys. Acta* 1361: 29-41.

Iwanaga, S., T. Muta, T. Shigenaga, N. Seki, K. Kawano, T. Katsu, and S. Kawabata. 1994. Structure-function relationships of tachyplesins and their analogues. *Ciba. Found. Symp.* 186: 160-174.

Jia, X., A. Patrzykat, R.H. Devlin, P.A. Ackerman, G.K. Iwama, and R.E. Hancock. 2000. Antimicrobial peptides protect coho salmon from *Vibrio anguillarum* infections. *Appl. Environ. Microbiol.* 66: 1928-32.

Kelly, D., W.R. Wolters, and J.M. Jaynes. 1990. Effects of lytic peptides on selected fish bacterial pathogens. *J. Fish Dis.* 13: 317-321.

Kelly, D.G., W.R. Wolters, J.M. Jaynes, and J.C. Newton. 1994. Enhanced disease resistance to enteric septicemia in channel catfish, *Ictalurus punctatus*, administered lytic peptide. *J. Appl. Aquaculture* 3: 25-34.

Kim, H.S., C.B. Park, M.S. Kim, and S.C. Kim. 1996. cDNA cloning and characterization of buforin I, an antimicrobial peptide: a cleavage product of histone H2A. *Biochem. Biophys. Res. Commun.*. 229: 381-387.

Lazarovici, P., N. Primor, and L.M. Loew. 1986. Purification and pore-forming activity of two hydrophobic polypeptides from the secretion of the Red Sea Moses sole (*Pardachirus marmoratus*). *J. Biol. Chem.* 261: 16704-16713.

Lee, I.H., Y. Cho, and R.I. Lehrer. 1997. Styelins, broad-spectrum antimicrobial peptides from the solitary tunicate, *Styela clava. Comp. Biochem. Physiol.* 118B : 515-21.

Lemaitre, C., N. Orange, P. Saglio, N. Saint, J. Gagnon, and G. Molle. 1996. Characterization and ion channel activities of novel antibacterial proteins from the skin mucosa of carp (*Cyprinus carpio*). *Eur. J. Biochem.* 240: 143-149.

Mitta, G., F. Vandenbulcke, and P. Roch. 2000. Original involvement of antimicrobial peptides in mussel innate immunity. *FEBS Lett.* 486: 185-190.

Mitta, G., F. Vandenbulcke, T. Noel, B. Romestand, J.C. Beauvillain, M. Salzet, and P. Roch. 2000. Differential distribution and defence involvement of antimicrobial peptides in mussel. *J. Cell Sci.* 113:2759-2769.

Mitta, G., F. Vandenbulcke, F. Hubert, M. Salzet, and P. Roch. 2000. Involvement of mytilins in mussel antimicrobial defense. *J. Biol. Chem.* 275: 12954-12962.

Mitta, G., F. Hubert, E.A. Dyrynda, P. Boudry, and P. Roch. 2000. Mytilin B and MGD2, two antimicrobial peptides of marine mussels: gene structure and expression analysis. *Dev. Comp. Immunol.* 24: 381-393.

Mitta, G., F. Vandenbulcke, F. Hubert and P. Roch. 1999. Mussel defensins are synthesised and processed in granulocytes then released into the plasma after bacterial challenge. *J. Cell. Sci.* 112: 4233-4242.

Mitta, G., F. Hubert, T. Noel, and P. Roch. 1999. Myticin, a novel cysteine-rich antimicrobial peptide isolated from haemocytes and plasma of the mussel *Mytilus galloprovincialis. Eur. J. Biochem.* 265: 71-78.

Miyata, T., F. Tokunaga, T. Yoneya, K. Yoshikawa, S. Iwanaga, M. Niwa, T. Takao, and Y. Shimonishi. 1989. Antimicrobial peptides, isolated from horseshoe crab hemocytes, tachyplesin II, and polyphemusins I and II: chemical structures and biological activity. *J. Biochem. (Tokyo)* 106: 663-668.

Nakamura, T., H. Furunaka, T. Miyata, F. Tokunaga, T. Muta, S. Iwanaga, M. Niwa, T. Takao, and Y. Shimonishi. 1988. Tachyplesin, a class of antimicrobial peptide from the hemocytes of the horseshoe crab *(Tachypleus tridentatus)*. Isolation and chemical structure. *J. Biol. Chem.* 263: 16709-16713.

Naylor, R.L., R.J. Goldburg, J.H. Primavera, N. Kautsky, M.C.M. Beveridge, J. Clay, C. Folke, J. Lubchenco, H. Mooney, and M. Troell. 2000. Effect of aquaculture on world fish supplies. *Nature* 405: 1017-1024.

Oren, Z. and Y. Shai. 1996. A class of highly potent antibacterial peptides derived from pardaxin, a pore-forming peptide isolated from Moses sole fish *Pardachirus marmoratus. Eur. J. Biochem.* 237: 303-310.

Osusky, M., G. Zhou, L. Osuska, R.E.W. Hancock, W.W. Kay, and S. Misra. 2000. Transgenic plants expressing cationic peptide chimeras exhibit broad-spectrum resistance to phytopathogens. *Nat. Biotechnol.* 18: 1162-1166.

Park, C.B., J.H. Lee, I.Y. Park, M.S. Kim, and S.C. Kim. 1997. A novel antimicrobial peptide from the loach, *Misgurnus anguillicaudatus. FEBS Lett.* 411: 173-178.

Park, I.Y., C.B. Park, M.S. Kim, and S.C. Kim. 1998. Parasin I, an antimicrobial peptide derived from histone H2A in the catfish, *Parasilurus asotus. FEBS Lett.* 437: 258-262.

Patrzykat, A., L. Zhang, V. Mendoza, G. K. Wama, and R.E.W. Hancock. 2001. Synergy of histone-derived peptides of Coho salmon with lysozyme and the flounder peptide pleurocidin. *Antimicrob. Agents Chemother.* 45: 1337-1342.

Patrzykat, A., C.L. Friedrich, V. Mendoza, and R.E.W. Hancock. Sub-lethal concentrations of pleurocidin-derived antimicrobial peptides inhibit macromolecular synthesis in *E. coli.* Submitted to *Biochemistry*.

Rapaport, D., S. Nir, and Y. Shai. 1994. Capacities of pardaxin analogues to induce fusion and leakage of negatively charged phospholipid vesicles are not necessarily correlated. *Biochemistry* 33: 12615-12624.

Rapaport, D., R. Peled, S. Nir, and Y. Shai. 1996. Reversible surface aggregation in pore formation by pardaxin. *Biophys J.* 70: 2502-2512.

Rapaport, D. and Y. Shai. 1991. Interaction of fluorescently labeled pardaxin and its analogues with lipid bilayers. *J. Biol. Chem.* 266: 23769-23775.

Rapaport, D. and Y. Shai. 1992. Aggregation and organization of pardaxin in phospholipid membranes. A fluorescence energy transfer study. *J. Biol. Chem.* 267: 6502-6509.

Robinette, D., S. Wada, T. Arroll, M.G. Levy, W.L. Miller, and E.J. Noga. 1998. Antimicrobial activity in the skin of the channel catfish *Ictalurus punctatus:* characterization of broad spectrum histone-like antimicrobial proteins. *Cell. Mol. Life Sci.* 54: 467-475.

Schnapp, D., G.D. Kemp, and V.J. Smith. 1996. Purification and characterization of a proline-rich antibacterial peptide, with sequence similarity to bactenecin-7, from the haemocytes of the shore crab, *Carcinus maenas. Eur. J. Biochem.* 240: 532-539.

Shai, Y. 1994. Pardaxin: channel formation by a shark repellant peptide from fish. *Toxicology* 87: 109-129.

Smith, V.J., J.M. Fernandes, S.J. Jones, G.D. Kemp, and M.F. Tatner. 2000. Antibacterial proteins in rainbow trout, *Oncorhynchus mykiss. Fish Shellfish Immunol.* 10: 243-260.

Tatner, M.F. 1996. Natural changes in the immune system of fish. Pages 255–288 in *The Fish Immune System,* G. Iwama and T.Nakanishi, eds. Academic Press, New York.

Thennarasu, S. and R. Nagaraj. 1995. Design of 16-residue peptides possessing antimicrobial and hemolytic activities or only antimicrobial activity from an inactive peptide. *Int. J. Pept. Protein Res.* 46: 480-486.

Thennarasu, S. and R. Nagaraj. 1996. Specific antimicrobial and hemolytic activities of 18-residue peptides derived from the amino terminal region of the toxin pardaxin. *Protein Eng.* 9: 1219-1224.

Thennarasu, S. and R. Nagaraj. 1997. Solution conformations of peptides representing the sequence of the toxin pardaxin and analogues in trifluoroethanol-water mixtures: analysis of CD spectra. *Biopolymers* 41: 635-645.

Videler, H., G.J Geertjes and J.J. Videler. 1999. Biochemical characteristics and antibiotic properties of the mucus envelope of the queen parrotfish. *J. Fish Biol.* 54: 1124-1127.

Wedemeyer, G. and A.J. Ross. 1969. Some metabolic effects of bacterial endotoxins in salmonid fishes. *J. Fish. Res. Board Can.* 26: 115-122.

Yang, Y.S., G. Mitta, A. Chavanieu, B. Calas, J.F. Sanchez, P. Roch, and A. Aumelas. 2000. Solution structure and activity of the synthetic four-disulfide bond Mediterranean mussel defensin (MGD-1). *Biochemistry* 39: 14436-14447.

Yoshida, K., Y. Mukai, T. Niidome, C. Takashi, Y. Tokunaga, T. Hatakeyama, and H. Aoyagi. 2001. Interaction of pleurocidin and its analogs with phospholipid membrane and their antibacterial activity. *J. Pept. Res.* 57: 119-126.

Zhang, L., M.G. Scott, H. Yan, L.D. Mayer, and R.E.W. Hancock. 2000. Interaction of polyphemusin I and structural analogs with bacterial membranes, lipopolysaccharide, and lipid monolayers. *Biochemistry* 39(47): 14504-14514.

Fish Protein Hydrolysates and their Potential Use in the Food Industry

Hordur G. Kristinsson[1] *and Barbara A. Rasco*[2]

[1]Department of Food Science and Human Nutrition, University of Florida, P.O. Box 110370, Gainesville, Florida 32611, U.S.A.
[2]Department of Food Science and Human Nutrition, Washington State University, P.O. Box 646376, Pullman, Washington 99164, U.S.A.

Introduction

Fish has an amino acid composition which makes it an excellent source of nutritive and easily digestible protein (Friedman 1996, Venugopal *et al.* 1996). Unfortunately, great amounts of protein rich byproduct materials from seafood processing plants are discarded without any attempt of recovery. In recent years strict environmental regulations have been imposed which no longer allow many fish processors to discard their offal, resulting in a high cost of refining the material before it is discarded or directing it into low grade fishmeal or plant fertilizer. To meet the need of the seafood processing industry alternative methods to make the best use of their byproducts have to be developed. However to get acceptance from industry these processes have to be more economically feasible than discarding the byproducts or using them for feed or fertilizer (Kristinsson and Rasco 2000a). In the latter half of the last century there were great research efforts to find means to utilize the vast amount of fish byproducts and underutilized species. Despite this effort we have not come very far, in part due to economic obstacles of processing and generally low acceptance of the final products (Kristinsson and Rasco 2000a). The issue of fish utilization is especially acute as we enter a new century. Fish capture is close 100,000 metric tons and including aquaculture harvest exceeds 120,000 tons (FAO 1998). This world catch is believed by many to be on the verge of or exceeding sustainable limits of our oceans with many common food fishes at the brink of being endangered. The rapid growth of aquaculture must also not be overlooked since it does and will more increasingly lead to large amounts of byproducts with high quality protein that may be used for human consumption.

Possibly more than 60% of fish tissue remaining after processing (species-dependent) is considered to be processing waste and not used as human food (Mackie 1982). This material is high in quality protein (^{3}10%) and other valuable compounds which could be utilized for human consumption. It has furthermore been estimated that about 30% of the world's catch is transformed into fishmeal (Rebeca *et al.* 1991) for animal feed and possibly more than 15% of the biomass harvested will not find any utilization (Kristinsson and Rasco 2000b). In light of an increasing world population, the danger of overfishing and the limited utilization of our fish harvest, there obviously is a great need to utilize processing byproducts and underutilized species with more intelligence and foresight. By applying enzyme technology to recover and modify fish proteins present in byproducts and in underutilized species it is possible to produce a broad spectrum of protein ingredients with a wide range of applications in the food industry (Kristinsson and Rasco 2000a,b). These ingredients, collectively called fish protein hydrolysates (FPH), are fish proteins that have been hydrolyzed by enzymes to a varying degree depending on enzyme type and reaction conditions. This promising approach could make more use of the fish industry byproducts and at the same time be employed on underutilized species, increasing the margin of profit for the fishing industry and creating a more environmentally friendly industry. Enzymatic modification of proteins is widely used and has a long history in the food industry (Mullally *et al.* 1994), primarily for vegetable and milk proteins. Much work on the hydrolysis of fish proteins was conducted in the 1960s and 1970's (Hoyle and Merritt 1994) and was directed into production of cheap nutritious protein sources for rapidly growing developing countries, or towards animal feed production, primarily through production of fish protein concentrates (FPC), which employs chemical hydrolysis. In the past 10 years or so only scattered reports have emerged in the literature on FPH, but some of the latest research publications have reported the potential of using hydrolysates in food formulations (e.g. Baek and Cadwallader 1995, Shahidi *et al.* 1995, Vieira *et al.* 1995, Onodenalore and Shahidi 1996, Kristinsson and Rasco 2000b,c,d). In this chapter we will address the present and current state of FPH, its production, and discuss its various properties and potential use as functional ingredients in food systems.

Biological Methods for Fish Protein Hydrolysis

Proteolytic modification of food proteins to improve palatability and storage stability of the available protein resources is an ancient technology (Adler-

Nissen 1986). Protein hydrolysates, defined as proteins which to a greater or lesser degree are chemically or enzymatically broken down into peptides of varying sizes (Skanderby 1994), find variety of uses in the food industry, e.g. as milk replacers, protein supplements, stabilizers in beverages and flavor enhancers in confectionery products. Several techniques exist for extracting and hydrolyzing protein from fish. These include harsh methods employing aqueous and organic solvents and chemical hydrolysis at elevated temperatures and extreme pH, which normally results in destruction of functional properties and may lead to the development of toxic byproducts (Kristinsson and Rasco 2000a). Methods employing chemical hydrolysis to produces fish protein concentrates (FPC) preceded enzymatic hydrolysis and were proposed as a means to increase protein availability and protein intake in developing countries. Biological processes using enzymes are becoming more frequently employed in industrial practices to make FPH since enzyme hydrolysis is a mild process that results in products of high functionality and nutritive value. The biological processes will be discussed below. For a good account of chemical methods for hydrolysis readers are referred to a recent review by Kristinsson and Rasco (2000a).

Enzymatic Protein Hydrolysis Using Added Enzymes

Using added enzymes to hydrolyze food proteins is a process of considerable importance that is used to modify the physicochemical, functional and organoleptic properties of the original protein often improving its nutritive value (Kristinsson and Rasco 2000b). Many different aquatic sources have been enzymatically produced into fish protein hydrolysates. These include Atlantic cod (*Gadus morhua*) (Liaset *et al.* 2000), Atlantic salmon (*Salmo salar*) (Kristinsson and Rasco 2000b,c,d, Liaset *et al.* 2000), capelin (*Mallotus villosus*) (Shahidi *et al.* 1995), crayfish (Baek and Cadwallader 1995), dogfish (*Squalus acanthias*) (Diniz and Martin 1996), hake (*Urophycis chuss*) (Cheftel *et al.* 1971, Hale 1972, Yanez *et al.* 1976), herring (*Clupus harengus*)(Hoyle and Merritt 1994), lobster (*Panulirus* spp.) (Vieira *et al.* 1995), Pacific whiting (*Merluccius productus*) (Benjakul and Morrissey 1997), sardine (*Sardina pilchardus*) (Quaglia and Orban 1987b, Quaglia and Orban 1990, Sugiyama *et al.* 1991), and shark (*Isurus oxyrinchus*) (Onodenalore and Shahidi 1996). From the list above one can see that many represent underutilized species, and those who do not, are connected to utilization of byproducts.

Enzymatic hydrolysis is a mild biological process which is favored over acid or alkali hydrolysis since does not produce toxic byproducts and gives a more functional and nutritive end product (Gonzalez-Tello *et al.* 1994, Kristinsson and Rasco 2000a). The unique specificity of action and function of enzymes under mild conditions makes no need for extreme pH and temperature to be involved in the process, which often have detrimental effects on the final product. Using added enzymes instead of chemicals or endogenous enzymes (i.e. enzymes naturally present in the material being hydrolyzed) offers many advantages since it allows good control of the hydrolysis and thereby the properties of the resultant products (Shahidi *et al.* 1995, Kristinsson and Rasco 2000b). Knowing the specificity and activity of proteolytic enzymes provides the possibility of controlling the extent and nature of protein hydrolysis. By selecting suitable enzymes, reaction conditions and times permits production of hydrolysates with different molecular structures and different functional properties that could find different applications in various food formulations (Quaglia and Orban 1990, Kristinsson and Rasco 2000a,b). Most published processes on enzymatic hydrolysis of fish proteins are done in research laboratories, and may have limited applications in the industry. Commercial production of fish protein hydrolysates is still limited on a world basis, but has reached a sizable level in a few countries, including Denmark, France, Iceland, Norway, Japan, U.S.A. and several Southeast-Asian countries. Although enzymatic hydrolysis is preferred over chemical hydrolysis it has some drawbacks such as: (a) potentially high cost of using large quantities of added commercial enzymes, (b) sometimes difficulty in controlling the extent of reaction (especially in mixed protease preparations) that can result in heterogeneous products consisting of fractions of varying molecular weight, (c) low yields of protein, and (d) the need to inactivate enzymes by high/low pH or heat treatment at the end of the reaction, which adds to the processing costs and may adversely affect some functional properties (Deeslie and Cheryan 1988, Kristinsson and Rasco 2000a). Also the enzymes employed in the process cannot be reused since they are inactivated (Adler-Nissen 1986).

Many things have to be considered when attempting to hydrolyze fish protein in a controlled manner. Substrate has to be carefully chosen and it is important to distinguish between lean and fatty species since the latter can lead to lipid oxidation problems in the final product creating unpleasant flavors and odors (Mackie 1982, Gildberg 1993). Aldol condensation of carbonyls produced from lipid oxidation may also react with basic protein side groups in the hydrolysate which can lead to formation of undesirable brown discoloration of the final product (Hoyle and Merritt 1994). The choice of enzyme is another very important factor since different enzymes have different specificity, and therefore produce products of different molecular makeup and functional properties. The choice is usually though determined by a combination of efficacy and economics (Lahl and Brown 1994, Kristinsson and Rasco 2000d). For example, when Kristinsson and Rasco (2000d) compared five different enzyme preparations on the same activity level on a substrate of salmon muscle proteins they found the least efficient hydrolyzing enzyme (Alcalase) to be most cost efficient. A wide variety of enzymes have been used to hydrolyze fish protein (Table 1). Pepsin has been shown to very efficiently hydrolyze fish protein (Hale 1969, Liu and Pigott 1981) but since it only works at low pH, it can adversely affect functional properties of the final product and its nutritive value (Tarky *et al.* 1973). Enzymes active in the neutral range have for this reason been favored in recent years. Alcalase, an alkaline enzyme with both high endo- and exopeptidase activity (i.e. cleaves both within the polypeptide backbone and terminal amino acids) has repeatedly been proven by many researchers to be one of the best enzymes used to produce functional FPH (Quaglia and Orban 1987b, Sugiyama *et al.* 1991, Diniz and Martin 1996, Benjakul and Morrissey 1997, Kristinsson and Rasco 2000b). Enzyme choice is also important for protein recovery, since the maximum recovery is desired for this process. Alcalase has shown to give good protein recovery, around 70% recovery for both sardine hydrolysis (Quaglia and Orban 1987b) and Pacific whiting solid waste hydrolysis (Benjakul and Morrissey 1997). The same enzyme also gave hydrolysates with high protein content of the final product, 72.4% for capelin hydrolysates (Shahidi *et al.* 1995), 78% for shark hydrolysates (Onodenalore and Shahidi 1996), 80% for Pacific whiting solid waste hydrolysates (Benjakul and Morrissey 1997) and over 88% for salmon protein hydrolysates (Kristinsson and Rasco 2000b). A similar high protein content as reported for the salmon hydrolysates was reported for hydrolysates made from a tropical fish, *Aristichthys noblis*, with a protease called P "Amano" 3 (87%) (Yu and Fazidah 1994). Other enzymes have resulted in high recovery and protein content. When five different enzyme preparations (Alcalase 2.4L, Corolase PN-L, Corolase 7089, Flavourzyme 1000L and a salmon pyloric caeca extract) were compared for their ability to produce salmon protein hydrolysates they all gave good protein recoveries and high proteins contents of the final product (Kristinsson and Rasco 2000b). Alcalase gave the highest protein content, over 88%, but the enzyme preparation Flavourzyme gave the highest protein recovery, with Alcalase and Corolase 7089 closely following (Kristinsson and Rasco 2000b). The protein recoveries are highly dependent on extent of hydrolysis, the higher the %DH the more recovery (Figure 1), while final protein content in the hydrolysate is less affected by %DH (Kristinsson and Rasco 2000b). This is likely due to the smaller size of the peptides at higher %DH which are more soluble and are not centrifuged out.

Prior to hydrolysis the starting material has to be thoroughly homogenized and diluted up to a point so enzymes can have easy access to proteins. The substrate is then usually adjusted to a pH and temperature representing the optimal reaction conditions of the enzyme. However specificity of enzymes (especially enzyme mixtures) may vary greatly depending on reaction conditions (Kristinsson and Rasco 2000c). Therefore employing reaction conditions different than the optimal conditions, which normally result in most efficient hydrolysis, may give rise to products of different properties. An understanding on how specificity of the enzymes

Table 1. Example of enzymes used to produce fish protein hydrolysates

Enzyme preparation	References
Alcalase 2.4L	Shahidi *et al.* 1995, Kristinsson and Rasco 2000b,d
Bromelain	Miller and Groninger 1976
Corolase PN-L	Kristinsson and Rasco 2000b,d
Corolase 7089	Kristinsson and Rasco 2000b,d
Flavourzyme 1000L	Kristinsson and Rasco 2000b,d
Neutrase	Shahidi *et al.* 1995, Benjakul and Morrissey 1997
Neutral Fish Protease Extract	Gildberg *et al.* 1989
Optimase	Baek and Cadwallader 1995
Pancreatin	Hale 1969
Papain	Quaglia and Orban 1987a,b
Pepsin	Hale 1969, Liu and Pigott 1981, Liaset *et al.* 2000
Salmon Pyloric Caeca Extract	Kristinsson and Rasco 2000b,c,d

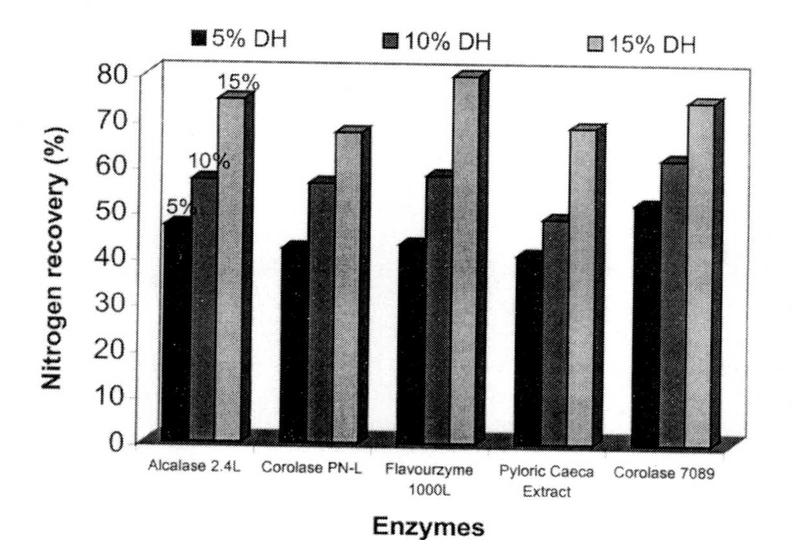

Figure 1. Influence of enzyme type used to make salmon protein hydrolysates and degree of hydrolysis on nitrogen recovery.

used in hydrolysis changes with changes in e.g. pH, ionic strength and temperature can be a very powerful tool to produce FPH with pre determined properties for different applications.

The hydrolysis is initiated by adding the enzyme preparation to the system and continually stirring at a constant rate, with pH and temperature being closely monitored and ideally kept constant. When the hydrolysis has reached its desired end point it is terminated by inactivating the enzymes by high temperature and/or reducing or increasing pH. Recovery of the hydrolyzed proteins is commonly by means of centrifugation or filtration to separate them from insoluble materials consisting of materials such as fat, scales, bones, connective tissue and unhydrolyzed protein. The soluble fraction containing the hydrolyzed protein can be further treated by passing it through charcoal to deodorize and decolorize it, and finally either freeze or spray drying it (Figure 2).

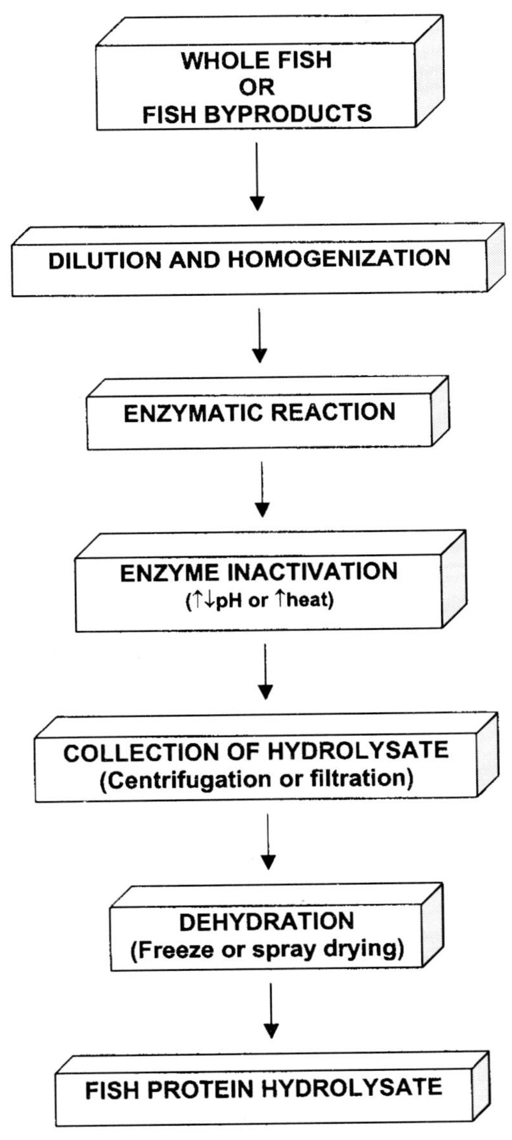

Figure 2. Simplified flowsheet for the enzymatic production of fish protein hydrolysates.

The enzymatic hydrolysis curve (Figure 3) is typically characterized by an initial rapid phase where a number of peptide bonds are being hydrolyzed, followed by a decrease in hydrolysis rate due to enzyme inactivation and/or substrate limitation and finally reaching a stationary phase where no or little hydrolysis takes place (Rebeca *et al.* 1991, Mullally *et al.* 1995, Shahidi *et al.* 1995, Kristinsson and Rasco 2000a,d). Controlling the extent of hydrolysis is very important if one wants to produce a product with specified molecular profile and functional properties (Kristinsson and Rasco 2000b,c,d). This is conveniently accomplished using the pH-stat method, which is based upon maintaining constant pH during the reaction. By pH-stat, the %DH is calculated from the volume and molarity of base or acid used to maintain constant pH, and defines the degree of hydrolysis as percent ratio of the numbers of peptide bonds broken (h) to the total numbers of bonds per unit weight (h_{tot}; meq/kg protein, calculated from the amino acid composition of the substrate) (Adler-Nissen 1986): $\%DH = (h/h_{tot}) \times 100$. %DH can also be expanded to:

$$\%DH = (B \times N_B \times 100)/(\alpha \times h_{tot} \times MP)$$

Where, B is the base consumption in ml (or acid in case of acid proteases), N_B is the normality of the base (or acid), α is the average degree of dissociation of the α-NH groups, and MP = mass of protein in grams (%N × 6.25). As the protein is being progressively more broken down by the enzyme the %DH increases. Unfortunately control over the hydrolysis

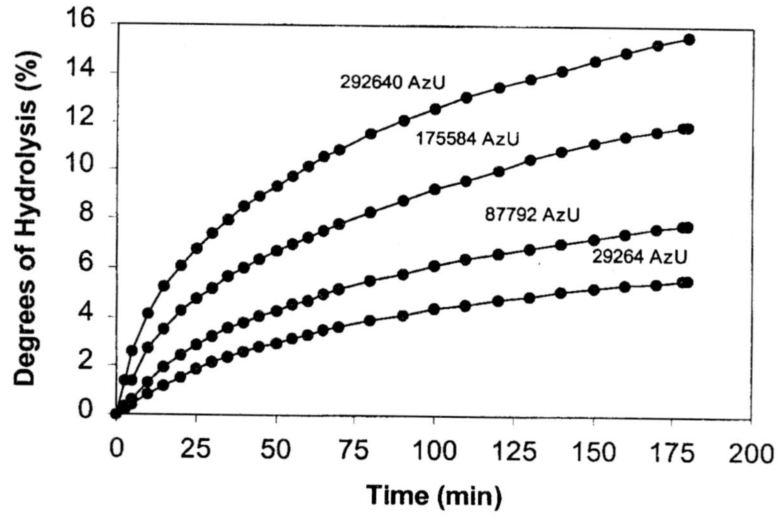

Figure 3. Enzymatic hydrolysis curve of salmon muscle proteins hydrolyzed with Alcalase 2.4L at different enzyme activities. AzU refers to Azocoll activity units.

process is often lacking in the literature and often researchers fail to specify the final %DH of their FPH preparations. Another limitation often encountered in the literature is that studies comparing different enzymes on the same substrates fail to use a standardized activity level and add different enzymes on the same weight basis. This is not a satisfactory comparison since different enzymes may have profoundly different activities and should be compared based on standardized activity (Kristinsson and Rasco 2000d). As an example, Kristinsson and Rasco (2000d) compared five different preparations of enzymes and found that on the same activity level (based on Azocoll activity units) they showed remarkably different reaction rates on salmon muscle proteins. Comparing enzymes based on weight or volume like many past studies do gives thus little insight on the efficacy of a particular enzyme preparation.

Enzymatic Hydrolysis Using Endogenous Enzymes — Autolytic Hydrolysis

Biochemical production of fish protein hydrolysates may be carried out by employing an autolytic process. The autolytic process depends on the action of endogenous digestive enzymes present in the substrate. The end product of autolytic hydrolysis is generally a fairly viscous liquid rich in free amino acids and small peptides. The digestive enzymes in question are primarily the serine proteases trypsin and chymotrypsin from the fish pyloric caeca, and the thiol protease pepsin from fish stomach. Lysosomal proteases, i.e. catheptic enzymes, present in fish muscle also contribute to proteolytic breakdown to some extent.

Using proteases present in the substrate, e.g. fish viscera, has the advantage over commercial enzyme preparations that enzyme costs can be greatly reduced. The drawback is however that the endogenous enzymes are a complex mixture of enzymes with different activity requirements. Another complication is that the presence of certain digestive enzymes and their concentration is highly seasonal and can vary tremendously within species as well as between species. These different activity requirements and variations make it very hard to control the hydrolytic process from one batch to another, so production of hydrolysates with specified molecular and functional properties becomes quite a difficult task. Despite these problems endogenous proteolytic enzymes are extensively used in some regions to produce hydrolyzed products for human consumption, specifically fish sauces.

Fish sauce is the major autolytically hydrolyzed fish product presently consumed in the world. Its production has thousands of years of tradition in Asia, and it is known to have been produced in Mediterranean countries in ancient times. Fish sauce is now mainly used as a condiment like the popular Nuoc-Mam produced in Vietnam, with an annual production in Southeast Asia of about 250,000 metric tons (Gildberg 1993). The substrate is often composed of small pelagic fishes that are immersed in a high salt (20–40%) solution at ambient temperatures (25–45°C). The endogenous proteases slowly break down the fish muscle for 6–12 months under anaerobic conditions. This slow but extensive breakdown results in a liquefied fish sauce composed predominantly of free amino acids, with up to 50% nitrogen recovery. Variations in enzyme profiles are not a major concern here since the goal is extensive hydrolysis. Although production of fish sauce does not improve the nutritive value of the protein, the keeping quality is greatly increased and organoleptic characteristics are generally improved (Van Veen and Steinkraus 1970). Microbial fermentation and hydrolysis of fish may also be used to make liquefied fish products. Normally the bacterial fermentation is initiated by mixing minced or chopped fish with a fermentable sugar which favors the growth of lactic acid bacteria, which is advantageous since it produces acid and antibiotics that together destroy competing spoilage bacteria (Raa and Gildberg 1982).

Research on fish protein hydrolysis using endogenous enzymes for human food applications is scarcely found in the literature. In the United States the sale of processed fish foods containing visceral material of any kind is prohibited by FDA regulations, which works against its development as a food source in the U.S. (Kristinsson and Rasco 2000a). Fish sauce is almost the only autolytically produced food of aquatic origin. In order to produce a functional protein hydrolysate with specific properties a good knowledge of the enzymes involved is crucial. Since endogenous enzymes in fish can be a highly variable mixture, the properties of functional protein hydrolysates so prepared may vary greatly under the same reaction conditions. Good protein recovery and excellent functional properties have however been reported by using endogenous enzymes (Kristinsson and Rasco 2000b). Kristinsson and Rasco (2000b,c,d) economically extracted a serine protease extract from *Atlantic salmon pyloric caeca* and used it to produce salmon FPH which gave better emulsifying properties than soy protein concentrate and FPH made from four commercially available enzymes, and gave a protein recovery close to 70% at 15% DH. Other studies using autolytic enzymes did not give as good results. Shahidi *et al.* (1995) hydrolyzed ground capelin (*Mallotus villosus*) by endogenous enzymes, and found that it enhanced the overall extraction of the fish protein at both acid and alkaline pH, since both acid and alkaline proteases are present in fish muscle and viscera. However the protein recovery of these autolytically produced hydrolysates gave far less protein recovery (22.9%) than FPH produced with Alcalase (70.6%) (Shahidi *et al.* 1995). Hale (1972) reported similar results with red hake when it was hydrolyzed with Alcalase or fish viscera. One of the reasons why these studies gave less recovery was that whole viscera were used and not a defined protease extract like Kristinsson and Rasco (2000b,c,d) used.

Functional Properties of Fish Protein Hydrolysates

Although protein recovery in fish hydrolysate production can be adequate and nutritional requirements of FPH are good, good functional properties of the hydrolysate are very important to its successful use in formulated foods. Since enzymatic hydrolysis of fish protein results in a decrease in the protein's molecular weight its characteristics are modified, sometimes improving the protein's function and bioavailability. Specific functional properties of FPH are important, particularly if the aim is to use them as ingredients in food products (Gildberg 1993, Kristinsson and Rasco 2000a,b). Specificity of enzymes is important because it strongly influences the molecular size and hydrophobicity of the hydrolysate (Gauthier *et al.* 1993, Kristinsson and Rasco 2000b). A broad enzyme specificity generally leads to a complex peptide profile with many small peptides, but limited specificity generally gives fewer and larger peptides often of better functionality. The extent of hydrolysis (%DH) is also important since uncontrolled or prolonged hydrolysis may result in small peptides, lacking functional properties such as emulsifying and foaming properties, but of good solubility, often with increased bitterness. If reaction conditions are properly controlled during hydrolysis and appropriate enzymes are selected one can potentially produce hydrolysates with different solubility, emulsifying characteristics, foaming properties and/or taste characteristics suitable for a wide array of products (Mullally *et al.* 1994, Kristinsson and Rasco 2000b).

Solubility

Solubility is one of the most important properties of protein hydrolysates, since many other properties are affected by solubility (Wilding *et al.* 1984). Intact fish myofibrillar proteins have the problem of lack of solubility in water (Spinelli *et al.* 1972a, Venugopal and Shahidi 1994) over the pH range of ca 4-10, which is in the range of most foods. Recent studies however show that fish muscle proteins can be highly soluble at very low ionic strength (<0.3mM NaCl) (Stefansson and Hultin 1994, Feng and Hultin 1997) or at extremes of pH (Hultin and Kelleher 2000, Kristinsson and Hultin unpublished results), but this is well out of the range of most food systems. Enzymatic hydrolysis is therefore an important technique to increase the solubility of fish muscle proteins, extending their use as food ingredients. Solubility generally is increased as the %DH increases since increased hydrolysis leads to smaller peptides and newly exposed amino and carboxyl groups that increase the FPH hydrophilicity (Gauthier *et al.* 1993, Mahmoud 1992). Sugiyama and coworkers (1991) showed that when sardine meal was treated with several alkaline, neutral and acid proteases the solubility was substantially increased, with the alkaline enzymes being the most effective. Another study on sardines found both Alcalase and papain to give hydrolysates of very high solubility (>90%), which was correlated with a decrease in molecular weight (Quaglia and Orban 1987a,b). Fish protein hydrolysates have the additionai advantage of being highly soluble over a wide range of pH (Shahidi *et al.* 1995, Vieira *et al.* 1995, Kristinsson and Rasco 2000b), an important property for a food protein ingredient. For example, FPH produced from Atlantic salmon with Alcalase exhibited 96-100% solubility in 100 mM NaCl from pH 2-11 (Kristinsson and Rasco 2000b). Intact fish proteins on the other hand exhibit low solubility in the pH range and ionic strength conditions of most food systems (except at high ionic strength, e.g. in meat batters) which puts limitations on their use. The good solubility and nutritive value of enzymatically hydrolyzed fish protein has made them a feasible choice to produce milk replacers for weanling animals (Yanez *et al.* 1976, Hale and Bauersfeld 1978, Rebeca *et al.* 1991) which is presently being pursued in Japan ("bio-fish flour") and France (Gildberg 1993). This property of FPH could also make it a promising candidate for nutritional drink formulas, provided that flavor is not adversely affected. The soluble FPH due to their good amino acid balance have also been studied as possible protein ingredients to supplement cereal protein for human consumption (Yanez *et al.* 1976) and to be used in the food industry for bakery products (Rebeca *et al.* 1991).

Water-Holding Capacity

One of the primary goals for many food producers is to be able to control the water balance of their food systems, to increase yield and profits. A number of commercially available hydrolyzed food proteins are used for this purpose, a good example being the addition of soy protein concentrates or isolates into meat patties. Hydrolysis modifies the ability of proteins to absorb and bind to water molecules. The increase of terminal carboxyl and amino groups produces a substantial effect on the amount of adsorbed water and the strength of the sorption bond making them highly hydroscopic. Due to this, FPH have been shown to have a good ability to bind and hold water, which is a particularly useful property for many food formulations. Fish protein hydrolysates from Atlantic salmon muscle proteins produced with four commercial proteases and an pyloric caeca extract were significantly better in preventing drip loss when frozen minced salmon muscle was thawed compared to egg albumin and soy protein when added at 1.5% (w/w) basis (Kristinsson and Rasco 2000b). For example the most successful treatment, Alcalse produced FPH at 10% DH only resulted in a 0.96% weight loss while the control with no addition gave close to 3% weight loss. The free small FPH peptides in the water phase of the salmon mince could increase its osmotic potential and thus work against water moving out of the gel thereby explaining the less drip loss. It was also noticed that the mince patties with the least drip loss had a gel-like appearance, which lead to the idea that the hydrolysates may possibly be binding to the myofibrillar protein entrapping water and at the same time promoting gelation of the mince. The effect on cooking was however not evaluated in this study. Other studies have found a beneficial effect of FPH on water binding. Shark muscle protein hydrolyzed with Alcalase and then added to comminuted pork increased its cooking yield proportionally with increased FPH addition (Onodenalore and Shahidi 1996). Using the same process to make capelin FPH and adding it to comminuted pork gave similar results (Shahidi *et al.* 1995), suggesting strong water binding. A significant increase in cooking yield was also observed for a sardine hydrolysate when it was added to hamburgers (Vareltzis *et al.* 1990). Water control is very important in the seafood industry since more water retained or added to the product means more profit and often better properties. Fish protein hydrolysates provide a great opportunity here for the seafood industry to properly control the water balance of fishery products. The wide use of polyphosphates for this purpose is not looked favorably upon in many countries, and some people are adversely affected by these compounds. The use of FPH for water balance is more favorable both since it is a natural product and due to its aquatic food origin would pose a lesser labeling problem.

Emulsifying Properties

Hydrolysates with good emulsifying properties may find use as ingredients to aid in the formation and stabilization of oil-in-water or water-in-oil emulsions, e.g. in margarine, dressing and meat batters, to name a few examples. Emulsifying properties of hydrolyzed proteins are influenced by the extent of hydrolysis. Increase in hydrolysis of fish proteins generally results in a decrease in their emulsifying properties (Kristinsson and Rasco 2000b). This is because the small peptides generated from extensive hydrolysis have less surface activity and less hydrophobicity than larger peptides from limited hydrolysis. Even though they may adsorb rapidly at the oil-water interface, because of their high solubility, they are less efficient in reducing the interfacial tension producing a good and stable emulsion (Kristinsson and Rasco 2000b). Protease specificity is important here since it can greatly influence the molecular size, topology and hydrophobicity of the resulting polypeptides, all which influence emulsifying properties of the FPH. Using enzymes with more narrow specificity and low degree of hydrolysis is generally recommended. The ability of FPH to form and stabilize emulsions has ranged from poor to excellent if some recent reports are looked at. Fish protein hydrolysates from Atlantic salmon muscle (Kristinsson and Rasco 2000b) had good emulsifying capacity compared to soy protein concentrate, but did not compare with egg albumin. The same was found for their ability to stabilize emulsions. The emulsifying properties did greatly depend on the specificity of the enzyme used to make them; the FPH with larger peptides were better emulsifiers. The emulsifying properties dropped as %DH increased due to breakdown of the larger peptides (Kristinsson and Rasco 2000b). Interestingly using a serine protease mixture from salmon pyloric caeca to produce the FPH gave the best properties which was attributed to the narrow specificity of the extract which gave larger polypeptides than the other enzymes with better ability to form and stabilize oil-in-water emulsions (Kristinsson and Rasco 2000b). These results are outlined in Figure 4. The same negative impact increase in %DH has on emulsifying properties was reported earlier for sardine hydrolysates (Quaglia and Orban 1990). To possess good emulsifying and interfacial properties it has been suggested that peptides should not be smaller than 20 residues (Lee *et al.* 1987). Liceaga-Gesualdo and Li-Chan (1999) recently reported increase in emulsion formation and stability when herring was hydrolyzed with Alcalase compared to unhydrolyzed proteins. Spinelli and

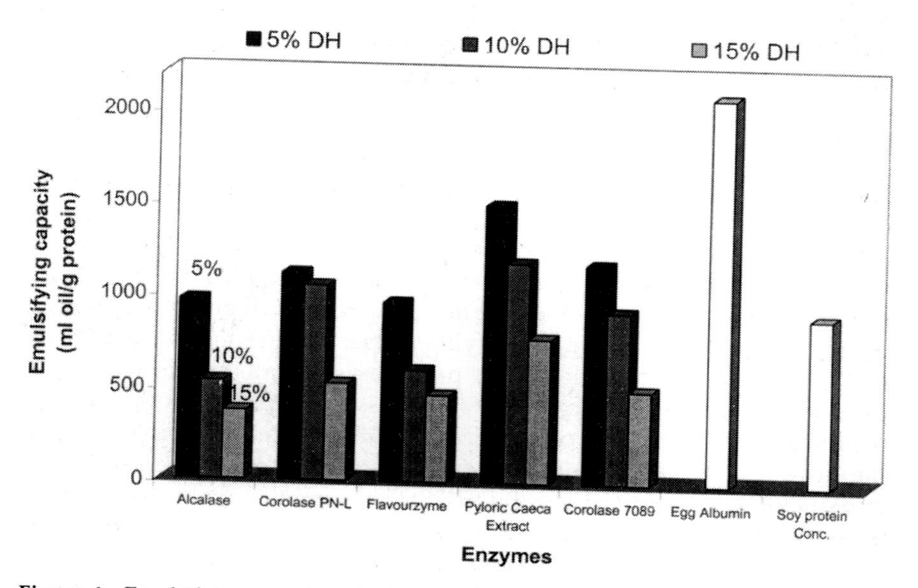

Figure 4. Emulsifying capacity of salmon protein hydrolysates at different degrees of hydrolysis made using different enzymes, compared to egg albumin and soy protein concentrate.

coworkers (1972a,b) reported similar findings earlier, that FPH from rockfish did have improved emulsifying capacity compared to the parent myofibrillar proteins. However the same fish hydrolyzed with bromelain gave a FPH with poor emulsifying capacity and stability (Miller and Groninger 1976). Other studies have reported poor emulsifying properties of FPH. When an undefined fungal protease was used to produce FPH from lobster processing wastes (Vieira *et al.* 1995) the product generated had very poor emulsifying ability. Also, relatively poor emulsifying capacity and stability was reported for capelin (Shahidi *et al.* 1995) and shark protein hydrolysates (Onodenalore and Shahidi 1996). The problem with these studies is that the %DH of the FPH was not reported, which makes interpretations of their results impossible. Another complication with all of the above studies and other functional properties are that different methods are used to study them, making it very difficult to compare them. Chemical modifications of FPH with poor emulsifying properties, such as acylation or succinylation, can be used to greatly improve emulsifying properties (Sikorski and Naczk 1981, Gildberg 1993, Groninger and Miller 1979) but these modifications are not allowed in food products due to safety concerns.

Foaming Properties

The ability of proteins to form and stabilize foams is an important property for some food applications (e.g. egg proteins and meringues; milk proteins and whipped cream). Fish protein hydrolysates could find applications to aid in foaming and foam stabilization, e.g. in fish based soufflés and patés. The factors influencing foaming properties of proteins and protein hydrolysates are similar to those influencing emulsifying properties. Few studies have been conducted on FPH's foaming properties. Capelin hydrolysates exhibited reasonable foaming properties (whippability and foam stability) when hydrolyzed by Alcalase to 12% DH (Shahidi *et al.* 1995) and shark hydrolysates were even better (Onodenalore and Shahidi 1996). However the above hydrolysates did not compare well to published results on commonly used whey protein hydrolysates (Althouse *et al.* 1995). The decrease in molecular weight, hence decrease in peptide length as hydrolysis progresses leads to a decrease in foaming ability since the small peptides don't have the ability to stabilize the air cells of the foam (Kuehler and Stine 1974, Liceaga-Gesualdo and Li-Chan 1999). Therefore low %DH are preferred to obtain FPH with good foaming properties. This was clearly illustrated recently with herring FPH where it initially formed good emulsions which rapidly broke down, while the control unhydrolyzed proteins exhibited slightly lower foamability but far better stability (Liceaga-Gesualdo and Li-Chan 1999). As for the emulsifying properties the foaming properties can be greatly improved by chemically modifying the FPH. For example, foams from enzyme modified succinylated FPH were more stable than foams from modified soy protein and egg white, but with lower foam volume (Groninger and Miller 1975). The succinylated FPH was incorporated into a dessert topping, a soufflé and chilled and frozen desserts with great results. Foams from succinylated FPH have also been successfully used in whipped gelatin desserts where they formed a stable foam with smooth texture (Ostrander *et al.* 1977). However according to present U.S. regulations these modifications are not allowable in food systems.

Fat Binding

Binding and stabilization of fat is an important functional property in many products such as sausages and confectioneries. This functional property of FPH has been somewhat neglected and only a handful of studies are found in the literature. Work on FPH made from capelin (Shahidi *et al.* 1995) and shark (Onodenalore and Shahidi 1996) showed some ability the hydrolysate to bind fat, but no units were reported to express extent of fat absorption. Kristinsson and Rasco (2000b) studied the effect of five different protease preparations and three %DH (5,10 and 15%) on salmon FPH's ability to bind fat. Fat binding varied depending on protein used for the hydrolysis and dropped as hydrolysis increased (Figure 5). The FPH were far better than egg albumin and soy protein concentrate which both are common food protein ingredients, which points to the powerful potential FPH may have in products were fat binding and stabilization is desired.

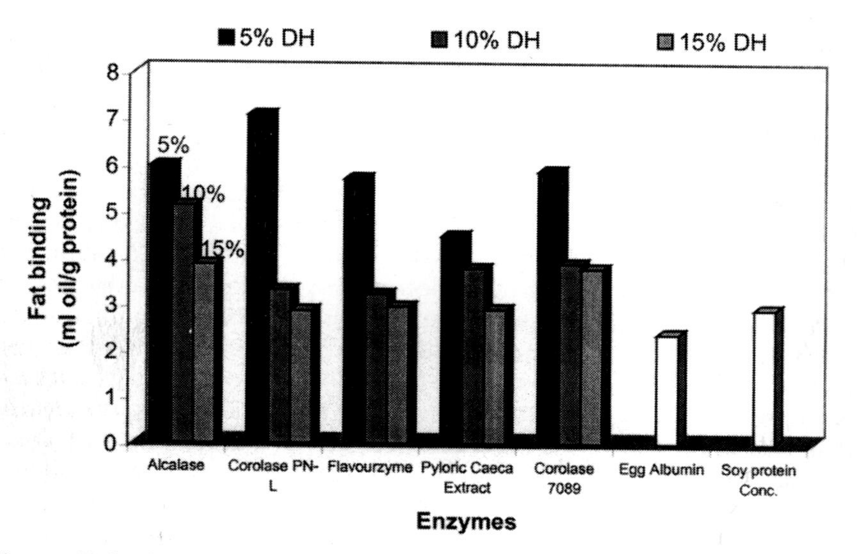

Figure 5. Fat binding capacity of salmon protein hydrolysates at different degrees of hydrolysis made using different enzymes, compared to egg albumin and soy protein concentrate.

Flavor

Even though enzymatic hydrolysis of fish proteins can lead to desirable functional properties of FPH, it has the disadvantage of generating bitterness, which is one major reason for its slow acceptance as a food ingredient. As hydrolysis progresses more and more exposed hydrophobic residues will be able to interact with the taste buds and result in increased bitterness. However if the protein is extensively hydrolyzed into small peptides and free amino acids (like in fish sauce) bitterness is greatly reduced. This is because larger peptides from less extensive hydrolysis with hydrophobic residues are by far more bitter compared to a mixture of the same amino acids. As %DH increased the bitterness intensity of FPH from *Aristichthys noblis* increased (Yu and Fazidah 1994). On the other hand very extensive hydrolysis of lobster processing waste gave a product of superior quality and no bitterness, possibly due to large amounts of free amino acids (Vieira *et al.* 1995) and flavor enhancing nucleotides which lobster is rich of. Although extensively hydrolyzed FPH can provide good flavor with little bitterness it lacks many of the previously discussed functional properties.

The choice of enzyme is important to avoid bitterness. Choosing an enzyme preparation such as Alcalase with exopeptidease activity and high preference for cleaving off hydrophobic amino acids is recommended (Petersen 1981, Adler-Nissen 1986). Flavourzyme from the same producers as Alcalase (Novo Nordisk) is another promising enzyme to limit the development of bitterness and at the same time get good functional properties (Kristinsson and Rasco 2000a,b). Hoyle and Merritt (1994) found that herring FPH made with papain gave higher bitterness scores than FPH made with Alcalase, even though the papain samples were at a lower %DH. When the same substrate was defatted and then used to make FPH much lower bitterness scores developed. This points to the lipids as a source of much of the bitterness (i.e. oxidation products). A recent study by Liu and coworkers (2000) on autolytic hydrolysis of frigate mackerel byproducts also pointed to the correlation of lipid oxidation and bitterness development during hydrolysis. This underlines the importance of substrate selection and quality of the starting material for FPH processes. Lalasidis and coworkers (1978) hydrolyzed cod filleting offal which gave a product of bitter taste, which was eliminated when the product was further treated with pancreatine, which is rich in exopeptidase activity. This emphasizes that researchers need to understand the specificity of their preparations to be able to select proteases which limit bitterness. However for an enzyme preparation to be effective in hydrolyzing fish proteins and at the same time limiting bitterness it has to have both exopeptidase and endoproteinase activity (Kristinsson and Rasco 2000b).

Fish protein hydrolysates have been incorporated into foods to evaluate their sensory acceptability. Chemically modified fish protein hydrolysates from rockfish were successfully used in dessert topping, soufflé, chilled and frozen desserts and whipped soufflés from a sensory standpoint (Groninger and Miller 1975, Ostrander 1977). Hydrolysates from *Oreochromis mossambicus*, an Asian freshwater fish, were incorporated into crackers and were highly acceptable up to 10% addition after frying the crackers (Yu and Tan 1990). It must however be kept in mind that sensory standards may vary quite a bit between countries and cultures, some favoring a slight fishy and bitter taste while others find it highly objectionable.

Other uses that have been somewhat neglected are FPH's proven potential to be used as a seafood flavoring (Baek and Cadwallader 1995) or flavor enhancer. This property is possibly the most feasible one for FPH from an application standpoint. Fujimaki and coworkers (1973) reported that fish protein concentrate treated with Pronase to make FPH had flavor potentiating activity much like that of monosodium glutamate (MSG) but accompanied with some bitterness. The peptides responsible for this MSG-like flavor were found to be three dipeptides and five tripeptides (Nogushi *et al.* 1975). Basic tripeptides containing asparagine and lysine as the second and C-terminal residues, respectively, with the N-terminal residue leucine or glycine have been found to be responsible for bitter taste in fish protein hydrolysates (Hevia and Olcott 1977). Proper selection of enzymes and an understanding on their behavior could aid researchers and industry to work against formation of FPH containing bitter peptides. Baek and Cadwallader (1995) concluded that the enzyme Optimase not only was one of the most economical enzyme to use to make hydrolysates from crayfish processing byproducts, but that it had the potential to be used to produce quality seafood flavor extracts. A recent company in Iceland (NothIce Ltd) has developed a product line of seafood extracts of superior flavor from shrimp, lobster and pollock using a proprietary cod serine protease preparation, which has unique specificity and ability to extract and hydrolyze aquatic proteins at low temperatures.

Several post-processing techniques have been suggested to mask or reduce bitterness of hydrolysates, but few applied to FPH. Treating FPH with activated carbon may partly remove bitter peptides (Shahidi *et al.* 1995). Treating FPH with exopeptidases may also lower the bitterness of the product (Lalasidis *et al.* 1978). Bitter peptides may also be extracted with solvents (Lalasidis *et al.* 1978) and FPH debittered by using ethyl alcohol (Chakrabarti, 1983). One interesting approach is to make plastein out of FPH, a process that reverses hydrolysis with proteases such as pepsin and papain, i.e. by rejoining the hydrolyzed fragments (Montecalvo *et al.* 1984, Gildberg 1993). During a polycondensation of the hydrolyzed units new polypeptides are formed, and they aggregate via hydrophobic associations, therefore masking bitterness (hydrophobic residues) and giving the product unique functional properties (Kinsella 1976, Gildberg 1993, Kristinsson and Rasco 2000a). The gel-like product may find use as fillers in a variety of products (Lanier 1994). This reaction has also been shown to be useful to recover protein from extensively autolytically produced fish silage (Ragunath and McCurdy 1991).

Studies on bitterness accompanied fish protein hydrolysis have been surprisingly limited compared to the problem it poses. While nutritional and safety criteria are met for FPH and functional properties are good, their sensory properties are extremely important for their successful adaptation and acceptance by the food industry and the consumer (Kristinsson and Rasco 2000a). More studies are needed on bitterness development to aid in the penetration of FPH into the food industry.

Antioxidant Properties

A little studied property of FPH is their antioxidant potential. Lipid oxidation is a major cause of quality deterioration in seafood, especially those with high concentrations of lipids (Hultin 1994). Fish protein hydrolysates could find a use in these systems to minimize oxidation. Hatate and coworkers (1990) found that sardine myofibril protein hydrolysates alone showed antioxidant activity to some extent but a very powerful synergistic effect when added to several commercial antioxidants. Shahidi and coworkers (1995) found that an addition of capelin FPH at 0.5–3.0% addition level in comminuted pork reduced TBARS formation by 17.7–60.4%, which indicated a strong antioxidant effect, which was speculated to be due to chelation. These effects are likely highly dependent upon the amino acid composition and molecular size of the FPH peptides. Chuang and coworkers (2000) showed that mackerel FPH had a powerful antioxidant effect by decreasing both lipoxygenase and hemoglobin catalyzed lipid oxidation which may have been due to histidine related dipeptides such as anserine and carnosine which the mackerel hydrolysate is rich of. The antioxidant mechanism is likely via scavenging of free radicals (Chuang *et al.* 2000). More studies on FPH's potential to prevent oxidation and their antioxidative function in complex food systems is highly recommended.

Physiological Effects

Studies have pointed to various physiological effects of FPH. Oxidative processes in-vivo lead to various diseases and FPH's potential as an antioxidant has been documented as outlined above. For example, lipoxygenases have been implicated in low density lipoprotein oxidation which is connected to hypertension and arteriosclerosis, and FPH has been shown to inhibit this enzymes activity (Chuang *et al.* 2000). Angiotensin is another compound connected to hypertension (Chuang *et al.* 2000) and FPH from a variety of species have been found to have inhibitory effects on an enzyme which converts angiotensin and reduces hypertension (Kohama *et al.* 1991, Ukeda *et al.* 1992, Matsumura *et al.* 1993, Wako *et al.* 1996). A commercial cod and mackerel hydrolysate (PC60 and its derivative Stabilium 200) showed another physiological effect, i.e. ability to reduce anxiety in humans (Dorman *et al.* 1995) and improve learning performance and memory in humans (Le Poncin,1996a) and rats (Le Poncin 1996b). Furthermore Bernet and coworkers (2000) reported the ability of the PC60 hydrolysates to reduce stress in laboratory rats, in a similar manner as the known drug diazepam (valium) does. Fish protein hydrolysates have also been found to enhance flow of red blood cells (Chuang *et al.* 2000), and therefore could enhance blood flow in-vivo. In light of the popularity of functional foods FPH may find a niche in this market as a nutritional supplement or nutraceutical due to it positive effects in-vivo.

Fish protein hydrolysates also have a proven ability to enhance growth of living organisms. Animals respond very well to hydrolysates due to its good amino acid balance and high protein content and FPH (mostly autolytically produced) is widely used as feed for piglets, minks, pigs and poultry and increasingly as pellets for pet food and aquaculture (Dong *et al.* 1992, Gildberg 1993, Kristinsson and Rasco 2000a). However bitterness problems can make the product highly unpalatable for animals fed feeds rich in FPH. Hydrolysates may also find other uses in animal feed than as a protein source. For example, Gildberg and coworkers (1996) reported that selected fractions (500-3000 daltons) of cod stomach FPH could be used as an adjuvant in fish vaccine and as an immune stimulant for aquacultured fish.

Increases in crop growth and yield when FPH is used as a fertilizer is well established (Kristinsson and Rasco 2000a). Novel applications taking advantage of this effect of FPH has been used to stimulate development of commercially important plant varieties. Eguchi and coworkers (1997) examined the effect of using FPH to stimulate somatic embryogenesis in anise (*Pimpinella anisum*) as compared to proline a known stimulator. FPH could well become a proline and amino acid substitute in plant tissue culture applications and can find use in value-added applications in the plant propagation industry (Milazzo *et al.* 1999). Andarwulan and Shetty (1999) found that when mackerel FPH was added to oregano cultures the plants produced more rosemaric acid and phenolics. Andarwulan and Shetty (2000) later found that FPH from mackerel stimulated the growth of Epoxy-pseudoisoeugenol-(2-methylbutyrate) (EPB) in anise root cultures. This phenolic compound has the potential to regulate nutraceutical type phytochemicals in plants. Fish protein hydrolysates contain a good amount of glutamic acid and proline, both which can be used to stimulate the pentose phosphate pathway and phenolic synthesis in plants (Andarwulan and

Shetty 1999). Using FPH can therefore stimulate expression of many commercially valuable phytochemicals which could find use as nutraceuticals in various food products or as antioxidants to extend shelf life of products, a new arena for FPH and possibly a very valuable one.

The increase in the biotechnology industry has crated a great need for a variety of different growth media for microorganisms. Fish protein hydrolysates show a potential here since extensively hydrolyzed FPH have been found to be an excellent growth media for a variety of microorganisms (Gildberg *et al.* 1989, de la Broise *et al.* 1998). The advent of Mad Cow Disease in Europe may make FPH a very attractive alternative to beef peptone in the future.

Conclusion

In this brief account on fish protein hydrolysates it can be seen that the potential exist to produce FPH of good functionality which could find a variety of uses in food products and for other applications. The use of FPH as a functional food ingredient will be more readily accepted by industry and consumers as more is known on how to economically produce it into a functional ingredient and at the same time limit the development of bitter taste. There exists a wonderful opportunity for this to happen. This is because of hardened regulations on byproduct disposal and the abundance of underutilized species which may well be the frontiers of fisheries in the future. The conversion of this protein rich material into valuable functional food protein ingredients for human consumption instead of converting it into animal feed or fertilizer or dumping it can be a great stimulant for the ever so struggling fishery industry and surely be a step forward environmentally.

References

Adler-Nissen, J. 1986. *Enzymic Hydrolysis of Food Proteins*, Elsevier Applied Science Publishers, Barking, UK.

Althouse, P.J., P. Dinakar and A. Kilara. 1995. Screening of proteolytic enzymes to enhance foaming of whey protein isolates, *J. Food Sci.* 60: 1110-1112.

Andrawulan, N. and K. Shetty. 1999. Influence of acetyl salicylic acid in combination with fish protein hydrolysates on hyperhydricity reduction and phenolic synthesis in oregano (*Origanum vulgare*) tissue cultures. *J. Food Biochem.* 23: 619-635.

Andrawulan, N. and K. Shetty. 2000. Stimulation of novel phenolic metabolite, epoxy-pseudoisoeugenol-(2-methylbutyrate) (EPB), in transformed anise (*Pimpinella anisum* L.) root cultures by fish protein hydrolysates. *Food Biotechnol.* 14: 1-20.

Baek, H.H. and K.R. Cadwallader. 1995. Enzymatic hydrolysis of crayfish processing by-products. *J. Food Sci.* 60: 929-934.

Benjakul, B. and M.T. Morrissey. 1997. Protein hydrolysates from Pacific whiting solid wastes. *J. Agric. Food Chem.* 45: 3423-3430.

Bernet, F., V. Montel, B. Noel, and J.P. Dupouy. 2000. Diazepam-like effects of a fish protein hydrolysate (Gabolysat PC60) on stress responsiveness of the rat pituitary-adrenal system and sympathoadrenal activity. *Psychopharmacology.* 149: 34-40.

de la Broise, D., G. Dauer, A. Gildberg, and F. Guerard. 1998. Evidence of positive effect of peptone hydrolysis rate on *Escherichia coli* culture kinetics. *J. Mar. Biotechnol.* 6: 111-115.

Chakrabarti, R. 1983. A method of debittering fish protein hydrolysate. *J. Food Sci. Technol.* 20 : 154-158.

Cheftel, C., M. Ahern, D.I.C. Wang, and S.R. Tannenbaum. 1971. Enzymatic solubilization of fish protein concentrate: Batch studies applicable to continuous enzyme recycling processes. *J. Agr. Food Chem.* 19: 155-161.

Chuang, W-L., B. Sun Pan, and J-S. Tsai. 2000. Inhibition of lipoxygenase and blood thinning effects of mackerel protein hydrolysate. *J. Food Biochem.* 24: 333-343.

Deeslie, W.D. and M. Cheryan. 1988. Functional properties of soy protein hydrolysates from a continuous ultrafiltration reactor. *J. Agric. Food Chem.* 36 : 26-31.

Diniz, F.M. and D.M. Martin. 1996. Use of response surface methodology to describe the combined effects of pH, temperature and E/S ratio on the hydrolysis of dogfish (*Squalus acanthias*) muscle. *Int. J. Food Sci. Tech.* 31: 419-426.

Dong, F.M., W.T. Fairgrieve, D.I. Skonberg, and B.A. Rasco. 1993. Preparation and nutrient analyses of lactic acid bacteria ensiled salmon viscera. *Aquaculture* 109: 351-366.

Dorman, T., L. Bernard, P. Glaze, J. Hogan, R. Skinner, D. Nelson, L. Bowker, and D. Head. 1995. The effectiveness of *Garum armoricum* (stabilium) in reducing anxiety in college students. *J. Adv. Med.* 8: 193-200.

Eguschi, Y., J.S. Bela, and K. Shetty. 1997. Simulation of somatic embryogenesis in Anise (*Pimpinella anisium*) using fish protein hydrolysates and proline. *J. Herbs Spices* 5: 61-68.

FAO. 1998. The State of World Fisheries and Aquaculture. Food and Agricultural Organization, Rome, Italy.

Feng, Y. and H.O. Hultin. 1997. Solubility of the proteins of mackerel light muscle at low ionic strength. *J. Food Biochem.* 21: 479-496.

Friedman, K. 1996. Nutritional value of proteins from different food sources: A review. *J. Agric. Food Chem.* 44 : 6-29.

Fujimaki, M., S. Arai, M. Yamashita, H. Kato, and M. Nogushi. 1973. Taste peptide fractionation from a fish protein hydrolysate. *Agric. Biol. Chem.* 37: 2891-2895.

Gauthier, S.F., P. Paquin, Y. Pouliot, and S. Turgeon. 1993. Surface activity and related functional properties of peptides obtained from whey proteins. *J. Dairy Sci.* 76: 321-328.

Gildberg, A., I. Batista, and E. Strom. 1989. Preparation and characterization of peptone obtained by a two-step enzymatic hydrolysis of whole fish. *Biotech. Appl. Biochem.* 11: 413-423.

Gildberg, A. 1993. Enzymic processing of marine raw materials. *Process Biochem.* 28: 1-15.

Gildberg, A., J. Bogwald, A. Johansen, and E. Stenberg. 1996. Isolation of acid peptide fractions from a fish protein hydrolysate with strong stimulatory effect on Atlantic salmon (*Salmo salar*) head kidney leucocytes. *Comp. Biochem. Physiol. B* 114: 97-101.

Gonzalez-Tello, P., F. Camacho, E. Jurado, M.P. Paez, and E.M. Guadix. 1994. Enzymatic hydrolysis of whey proteins: II. Molecular-weight range. *Biotechnol. Bioeng.* 44: 529-532.

Groninger, H.S. and R. Miller. 1975. Preparation and aeration properties of an enzyme-modified succinylated fish protein. *J. Food Sci.* 40: 327-330.

Groninger, H.S. and R. Miller. 1979. Some chemical and nutritional properties of acylated fish proteins. *J. Agric. Food Chem.* 27: 949-955.

Hale, M.B. 1969. Relative activities of commercially-available enzymes in the hydrolysis of fish proteins. *Food Technol.* 23: 107-110.

Hale, M.B. 1972. Making fish protein concentrate by enzymatic hydrolysis, Pages 1-31 in *NOAA Technical Report NMFS SSRF-675*, US Department of Commerce, Seattle, WA, USA.

Hale, M.B. and P.E. Bauersfeld, Jr. 1978. Preparation of a menhaden hydrolysate for possible use in a milk replacer. *Mar. Fish. Rev.* 40: 14-17.

Hatate, H., Y. Numata, and M. Kochi. 1990. Synergistic effect of sardine myofibril protein hydrolyzates with antioxidant. *Nippon Suisan Gakkaishi* 56: 1011.

Hevia, P. and H.S. Olcott. 1977. Flavour of enzyme-solubilized fish protein concentrate fractions. *J. Agric. Food Chem.* 25: 772-775.

Hoyle, N. and J.H. Merritt. 1994. Quality of fish protein hydrolysates from herring (*Clupea harengus*). *J. Food Sci.* 59: 76-79.

Hultin, H.O. 1994. Oxidation of lipids in seafood. Pages 49–74 in *Seafoods: Chemistry, Processing Technology and Quality*, F. Shahidi and J.R. Botta, eds., Blakie Academic and Professional, Glasgow.

Hultin, H.O. and S.D. Kelleher. 2000. Surimi processing from dark muscle fish. Pages 59–77 in *Surimi and Surimi Seafood*, J.W. Park, ed., Marcel Dekker, Inc., New York.

Kinsella, J.E. 1976. Functional properties of proteins in foods: a survey. *Crit. Rev. Food Sci. and Nutr.* 8: 219-280.

Kohama, Y., H. Oka, Y. Kayamori, K. Tsujikawa, T. Mimura, Y. Nagase, and M. Satake. 1991. Potent synthetic analogues of angiotensin-converting enzyme inhibitor derived from tuna muscle. *Agric. Biol. Chem.* 55: 2169-2170.

Kristinsson, H.G. and B.A. Rasco. 2000a. Fish protein hydrolysates: production, biochemical and functional properties. *CRC Crit. Rev. Food Sci. Nutr.* 32: 1-39.

Kristinsson, H.G. and B.A. 2000b. Rasco. Biochemical and functional properties of Atlantic salmon (*Salmo salar)* muscle proteins hydrolyzed with various alkaline proteases. *J. Agric. Food Chem.* 48: 657-666.

Kristinsson, H.G. and B.A. Rasco. 2000c. Hydrolysis of salmon muscle proteins by an enzyme mixture extracted from Atlantic salmon (*Salmo salar*) pyloric caeca. *J. Food Biochem.* 24: 177-187.

Kristinsson, H.G. and B.A. Rasco. 2000d. Kinetics of the hydrolysis of Atlantic salmon (*Salmo salar*) muscle proteins by alkaline proteases and a visceral serine protease mixture. *Proc. Biochem.* 36: 131-139.

Kuehler, C.A. and C.M. Stine. 1974. Effect of enzymatic hydrolysis on some functional properties of whey protein. *J. Food Sci.* 39: 379-382.

Lahl, W.J. and S.D. Braun. 1994. Enzymatic production of protein hydrolysates for food use, *Food Tech.* 58: 68-71.

Lalasidis, G., S. Bostrom, and L-B. Sjoberg. 1978. Low molecular weight enzymatic fish protein hydrolysates: Chemical composition and nutritive value. *J. Agric. Food Chem.* 26: 751-756.

Lanier, T.C. 1994. Functional food protein ingredients from fish. Pages 127-159 in *Seafood Proteins*, Z.E. Sikorski, B. Sun Pan and F. Shahidi, eds., Chapman and Hall, New York.

Le Poncin, M. 1996a. Experimental study: stress and memory. *Eur. Neuropsychopharmacol.* 6: 110-P10-2.

Le Poncin, M. 1996b. Nutrient presentation of cognitive and memory performances. *Eur. Neuropsychopharmacol.* 6: 187-P19-4.

Lee, S. W. M. Shimizu, S. Kaminogawa, and K. Yamaguchi. 1987. Emulsifying properties of a mixture of peptides derived from the enzymatic hydrolysates of b-casein. *Agric. Biol. Chem.* 51: 161-165.

Liaset, B., E. Lied, and M. Espe. 2000. Enzymatic hydrolysis of by-products from the fish-filleting industry; chemical characterisation and nutritional evaluation. *J. Sci. Food Agric.* 80: 581-589.

Liceaga-Gesualdo, A.M. and E.C.Y. Li-Chan. 1999. Functional properties of fish protein hydrolysate from herring (*Clupea harengus*). *J. Food Sci.* 64: 1000-1004.

Liu, L.L. and G.M. Pigott. 1981. Preparation and use of inexpensive crude pepsin for enzyme hydrolysis of fish. *J. Food Sci.* 46: 1569-1572.

Liu, C., K. Morioka, Y. Itoh, and A. Obatake. 2000. Contributions of lipid oxidation to bitterness and loss of free amino acids in the autolytic extract from fish wastes: Effective utilization of fish wastes. *Fisheries Sci.* 66: 343-348.

Mackie, I.M. 1982. Fish protein hydrolysates. *Proc. Biochem.* 17: 26-32.

Mahmoud, M.I., W.T. Malone, and C.T. Cordle. 1992. Enzymatic hydrolysis of casein: effect of degree of hydrolysis on antigenicity and physical properties. *J. Food Sci.* 57: 1223-1229.

Matsumura, N., M. Fujii, Y. Takeda, and T. Shimizu. 1993. Isolation and characterization of angiotensin I-converting enzyme inhibitory peptides derived from bonito bowels. *Biosci. Biotechnol. Biochem.* 57: 1743-1744.

Milazzo, M.C., Z. Zheng, G. Kellett, K. Haynesworth, and K. Shetty. 1999. Stimulation of benzyladenine-induced *in vitro* shoot organogenesis and endogenous proline in melon (*Cucumis melo* L.) by fish protein hydrolysates in combination with proline analogues. *J. Agric. Food Chem.* 47: 1771-1775.

Miller, R. and H.S. Groninger. 1976. Functional properties of enzyme-modified acylated fish protein derivatives. *J. Food Sci.* 41: 268-272.

Montecalvo Jr., J., S.M. Constantinides, and C.S.T. Yang. 1984. Enzymatic modification of fish frame protein isolate. *J. Food Sci.* 49: 1305-1309.

Mullally, M.M., D.M. O'Callaghan, R.J. FitzGerald, W.J. Donnelly, and J.P. Dalton. 1994. Proteolytic and peptidolytic activities in commercial pancreatin protease preparations and their relationship to some whey protein hydrolysate characteristics. *J. Agric. Food Chem.* 42: 2973-2981.

Mullally, M.M., D.M. O'Callaghan, R.J. FitzGerald, W.J. Donnelly, and J.P. Dalton. 1995. Zymogen activation in pancreatic endoproteolytic preparations and influence on some whey protein characteristics. *J. Food Sci.* 60: 227-233.

Noguchi, M., S. Arai, M. Yamashita, H. Kato, and M. Fujimaki. 1975. Isolation and identification of acidic oligopeptides in a flavor potentiating fraction from a fish protein hydrolysate. *J. Agric. Food Chem.* 23: 49-53.

Onodenalore, A.C. and Shahidi, F. 1996. Protein dispersions and hydrolysates from shark (*Isurus oxyrinchus*). *J. Aquat. Food Prod. Technol.* 5: 43-59.

Ostrander, J.G., P.J. Nystrom, and C.S. Martinsen. 1977. Utilization of a fish protein isolate in whipped gelatine desserts. *J. Food Sci.* 42: 559-560.

Petersen, B.R. 1981. The impact of the enzymatic hydrolysis process on recovery and use of proteins. Pages 149-175 in *Enzymes and Food Processing*, Elsevier Applied Science Publishers, London, UK.

Quaglia, G.B. and E. Orban. 1987a. Enzymic solubilisation of proteins of sardine (*Sardina pilchardus*) by commercial proteases. *J. Sci. Food Agric.* 38: 263-269.

Quaglia, G.B. and E. Orban. 1987b. Influence of the degree of hydrolysis on the solubility of the protein hydrolsyates from sardine (*Sardina pilchardus*). *J. Sci. Food Agric.* 38: 271-276.

Quaglia, G.B. and E. Orban. 1990. Influence of enzymatic hydrolysis on structure and emulsifying properties of sardine (*Sardina pilchardus*) protein hydrolysates. *J. Food Sci.* 55: 1571-1573.

Raa, J. and A. Gildberg. 1982. Fish silage: a review. CRC Crit. Rev. Food Sci. Nutr. 14: 383-419.

Raghunath, M.R. and A.R. McCurdy. 1991. Synthesis of plastein from fish silage. *J. Sci. Food Agric.* 54: 655-663.

Rebeca, B.D., M.T. Pena-Vera, and M. Diaz-Castaneda. 1991. Production of fish protein hydrolysates with bacterial proteases; Yield and nutritional value. *J. Food Sci.* 56, 309-314.

Sikorski, Z.E. and M. Naczk. 1981. Modification of technological properties of fish protein concentrates. *Crit. Rev. Food Sci. Nutr.* 4: 201-230.

Shahidi, F., X-Q. Han, and J. Synowiecki. 1995. Production and characteristics of protein hydrolysates from capelin (*Mallotus villosus*). *Food Chem.* 53: 285-293.

Skanderby, M. 1994. Protein hydrolysates: their functionality and applications. *Food Technol. Int. Eur.* 10: 141-144.

Spinelli, J., B. Koury, and R. Miller. 1972a. Approaches to the utilization of fish for the preparation of protein isolates; Isolation and properties of myofibrillar and sarcoplasmic fish protein. *J. Food Sci. 37*: 599-603.

Spinelli, J., B. Koury, and R. Miller. 1972b. Approaches to the utilization of fish for the preparation of protein isolates; enzymic modifications of myofibrillar fish proteins. *J. Food Sci. 37*: 604-608.

Stefansson, G. and H.O. Hultin. 1994. On the solubility of cod muscle protein in water. *J. Agric. Food. Chem.* 42: 2656-2664.

Sugiyama, K., M. Egawa, H. Onzuka, and K. Oba. 1991. Characteristics of sardine muscle hydrolysates prepared by various enzymic treatments. *Nippon Suisan Gakkaishi*. 57: 475-479.

Tarky, W., O.P. Agarwala, and G.M. Pigott. 1973. Protein hydrolysate from fish waste. *J. Food Sci.* 38: 917-918.

Ukeda, H., H. Matsuda, K. Osjima, H. Matufuji, T. Matsui, and Y. Osjima. 1992. Peptides from peptic hyrolysate of heated sardine meat that inhibit angiotensin I converting enzyme. *Nippon Nogeikagaku Kaishi* 65: 1223-1228.

Vareltzis, K., N. Soultos, F. Zetou, and F. Tsiaras. 1990. Proximate composition and quality of a hamburger type product made from minced beef and fish protein concentrate. *Lebensm. Wiss. Technol.* 23: 112-115.

van Veen, A.G. and K.H. Steinkraus. 1970. Nutritive value and wholesomeness of fermented foods. *J. Agric. Food Chem.* 18: 576-578.

Venugopal, V. and F. Shahidi. 1994. Thermostable water dispersions of myofibrillar proteins from Atlantic Mackerel (*Scomber scombrus*). *J. Food Sci.* 59: 265-268.

Venugopal, V., S.P. Chawla, and P.M. Nair. 1996. Spray dried protein powder from threadfin beam: preparation, properties and comparison with FPC type-B. *J. Muscle Foods* 7: 55-72.

Viera, G.H.F., A.M. Martin, S. Saker-Sampaiao, S. Omar, and R.C.F. Goncalves. 1995. Studies on the enzymatic hydrolysis of Brazilian lobster (*Panulirus* spp.) processing wastes. *J. Sci. Food Agric.* 69: 61-65.

Wako, Y., S. Ishikawa, and K. Muramoto. 1996. Angiotensin I-converting enzyme inhibitors in autolysates of squid liver and mantle muscle. *Biosci. Biotechnol. Biochem.* 60: 1353-1355.

Wilding, P., P.J. Lilliford, and J.M. Regenstein. 1984. Functional properties of proteins in foods. *J. Chem. Tech. Biotechnol.* 34B: 182-190.

Yanez, E., D. Ballester, and F. Monckeberg. 1976. Enzymatic fish protein hydrolyzate: chemical composition, nutritive value and use as a supplement to cereal protein. *J. Food Sci.* 41: 1289-1292.

Yu, S.Y. and L.K. Tan. 1990. Acceptability of crackers ('Keropok') with fish protein hydrolysates. *Int. J. Food Sci. Technol.* 25: 204-210.

Yu, S.Y. and S. Fazidah. 1994. Enzymic hydrolysis of proteins from *Aristichthys noblis* by protease P"Amano"3. *Trop. Sci.* 34: 381-391.

Shelf Life Extension and Value Addition of Fishery Products: A Critical Evaluation

Vazhiyil Venugopal

Food Technology Division, Bhabha Atomic Research Centre, Mumbai 400 085, India.

Introduction

Global fish production is showing stagnation at around 120 million tonnes, essentially due to overfishing and the resulting depletion of stocks. About 30% of the total catch is lost due to spoilage because of its inherent characteristics and also due to the lack of required infrastructure for its preservation. While several species of by-catch, non-conventional and under-utilised species are equally important from a nutritional point of view, enough attention is not being paid for their conservation because of poor consumer interests in these species. At the same time, interests in fish are increasing throughout the world because of the rising population and awareness of the health benefits of fish consumption. Aquaculture is gaining rapid acceptance as a way to supplement the demand for selected fish. The current emphasis, however, needs to be directed towards increased utilisation of the available catch rather than increasing production.

The two major problems that confront commercial fisheries are high perishability of the commodity and potential health hazards due to possible presence of pathogenic microorganisms, parasites, and toxins. Perishability of the commodity is caused by both autolytic enzymes as well as enzymes of spoilage causing microorganisms that contaminate the fishery products. Unlike red meat, the intrinsic characteristics of fish muscle, such as high moisture and low collagen contents, make it more sensitive to microbial spoilage. Since fishing voyages usually take several days, it is essential that fish quality is not compromised until it reaches the processing centres. Retention of quality is also important in distribution among consumers in the interior parts of the countries, where fish are used fresh or processed in traditional ways. All over the world consumers as well as regulatory authorities demand that fishery products intended for human consumption should be fresh as well as safe against any health hazards. It is important for any seafood processing industry to process fish while it is still firm and intact. Importers of seafood will not accept products which are rancid, discoloured, stale, and possess off-odours and off-flavours and also if the products pose any health hazards (Martinez 1997).

While chilling is the age old method to control spoilage, the chilled storage life of fish is limited to a few days. Combination of additional processes in conjunction with chilling can further extend the refrigerated shelf life of many products up to several weeks. These include treatment with chemicals, modified atmosphere, and low dose radiation, in combination with chilling. These can lead to better marketability, nutritional benefits and economic gains from the catch. Efforts in this direction can also help better conservation of aquacultured fishery products. This article will discuss the different techniques for fish preservation and value addition of fishery products.

Spoilage of Fish

Biochemical Changes in the Muscle

Fish is prone to rapid spoilage due to autolytic changes, bacterial activity and other reactions such as lipid oxidation. Immediately after death, fish muscle is relaxed and elastic. Within a few hrs, *rigor mortis* sets in, when the whole body becomes inflexible and hard. The onset of *rigor* depends upon the temperature, particularly on the difference of temperature between that of water and storage. When the difference is large, the time from death to onset of *rigor* is short and *vice versa*. During *rigor mortis*, aerobic respiration ceases and the anaerobic oxidation of glucose leads to accumulation of lactic acid which results in a drop in the muscle pH from about 6.8 to 6.5. However, most teleost fish and crustaceans contain little carbohydrate, whereas its content is higher in bivalve molluscan shellfish. The final pH depends upon the species and composition of the organism. During *rigor,* depletion of adenosine triphosphate (ATP) (to a content of about one micromole per g) results in stiffening of the muscle due to the irreversible association of myosin and actin molecules in the absence of the nucleotide. Resolution of the *rigor* is a slow process essentially due to the low pH-favoured hydrolysis of actomyosin by acid proteases such as cathepsins present in the muscle. *Rigor mortis* of fish has technological significance since the quality of fillets is influenced by the process. Ideally, fish should be filleted *post-rigor*. Fillets prepared in *rigor* will be stiff with poor yields, which can cause gaping, when handled roughly (FAO 1995, Connell 1995). If the fillets are removed from the bone *pre-rigor*, the muscle can contract freely and the fillets will shorten following the onset of *rigor*. For example, studies on salt absorption behaviour of salmon and cod conducted by Sorensen *et al.* (1997) showed that *post-rigor* fish, when immersed in saturated brine, absorbed 3% salt with a 6.5% increase in weight within a period of one hr. On the other hand, *pre-rigor* fillets required 3 hr to absorb same amount of salt, but was associated with a 7% loss in weight.

Immediately after death, the ATP present in the muscle is degraded by autolytic enzymes by the sequence shown below:

$$\text{ATP} \rightarrow \text{ADP} \rightarrow \text{AMP} \rightarrow \text{IMP} \rightarrow \text{HxR} \rightarrow \text{Hx} \rightarrow \text{X} \rightarrow \text{U}$$

The first five steps of the reaction sequence proceed at a relatively faster rate. The oxidation of hypoxanthine (Hx) to xanthine (X) and ultimately to uric acid (U) is much slower and is the result of microbial enzyme activity. A strong correlation has been observed between nucleotide catabolism and loss of freshness of fish (Lakshmanan and Gopakumar 1999, Ashie 1996, Saito *et al.* 1959).

Saito *et al.* (1959) proposed that the freshness of fish could be determined in terms of 'K-value', by the content of the autolytic degradation products of ATP, as given below:

$$\text{K-value} = \frac{[\text{HxR}] + [\text{Hx}]}{[\text{ATP}] + [\text{ADP}] + [\text{AMP}] + [\text{IMP}] + [\text{Ino}] + [\text{Hx}]}$$

where [ATP], [ADP], [AMP], [IMP], [HxR] and [Hx] represent the relative concentrations of adenosine triphosphate, adenosine diphosphate, adenosine monophosphate, inosine monophosphate, inosine and hypoxanthine, respectively, at a specific time during storage of the fish. The content of IMP is known to contribute to the fresh fish flavour, and its loss due to subsequent breakdown causes loss of freshness, associated with a higher K-value. Since the concentrations of ATP, ADP and AMP significantly change within the first day of death, a simplified K-value (see below) has also been employed.

$$\text{K-value} = \frac{\text{HxR} + \text{Hx}}{\text{IMP} + \text{HxR} + \text{Hx}}$$

The K-value has been found suitable for several species, although it has limitation for some fish such as cod. Instead of K-value, a modified K_p-value equal to the concentrations of hypoxanthine/adenine has been found suitable for freshness valuation of certain shellfish (FAO 1995). The enzymes involved in the oxidation of Hx and X are given below:

Nucleoside phosphorylase: Inosine + Pi $\rightarrow$ Hypoxanthine + ribose–Pi

Xanthine oxidase: Hypoxanthine + Oxygen → X anthine + hydrogen peroxide

Xanthine oxidase: Xanthine + oxygen → uric acid + hydrogen peroxide

Lipid oxidation is associated with early post-mortem changes in fish tissue. The process is initiated by the accumulation of active oxygen species, the activation of haemoproteins, an increase in free iron and the consumption of antioxidants (Hultin 1994). Lipid oxidation is comparatively more during frozen storage, rather than chilled (0° – 2°C) storage, and can be both non-enzymatic as well as enzymatic. Enzymes such as lipoxygenase, peroxidase and microsomal enzymes from animal tissues can potentially initiate lipid peroxidation producing hydroperoxides (German and Kinsella 1985, Hsiegh and Kinsella 1986, McDonald and Hultin 1987, Wasson 1992). The breakdown products of these hydroperoxides include aldehydes, ketones, and alcohols which may cause development of off-flavours. Fish lipids rich in n-3 polyunsaturated fatty acids are very susceptible to oxidation, giving rise to n-3 aldehydes which cause distinctive oxidative off-flavours. These compounds include *cis*-4-heptenal, *trans*-2-heptenal, *trans*-2,cis-4-heptadienal etc. and also 1,5-octadien-3-ol, 1-octen-3-ol and hexanal.

Lipoxygenase-dependent oxidative activity has been detected during chilled storage of fatty fish species, sardine (*Sardina pilchardus*) and herring (*Clupea harengus*). Lipoxygenases were concentrated in the skin tissue of fish, and were active for up to 48 h of chilled storage. The pro-oxidative activity due to haem proteins continued for longer than that due to lipoxygenase. The most abundant degradation products of the hydroperoxides formed from arachidonic and docosahexaenoic fatty acids were 12- and 16-hydroxy acids. Trends of fluorescent formation resulting from interaction between oxidation products and biological amino constituents were compared with the pro-oxidative activities to establish co-relations with quality loss during chilling (Medina *et al.* 1999). The extent of lipid oxidation can be reduced by glutathione peroxidase which reduces unstable lipid hydroperoxides to non-radical, stable products, which are inactive in the oxidative chain propagating mechanism (Hultin 1994). Other enzymes include superoxide dismutase and catalase, which remove superoxides from the peroxidation mechanism.

Microbial Spoilage

Fish are prone to contamination with a variety of microorganisms depending upon environmental habitats, handling conditions and temperature (Gram and Huss 1996). The low molecular weight peptides and free amino acids produced by the autolysis of proteins not only lower the commercial acceptability of fish, but also provide an ideal medium for growth of contaminant microorganisms. The bacterial flora of fish caught in clean waters include gram-negative organisms such as *Pseuodomonas, Moraxella, Acinetobacter, Shewanella putrefaciens, Flavobacterium, Vibrio* and *Aeromonas* spp. and gram-positive bacteria namely, *Bacillus, Clostridium, Micrococcus,* and *Lactobacillus* spp. In polluted waters a high number of Enterobacteriaceae may be found. The flora of tropical fish often carry a slightly higher load of gram positive and enteric bacteria. During storage, the bacteria will grow with a doubling time of approximately 20 min and one day at ambient and ice temperatures, respectively, to reach a value of 10^8 per gram within a period 24 h at ambient temperatures and 2 to 3 weeks, when iced. The composition of the microflora of several different species of cold water fish at spoilage are reported to be dominated by gram-negative bacteria, while gram-positive bacteria may dominate in warm water fish species (Gram and Huss 1996). Nevertheless, bacterial spoilage patterns of both cold and tropical water fish during ice storage are comparable. Delay in ice storage can cause a shift in the composition of the microflora.

Bacteriological growth profiles in fish during storage have been studied in detail by several authors (Gram and Huss 1996, Venugopal 1990, Kraft 1992, Liston 1980). Initially, aerobic microorganisms grow on the fish surface, and when the surface becomes covered and slime builds up, conditions become more favourable for the growth of anaerobes. Associated with the bacterial proliferation, several volatile flavour-affecting compounds are generated on the fish muscle. Some of the bacteria such as *Alteromonas, Vibrio, Proteus* etc. present on fish are able to carry out respiration by using trimethylamine-oxide (TMA-O), as an electron acceptor, which is abundantly present in marine fish. This results in accumulation of trimethylamine (TMA), which is one of the dominant compounds (about 10 to 25 mg TMA per 100 g muscle) of spoiling fish contributing to the characteristic spoiled fishy odour. TMA, in addition to ammonia and other volatile amines, is a major component of the total volatile basic nitrogen (TVBN) compounds during fish spoilage. TMA formation in many fish species during spoilage is paralleled by bacterial production of hypoxanthine. Many spoilage-causing bacteria such as *Shewanella putrefaciens* and *Vibrio* spp. also produce off-smelling volatile sulphur compounds such as hydrogen sulphide, methylmercaptan and dimethylsulpide, formed from sulphur containing amino acids such as methionine.

Shewanella putrefaciens and *Pseudomonas* spp. have been recognised as the specific spoilage causing bacteria of iced fresh fish regardless of the origin of the fish (Dainty 1996, Gram and Huss 1996). Modified atmosphere-stored marine fish from temperate waters are spoiled by the CO_2-resistant *Photobacterium phosphoreum*, whereas, gram-positive bacteria are likely spoilers of CO_2-packed fish from tropical waters. *Photobacterium phosphoreum* has also been identified as the main spoiler in vacuum-packed cod (Dalgaard *et al.* 1993), while in iced, gutted cod, the specific spoilage organism was *S. putrefaciens* (CAC 1996). Fish products with high salt contents may spoil due to growth of halophilic bacteria, anaerobic bacteria and/or yeast, whereas in lightly salted fish, spoilage could be because of lactic acid bacteria and certain Enterobacteriaceae (Gram and Huss 1996). At the advanced stage of microbial growth, extracellular proteases secreted by the microorganisms cause further degradation of the fish muscle. A general pattern of spoilage of fish is depicted in Figure 1.

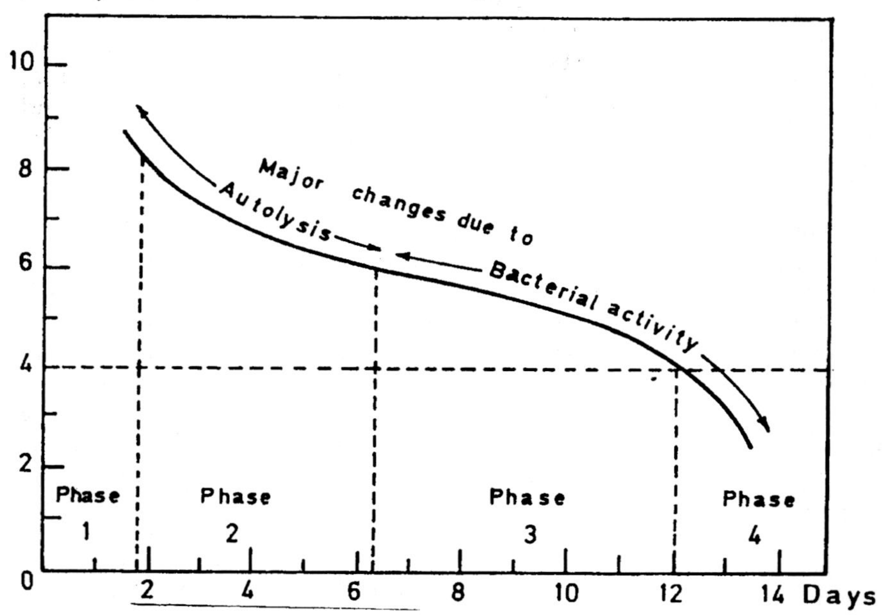

Figure 1. Typical spoilage profile of fish during ice storage.

Flavour Changes

The flavour of fish significantly influences its consumer acceptability. Fresh marine fish are nearly odourless, because they contain a small quantity of volatiles, whereas freshwater fish give off earthy-odour compounds, such as pyrrolidine, depending upon the nature of their habitats (Kawai 1996, Lindsay 1991). During storage, the action of exogenous enzymes in post-mortem tissue contributes to flavour changes. Fish chilled for one or two days may have better flavour due to their content of inosine monophosphate (IMP) formed through the enzymatic degradation of ATP. The flavour of cooked cod, for example, has the strongest intrinsic characteristics after 2 days in melting ice. The rapid oxidation of large amounts of unsaturated lipids present in fish is a major reason for changes in smell, taste, colour, texture and nutritional value. Enzymatic as well as non-enzymatic oxidation of polyunsaturated fatty acids of fish muscle lead to a variety of carbonyls, alcohols etc. that are responsible for flavour changes in fish, as mentioned above. Many fatty species such as salmon, ocean perch or halibut improve much in flavour, taste and texture, during the first 2 to 4 days of ice storage (Martinez *et al.* 1997).

Associated with the lipid oxidation compounds, microbial formation of several volatile nitrogenous compounds results in significant loss of fresh fish flavour. The chemical compounds produced by the action of microorganisms that affect sensory quality of fish include trimethylamine, hydrogen sulphide, dimethyl sulphide and methyl mercaptan, various amines and ammonia, indole, skatole, putrescine and cadaverine (Reineccius 1991). The factors influencing the flavour quality of farm-raised fish have been discussed by Johnson (1989) and Kawai (1996). In the case of elasmobranchs like sharks and rays, ammonia may form from the action of endogenous urease on urea in the flesh. Similarly, rapid formation of ammonia in shrimp is correlated with adenosine deaminase and AMP deaminase activity on nucleotides and arginase-catalysed formation of urea and its conversion to ammonia.

Changes in Texture

Tenderness is, perhaps, the most important quality parameter of muscle foods including fish. Fish is generally more soft than red meat because the former is characterised by a lesser amount of connective tissue and lower degree of its cross linking (Venugopal and Shahidi 1996). Therefore, tissue softening in fish is faster than in red meat. Tenderisation or flesh softening is associated with the disappearance of Z-disks in the muscle cell with the release of α-actinin, dissociation of the actomyosin complex, destruction of connectin, and general denaturation of connective tissue (Bremner 1992). Haard (1992a,b,c) observed that degradation of fish muscle caused by endogenous proteases is a primary cause of quality losses during cold storage. Muscle proteases including cathepsin D and cathepsin L, calcium-activated proteases (calpains), trypsin, chymotrypsin, alkaline proteases and collagenases have all been involved in the softening of fish tissue during storage (Sherekar *et al.* 1986, Bremner 1992). Doke *et al.* (1979) reported the role of hydrolytic enzymes in the autolytic spoilage of fishery products. The major source of hydrolytic activity in marine fish (Bombay duck) was concentrated essentially in the drip, while in the freshwater fish *Tilapia mossambica* it was in the skin. Removal of cathepsin D by removal of the drip or deskinning tilapia suppressed enzyme-mediated spoilage. It has been suggested that the proteins of muscle are largely unaffected during post-mortem storage, and that the softening of muscle is due not to the breakdown of myofibrils but to proteolytic digestion of minor cell components that link the major structural units together (Ando *et al.* 1991). Collagenases have been found to be involved in the development of mushiness in prawns stored on ice (Lindner *et al.* 1988).

Discolouration

Another major quality problem in seafood industry is the discolouration of items such as shrimps, lobsters etc. The pink/red colour of the skin of most seafoods fade during iced/chilled storage due to oxidation of the carotenoid pigments. The extent of the loss of colour is dependent upon the fish, availability of oxygen and the storage temperature (Haard 1992). Carotenoid fading may involve several mechanisms including (i) autooxidation of the conjugated double bonds by reaction with atmospheric oxygen; (ii) coupled oxidation with lipids involving free radicals released from fatty acids combining with carotenoids to form lipid hydroxyperoxides and carotenoid radicals; and (iii) enzyme activity (Simpson 1985). A lipoxygenase-like enzyme capable of catalysing oxidation of carotenoids like astaxanthin, tunaxanthin and β-carotene has been characterised from some fish species (Tsukuda 1970). Another enzyme involved in carotenoid oxidation has its activity influenced by hydrogen peroxide and halides, particularly bromides and iodides. The bleaching of β-carotene could be accomplished by one of these three possible mechanisms (Kanner and Kinsella, 1983).

Oxidation of muscle pigments, namely myoglobins is another reason for colour change in fish. Oxidation of bright red myoglobin to brown metmyoglobin can occur through both non-enzymatic and enzymatic routes. An enzyme, metmyoglobin reductase, has been isolated from the skeletal muscle of dolphins and bluefin tuna (Hsieh and Kinsella 1986). Development of black spot due to melanosis or enzymic browning in shrimp and lobster during storage is caused by the phenolases present in their tissues (Savagaon and Srinivasan 1978, Yan *et al.* 1989).

Food Safety Problems

Apart from spoilage-causing microorganisms, contamination of fish, shrimp, oysters, mussels, crabmeat etc. by pathogenic organisms, viruses and parasites has become a major issue in recent years because of their potential to cause food-borne diseases. Some of the common pathogens found in fishery products include *Salmonella* spp., *Staphylococcus aureus,* different species of *Clostridium botulinum, Bacillus cereus, Campylobacter jejuni, Escherichia coli* O157:H7, *Vibrio parahaemolyticus, Yersinia enterocolitica* and, *Listeria monocytogenes.* The incidence of many of the above pathogens in processed fishery items, including frozen products, has been of great concern with respect to their international trade. Furthermore, a variety of parasites, including flatworms, roundworms, tapeworms, and protozoa, infest the gills, viscera and skin of marine, freshwater, as well as farm-raised fish species which can cause diseases to consumers (Venugopal *et al.* 1999).

Several insects, such as flesh flies, beetles and mites, also infest fish, particularly during sun drying, the most destructive pest being the hide beetle, *Dermestes maculatus* Deg. Virus contamination is another common quality problem because of the habit of filtering large amounts of water by molluscan shellfish. From the sources of these contamination, public health issues associated with seafoods can be grouped as environment-, process-, distribution- and consumer-induced (Garret *et al.* 1997). In addition to the above contaminants, the presence of toxins, such as ciguatera, paralytic shellfish poisoning (PSP), diarrhetic shellfish poisoning (DSP), amnestic shellfish poisoning (ASP), histamine, etc., and environmental pollutants, such as mercury, pesticides residues etc., are other major problems faced by the seafood as well as aquaculture industry (Venugopal *et al.* 1999, WHO 1999). Understanding the profound influence of the above factors on the fishery products, the industry has adopted quality standards in order to safeguard the interests of the consumer.

Evaluation of Fish Freshness

Since fish freshness deteriorates fast, its evaluation is of utmost importance for the seafood industry for development of products of acceptable quality. The various freshness quality indices of seafoods have been discussed by Alur *et al.* (1997), Olafsdottir *et al.* (1997), Botta (1994) and Gill (1990). Sensory evaluation is the rapid method for quality evaluation and is defined as the scientific discipline used to evoke, measure, analyse and interpret characteristics of food as perceived by the senses of sight, smell, taste, touch and hearing (Olafsdottir *et al.* 1997). Sensory tests can be divided into three groups: discriminative, descriptive, and affective tests (FAO 1995). Discriminative testing is used to determine if a difference exists between samples (triangle test, ranking test). Descriptive tests are used to determine the nature and intensity of the differences (profiling and quality tests). The affective test (market test) is based on a measure of preference or acceptance by the consumer.

The most common descriptive tests are structured scaling for quality assessment and profiling for a detailed assessment of one or more attributes. In Europe, the most commonly used method for the quality assessment of raw fish in the inspection service and in the industry is the European Union Scheme (EC 1996). The quality index method (QIM) is a descriptive method, which has been suggested for raw fish based on sensory parameters. In this method, the descriptions of the individual grades are precise, objective, independent and primary rather than a cluster of terms. The scores for all of the characteristics are then added to give an overall sensory score (quality index), which can also be used to predict storage life. The method can be used to estimate the influence of time, temperature, and handling on raw as well as cooked quality of fish species including sardines, anchovies, saithe, place, and cod. Computer programs based on these evaluations have also been developed to predict the overall quality and hence, the value of these fish species at any given point of ice storage (Neilson 1993). A descriptive, 10-point hedonic scale can also be used to determine the chilled storage life of fresh or cooked fish, where a score of '10' indicates the maximum freshness characteristic of the fish, and '1' indicates a strong putrid odour. A score of '5' is indicative of borderline acceptability (FAO 1995, Bilinsky *et al.* 1983).

Microbiological and biochemical methods have also been used to determine fish quality. Since the activity of microorganisms is the main factor limiting the shelf life of fresh fish, an estimation of the total viable counts (TVC) or total plate count (TPC) in terms of colony forming units (cfu) helps quality assessment. However, the value is seldom a good indicator of the sensory quality or expected shelf life of the product. Newly caught fish have a TVC of 10^2 to 10^6 colony forming units (cfu) per g. At the point of sensory rejection, the TVC of fish products are typically 10^7 to 10^8 cfu per g. Nevertheless, TVC counts are used in standards, guidelines and specifications that often use much lower TVC as indices of acceptability. Biochemical methods of measurement of odour-bearing compounds can be used for freshness evaluation. Classical chemical methods for the analysis of total volatile bases (TVB) and trimethylamine (TMA) have been used for the determination of fish freshness in the industry. TVB values less than 30 mg N per 100 g are indicative of excellent quality, while values in excess of 45 mg N per 100 g indicate an unacceptable product. There is a good correlation between TVB values and sensory analysis (Woyewoda and Ke 1980). Trimethylamine oxide (TMAO) found in appreciable quantities in marine fish is broken down to trimethylamine by the bacterial enzyme trimethylamine oxide reductase. The content of TMA has been related to loss of freshness in marine fish.

Fish lipids are prone to hydrolysis and oxidation during storage, and the products formed affect its quality. Fatty fish are highly susceptible to such quality changes. Formation of free fatty acids is more in ungutted than in gutted fish, probably due to the involvement of digestive enzymes, particularly lipases. Autooxidation of fish lipids is initiated by removal of a proton from the central carbon of the unsaturated fatty acid, usually a pentadiene moiety of the fatty acid, and formation of a lipid radical. The latter reacts very quickly with atmospheric oxygen making a peroxy-radical ($LOO^{\cdot}$). The chain reaction involving the peroxy-radical results in formation of hydroperoxides, which are readily broken down, catalyzed by heavy metal ions, to secondary products including aldehydes, ketones, alcohols, carboxylic acids and alkanes. Development of rancidity in fishery products as a result of autooxidation is estimated in terms of the content of malonaldehyde (Gray 1978, Sinhuber and Yu 1960, Tarladgis 1960). Malonaldehyde is the principal compound in oxidized lipids that react with 2-thiobarbituric acid (TBA) to give a red pigment, which is measured colourimetrically.

Recently there is much interest in the development of biosensors and 'electronic noses' for rapid measurement of odour compounds in fish. The potentials for various biosensors for fish quality evaluation has been discussed recently by Venugopal *et al.* (2000).

Changes in fish freshness can also be determined by measuring the electrical properties of fish muscle. Three different instruments are available to measure the changes in electrical properties: The Torrymeter (Distell Industris Ltd., Fauldhouse, West Lothian, UK), the Fishtester VI (Intelectron International Electronics, Hamburg, Germany) and the RT-Freshness Grader (RT Rafagnataekni, Reykjavik, Iceland), which all show good correlation with sensory scores of fish freshness when used within their applicable range of operation. These meters cannot be used with thawed fish or fish that have been stored in chilled seawater, and their use for fillets is limited to a few days.

Time-temperature indicators (TTIs) are devices that can be attached to foods to indicate the time-temperature history of the food, based on measurement of some biological, physical or chemical process that depends on time and temperature. The temperature record can be used to develop appropriate models to predict the shelf life of fish. There is good potential for use of such devices for fresh, processed and chill stored fish (Labuza and Fu 1995).

The sensory quality of some fish, like yellowfin, skipjack, and bonito are highly affected by the colour of the muscles. Brown discolouration due to oxidation of myoglobin, replacing the original pink or deep red colour, makes the fish unsuitable or less appealing for consumption (Sikorski and Pan 1994). There is need to develop instrumental methods for colour measurement of such fish. Spectroscopic methods have gained acceptance to determine whether a fish has been frozen and also to estimate the storage time of fish on ice. Recently, the use of near-infrared spectroscopy has shown promise of being useful to assess the storage time of fresh fish (Olafsdottir *et al.* 1996). L*a*b* is an international standard for colour measurements, adopted by the Commission Internale d'Eclairage (CIE) in 1976. L* is the luminance or lightness component, which ranges from 0 to 100, and a* (from green to red) and, b* (from blue to yellow) are the two chromatic components which range from –120 to +120. The principles of colour measurement have been discussed by Papadakis *et al.* (2000) and Cydesdale (1978). Table 1 shows some of the methodologies employed for fish quality evaluation.

Table 1. Conventional methods for quality evaluation of fishery products

Method	References	Remark
Sensory evaluation	FAO 1995, Connell 1995	Depends upon sight, smell, taste, touch and hearing as judged by experienced persons
Chemical methods: Total volatile basic nitrogen	Convey microdiffusion method Farber and Ferro 1956, Gill 1990	Measurement of trimethylamine, ammonia and other volatile basic nitrogen by titration
Trimethylamine	Wong and Gill 1987	Poor correlation with total bacterial counts
Ammonia	Leblank and Gill 1984	Indicates only advanced spoilage of finfish
Volatile acids	Venugopal *et al.* 1981	Good correlation with bacterial spoilage
Nucleotide catabolites	Saito *et al.* 1959. HPLC method is used	Degradation products of ATP. Reliable quality indices of several fish/shellfish, based on K-value.
Biogenic amines	HPLC method is used Yen and Hsieh 1991	The amines are thermostable and hence cooked fish can be analysed
H_2S, CH_3SH, $(CH_3)_2S$	Determined by gas chromatography	Indicates advanced degree of spoilage
Ethanol	Kelleher and Zall 1983	Good quality index for several fish such as salmon, raw tuna, redfish, pollock, flounder and cod
2-Thiobarbituric acid (TBA) value	Tarladgis *et al.* 1960, Sinnhuber and Yu 1958, Gray 1978	Good index of oxidative rancidity in fish lipids. Reasonable correlation with sensory properties.
Instrumental methods	Pau and Olafsson 1991 Commercial pH and texturometers, moisture and fat analyzers	Electrical properties, pH, vision properties of fish muscle. Presence of parasites and blood clots in fish fillets are determined by computer-aided vision techniques
Microbiological methods	Total plate count Liston 1980	Indicative of microbial spoilage of fresh fish

Spoilage Rate of Fish

Spoilage of fish is directly influenced by temperature. Reactions by muscle as well as microbial enzymes are directly related to temperature. Therefore, the spoilage rate of fish has been found to be linearly related to storage temperature. Chilling fish immediately after they have been caught retards the rate of spoilage, since many bacteria are unable to grow at temperatures below 10°C and even psychrotrophic organisms grow very slowly when the temperature approaches 0°C. It is known that a rise in temperature of refrigerated fish from 0° to 3°C doubles spoilage, while an increase to 10°C enhances spoilage by a factor of 5 to 6. The effect of temperature on the rate of chemical reactions is often described by the Arrhenius Equation. This equation, however, has been shown not to be accurate when used to study the effect of a wide range of temperatures on the growth of microorganisms and spoilage of foods.

The *relative rate of spoilage* at any temperature 't' can be given as the ratio of the spoilage rate at temperature 't' to that 0°C. Spoilage rate can be determined by the slopes of the plots of sensory, bacteriological or chemical quality indices of fish plotted against storage time at each temperature, since these plots are essentially linear. Using the volatile acid content as an index of spoilage, Venugopal *et al.* (1981) determined the spoilage rate of Indian mackerel. Irradiation of the fish at a dose of 1.5 kGy reduced the spoilage rate and extended its refrigerated shelf life.

The relationship between spoilage rate and temperature is given by the Spencer and Baines equation, namely,

$$k = k_0 (1 + Ct)$$

where, k = spoilage rate at temperature, t (degree Celsius), k_0 = spoilage rate at 0°C, and C = linear temperature response. The relative rate concept has made it possible to quantify and mathematically describe the effect of temperature on the rate of spoilage of various types of fish products. In the temperature range of 0 to 8°C, the relative rate of spoilage of fish may be computed as:

$$k/k_0 = 0.24\ T$$

where k and k_0 represent the rate of spoilage, in spoilage units per day, at temperature T and 0°C.

Ratkowsky *et al.* (1982) suggested a 2-parameter square root model for the effect of sub-optimal temperature on growth of microorganisms, given as,

$$\sqrt{\mu_{max}} = b\ (T - T_{min})$$

where T is the absolute temperature (Kelvin) and T_{min} is a parameter expressing the theoretical minimum temperature of growth. The square root of the microbial growth rates plotted against the temperature form a straight line from which T_{min} is determined. Based on the T_{min}, a spoilage model has been developed on the assumption that the relative microbial growth rate would be similar to relative rate of spoilage. Based on this, the square root of the relative spoilage rate has been found to equal 0.1 ´ t°C +1. The relative spoilage rates for tropical fish are more than twice as high as estimated for temperate fish species. For these fish, the logarithm of the relative spoilage rate has been found to be 0.12 x t°C.

At a constant storage temperature, measurements of quality will change linearly from an initial to a final level when the product is no longer acceptable. Shelf life at a given temperature and given initial quality is determined by the equation:

$$\text{Shelf life} = \frac{\text{Final} - \text{Initial level of a quality index}}{\text{Spoilage rate in terms of the index under specified conditions}}$$

Knowing the shelf life at a given temperature, the shelf life at any other temperature can be determined from a temperature spoilage model, as mentioned above.

The effect of time/temperature storage conditions on product shelf life has been shown to be cumulative (Charm *et al.* 1972). Ronsivalli *et al.* (1973) developed a slide rule for predicting the shelf life of whole and fillets of cod and haddock based on the Spencer and Baines equation. The slide rule has three components, namely the day/temperature scale, the shelf life scale and the index. By knowing the storage temperature history of the fish items, the remaining shelf life could be predicted. The temperature spoilage models allow time/temperature function integration to be used for product evaluation, distribution and storage and also the shelf life of various fish products. An electronic time/temperature function integrator for shelf life prediction was developed based on the Ratkowsky equation (Owen and Nesbitt 1984).

Methods for Extension of Shelf Life of Fishery Products

Chilling

The common preservation method of fresh fish in tropical and temperate climates is by chilling it to about 0°C. Such chilled raw, unpacked, wet fish is usually regarded as fresh by the consumer. The extension of shelf life by chilling is essentially due to the reduction in the growth rate of microorganisms that contaminate seafoods from various sources (Olley and Ratkovsky 1973, Shewan 1971, Venugopal 1990). The fish catch prior to chilling may be washed to remove contaminating microorganisms, slime, and blood. Spray washing has been recommended for rapid cleaning of fish (Kosak and Toledo 1981). Although the ideal temperature of chilled fish is 0-2°C, even at this temperature some metabolic activity of contaminant microorganisms persists causing the fish to spoil (Venugopal 1990). A chill temperature of below 0°C, ideally, –2°C, has, therefore, been advocated. Chilling at temperatures of –2° to –4°C extends the shelf life of fish from 2 weeks at 0°C to about 4 to 5 weeks by effectively retarding bacterial spoilage, although it may cause some loss of sensory quality (Sikorski and Pan 1994). A reduction in the storage temperature from 5° to 0.6°C doubled the shelf life of several varieties of fish (Nickerson *et al.* 1983). Whole fatty fish such as herring remain acceptable at 20°C even after 6 to 12 hr of harvest and have typical shelf lives (5 to 7 days at 0°C, 4 to 6 weeks at –10°C and 6 to 9 months at –30°C) (Hardy 1986). The FAO (1973) recommends that the fish be cooled to the temperature of melting ice (0°C) as quickly as possible. Therefore, chilling on board immediately after catch is recommended. However, non-commercial, under-utilised fish are likely to be neglected in this respect.

The most common chilling media are wet ice, mixtures of ice and seawater, or ice-cold seawater. The normal storage life of cold-water fish chilled at 0°C immediately after post-*mortem* is 1 to 2 weeks, while fish from warm tropical water remain somewhat longer at 0°C. Chilled storage life depends on several factors such as fish composition, lipid content, contamination and type of microflora. Although rancidity development in lean fish is seldom a cause of quality deterioration, rancid off-flavours may be noticed in fish such as mullet and bulefish after 4 to 5 days on ice (Sikorski and Pan 1994).

The quantity of ice used to chill a batch of fish can be calculated as a function of both the quantity as well as temperature of the batch. Thus, the amount of ice required in kg is equal to:

$$C_{pf}\ ´\ T_f\ ´\ M_f /80,$$

where, C_{pf} = specific heat of fish (kcal/kg) C, approximately 0.80 for lean fish, 0.78 for medium fatty fish and 0.75 for fatty fish, T_f = fish temperature, usually taken as seawater temperature, M_f = mass of fish and 80 is the latent heat of fusion of ice (kcal/kg). If all the factors are put together, the amount of ice required to cool lean fish to 0°C would be approximately equal to T_f ´ M_f /100. Therefore, the ice necessary to cool 1 kg of lean fish from 25°C to 0°C will be 0.25 kg. While cooling fish to 0°C, some thermal loss of ice also takes place which results in its melting. Thermal loss of ice depends mainly on the external temperature and the type of container in which the iced fish is stored. Thermal loss of ice (L) can be expressed as L = k ´ T_e ´ t, where k = specific ice melting rate of the box/container, T_e = average external temperature, and t = time elapsed in h since icing. The k value can be determined experimentally in boxes and insulated containers. The values of k (kg of ice per day) for plastic box (polyethylene, 40 kg) and insulated container (Metabox 70 DK) have been found to be 0.22 and 0.108, respectively (Boeri *et al.* 1985, Lupin 1985). Ice stored at ambient temperature contains 12–20% water on its surface depending upon the nature of the ice (flake, crushed block or chips) which has a negligible cooling effect on fish (useful only to improve heat transfer and to keep the fish moist) (FAO 1993). To compensate for water loss from melting ice and bad handling, it is advisable to add about 12 to 20% extra ice to the fish (Zugarramurdi *et al.* 1995).

Ice may be used in the form of blocks, plates, tubes, shells, soft and flakes (FAO 1993, Rohr 1995). Of these, flake ice is the most popular for industrial use because of its advantages related to cooling, efficiency and economics of production. It is also dry and hence, will not stick together or form blocks when stored for long periods of time at sub-zero temperatures. Flake ice cools fish efficiently and faster due to a large heat exchange surface which can absorb approximately 83 kcal per kg. The ideal fresh water ice has a temperature of 0.5°C; it stays loose and workable and can be stored without refrigeration. Flake ice may be made from potable water or seawater.

It is generally accepted that the storage life, in ice, of tropical fish species is slightly longer than that of temperate species (Connell 1995, Poulter and Nicolaides 1985). Smaller fish generally have a shorter shelf life on ice. On average, the storage life of small fatty fish is 5 to 8 days, and lean white fish caught in cold waters up to 14 days; up to 21 days for big fish such as snappers and halibut, and even up to 30 days for some species from tropical waters such as mullets and breams. Mullets from warmer waters, chilled in ice immediately after capture, have a longer shelf life than those from colder waters. The storage life of dressed gray fish in ice is reported to be 6 to 8 days (ASHRAE 1987, Jhaveri and Constantinides 1982), 18 days for white flounder (Shaw *et al.* 1977), and 5 days for whole sardine (Nunes, 1990). The ice storage shelf life (days) of some marine species of interest are cod (15 to 16), hake (8 to 10), shark (8), herring (4 to 5), mackerel (4 to 5), tuna (21 to 22) and halibut (21 to 22) (Rodrick and Dixon 1994). Freshly chilled sardines (*Sardina pilchardus*) have a storage life of 9 days (Marrakchi *et al.* 1990), while that of the oil sardine (*Sardina longiceps*) is 5 to 6 days. During ice storage of elasmobranchs, such as sharks and rays, formation of ammonia from urea is the major quality index associated with pH increase (Pastoria and Sampedro 1994, Ghadi and Ninjoor 1989). Atlantic mackerel can be stored fresh in ice up to 9 days, after which accumulation of bacteria, trimethylamine and histamine occurs rapidly (Jhaveri *et al.* 1982).

Extensive investigations have shown that several Indian marine fish can attain an acceptable shelf life in ice up to 14 days, depending upon the species. The fish include the oil sardine (5–6 days), mackerel (6 days), threadfin bream (7 days), and seer (12–14 days) (Gopakumar 1990). While several large demersal fish species are marketed as whole, gutted, steaks, and bone-free fillets, small pelagics are sold as gutted or as fillets. Pelagic fish such as herring and scad can be filleted, which may have a large number of fine bones distributed throughout the flesh. Technological developments have made it possible to combine small fillets to give a large composite fillet without loss of desired textural qualities. These fillets and steaks may also be smoked prior to chilling or freezing.

Freshwater fish generally have a longer shelf life in ice as observed in the case of milk fish (14 days), tilapia (12–13 days), carps (16–21 days), mullet (8 days), and catfish (16–21 days) (Gopakumar 1990). Whole freshwater rainbow trout can be stored up to 12 days. However, longer refrigerated shelf lives in the range of 25 to 40 days have been observed in the case of several other freshwater fish species. Both K and H_x contents are useful indicators of the freshness of fish, regardless whether they have been gutted or not (Rodriguez *et al.* 1999).

Bleeding and gutting immediately after catch can help in the extension of chilled storage life. For example, bleeding and gutting extended the shelf life of dogfish (Ravesi *et al.* 1985). Heading and evisceration of the Atlantic croaker and grey trout helped retain their top quality during ice storage 7 to 10 days longer than that of uneviscerated fish (Stanier 1981). Belly bursting of fish such as anchovy, sardine and mackerel may be a problem during their chilled storage, which could be controlled if the pH of the medium was reduced to 5 by the addition of lactic or acetic acid (Martinez and Gildberg 1988). Evisceration (Bilinsky *et al.* 1983) and bleeding (Ward 1994) have been recommended prior to storage aboard the fishing vessel, in order to enhance shelf life. Immediate chilling of the fish on board affords good quality products (Hansen and Jensen 1982). Delay in icing is of particular concern with fish of the Scrombridae family (tuna, mackerel and bonita) because of the possibility of histamine poisoning, a chemical intoxication resulting from decarboxylation of histidine (Gopakumar 1990, Ward 1994).

Shrimp, unlike other crustaceans and fin fishes, die very soon upon capture, thereby leading to early deterioration. Bacterial spoilage and melanosis (blackening) are the two major causes for spoilage of shellfish. Furthermore, the cephalothorax of shrimp containing digestive enzymes undergoes immediate spoilage and results in loosening as well as blackening of the heads. Therefore, rapid chilling with extensive use of ice is necessary to control spoilage (Chandrasekharan 1994, Govindan 1985).

Chilling of fish on board may be achieved using ice or refrigerated seawater. Ice has a high cooling capacity, but is inefficient if not distributed uniformly. Icing of fish is not without disadvantages, such as bruising of the flesh with concurrent leaching of flavour components, water soluble proteins and vitamins. As an alternative to icing, mechanically refrigerated seawater (RSW) is being employed by the industry. In RSW chilling systems, seawater is chilled in advance to 0°C in tanks and then the fish are dipped into the water. Some water is also cooled to –1° to –2°C by passing over a heat exchanger of the mechanical cooling unit. Because of better contact and the buoyancy effect of water, RSW systems are more efficient in cooling with little pressure damage. The RSW systems are complicated and require sophisticated equipment. Furthermore, because of the circulating nature of RSW, the spoilage causing organisms may spread very rapidly and accelerate deterioration. Injection of carbon dioxide into the RSW system has been shown to delay spoilage by lowering the pH and hence bacterial activity. However, it may have some detrimental effects on fish quality such as flesh discolouration (Ashie *et al.* 1996). The RSW system may also turn the eye colour of some freshwater fish opaque (Sikorski and Pan 1994).

These problems have been reduced in the less expensive 'champagne system' in which the fish are put into insulated tanks up to 65% capacity. The tanks are equipped with perforated pipes for air percolation in the bottom and connected to an air compressor placed on deck. The seawater in the tank is cooled by ice which is mixed with the fish by blowing air through the pipes. Air bubbling is stopped when the temperature reaches 0°C, which may take less than 3 hr. The chilled seawater champagne system has also been used for container storage. These containers may be made from galvanized plate iron equipped with bottom air pipes as well as with lids, and insulated with polyurethane foam coated with fiberglass.

Mechanical handling systems for small pelagic fish on board have been described. These include mechanical graders and gutting machines. Traditionally three methods of fish storage on board are used, namely bulking, shelving and boxing. Traditionally, fish are stored in bulk on ice on board the vessel. If the stacks are more than 5 ft deep, the fish near the bottom may be damaged. This can be prevented by short shelving, where fish and ice are held between shelves of corrugated aluminium, usually 18 to 22 inches apart (Ward 1994). Shelving is primarily used for gutted larger species which are placed on the ice with the gut cavity facing down. For storage, separate holds are usually provided in each trawler (Hansen 1996). Boxing offers the best solution to the problem of on-board storage of fish. Containerisation in box lots of 50 to 250 kg capacity lends itself to rapid and easy unloading of the catch from the vessel. Insulated containers have the added advantages of better fish handling and a smaller requirement of ice.

Traditionally, bulk-stored iced fish are shovelled from the holding pens into baskets which are lifted up to the deck by hand. Mechanisation in fish unloading can prevent fish quality deterioration by speeding up the delivery system. Fish stored in ice-cold seawater, in containers on board, present no unloading problems. These may be lifted by a crane and transported to the plant by truck. Unloading by using fish pumps is increasing world-wide. These pumps may be elevators, dry suction pumps and wet suction pumps, the latter two being widely used for handling fish for human consumption. Pumping is highly suitable for unloading small-size fish (menhaden, sardine, herring etc.). Many of the good wet suction pumps cause less damage to small fish than do the dry suction pumps. Most purse-seiners in northern Europe, particularly in the Scandinavian countries, which land herring and mackerel for the food industry employ scoops for unloading, at about 30-50 tons per hr. Automatic unloading pumps are also used for rapid transfer of RSW/CSW stored anchoveta, sardine, and jack mackerel for the processing plants.

It is essential to maintain hygienic conditions on board the fishing vessels in order to ensure the high quality of the catch. The possibility of contamination and the survival of several psychrotropic pathogenic bacteria, even at chill temperatures, have a detrimental effect on the hygienic quality of the catch. The FAO (1994), therefore, recommends observation of strict hygienic conditions in the handling of fish. Rational design of fishing vessels, use of easily cleanable materials which are resistant to bacterial attack (such as stainless steel and teflon), stricter sanitising procedures as well as new systems of handling the catch are among the major steps required to improve hygiene standards.

Management of the raw material during storage is a fundamental part of production planning in the food fish industry. The 'ice time', namely, the days of pre-processing storage of raw fish in ice, and the remaining shelf life and utility of the fish can be computed (Neilson 1993). Figure 2 shows a general pattern of utility of raw fish with respect to pre-processing ice storage. Considering a maximum yield of 35–50% of finished product from the raw material, the utility of raw fish does not change within the first 5 days on ice, and yields high value products up to 35%. After this period, the yield of high-value products decreases to give more medium-value

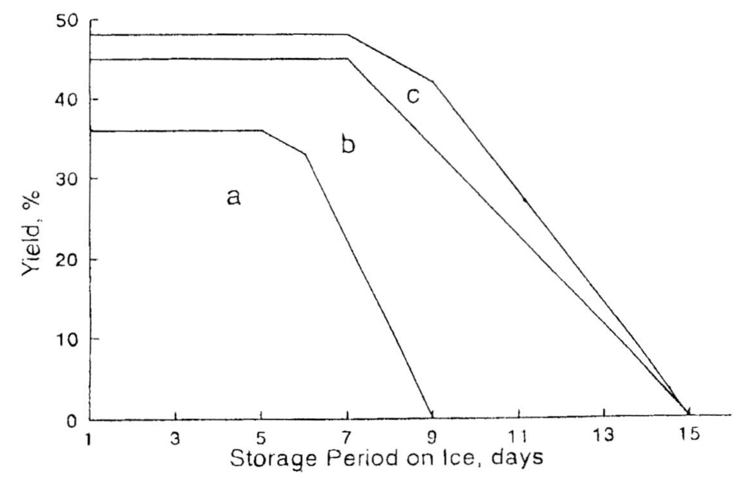

Figure 2. Yield of products of high (a), medium (b) and (c) poor quality as a function of ice storage of fish. (From Neilson, with permission.)

products. Beyond 9 days of ice storage, no high-value products could be present. In fish stored up to terminal shelf life in ice, only low-value products could be produced. Consequently, a change in utility will cause a change in the profit margin per kg of raw material (Neilson 1993).

In industrial plants, fish kept in ice flakes are generally stored in a chill room at 5°C, where fish is retained for about 12 hr. Reduction of the chill room temperature from 5° to 3°C will mean a reduction of about 40% in the amount of ice required, resulting in a considerable increase in fish storage capacity (Zugarramurdi *et al.* 1995).

The effect of handling has a significant influence on the yield of fish fillets and their quality. Himelbloom *et al.* (1994) examined fillet yield, defect levels, thaw drip, cooked texture, and the bacteriological content of rock sole (*Lepidopsetta bilineata*) handled and processed under a variety of conditions. Boxing, hand unloading, and hand filleting gave the highest quality fillets, while bulk stowed, wet pump-unloaded fish gave the lowest quality. The lowest bacterial counts were found in machine-filleted, wet-trimmed fish which also had a firmer cooked texture. Pre-rigor filleting of fish gave the highest yields with the least drip loss but underwent gaping and had the shortest shelf life. The lowest yield was obtained when fish were filleted after 3 days of ice storage (Curran *et al.* 1986). Shelf stability of fillets could be improved by a dip in carbonic acid and packaging. For example, cod fillets dipped in carbonic acid had a shelf life of 7–21 days. Low oxygen permeable packaging films slowed bacterial growth and extended the shelf life, but retained off odours and established low oxygen conditions in the packages, which may encourage growth of bacteria (Daniels *et al.* 1986).

Machines are available for faster processing of fish which include operations like gutting, washing, heading, dressing, splitting, filleting, skinning and deboning. The advantages of mechanised fish processing are a saving in labour costs, rapid handling of the catch, particularly in times of glut, and better quality of products. The machines, however, should suit the wide variety of fish species, the differences in size, shape, freshness, and texture and also the desired capacity of production.

Super chilling, i.e., chilling at temperatures below the freezing point, at –3° to –4°C, extends the shelf life of fish from 2 weeks at 0°C to about 4 to 5 weeks, by effectively retarding bacterial spoilage. The treatment is done by icing the catch in a fish room refrigerated to –2° to –4°C, by using ice of a lower melting point, by chilling in a seawater/ice slurry, or in refrigerated sea water. The sensory quality of the fish, however, decreases significantly due to toughening, excessive drip, and lipid hydrolysis (Sikorski and Pan 1994). The temperature of salt water ice is about –7°C. It cools the fish quickly and intensively, but the low temperature could at times burn the fish. Due to its salt content, seawater ice is softer than freshwater ice and tends to stick together during storage (Rohr 1995). Ice from seawater melts at about 2°C lower than that of fresh water ice, resulting in a colder product. Some species such as tuna, however, may not tolerate direct contact with salt water ice, since it results in skin damage of the fish. The marine environment being corrosive, on board ice may presents problems in the long run and therefore may require corrosion-free equipment (Bartholmey 1994).

Combination Methods Based on 'Hurdle Technology'

As discussed earlier, chilling of fish gives only a limited shelf life. The inadequacies of ice preservation on the storage and distribution of fresh fish in the expanding market place and the need for development of better methods were recognised a long time ago (Hoston and Slavin 1969). Use of certain processes in combination with chilling can significantly enhance the freshness of several fish varieties. The resulting 'combination treatments' may involve synergistic or cumulative action of the combination partners, leading to a lower level of treatment requirement for one or both the agents. This can result in an improvement in the sensory and bacteriological quality of the treated food. Preservative effects of a combination of treatment in controlling microbial growth and resulting spoilage is based on 'hurdle concept' and involves the creation of a series of hurdles in the foods for microbial growth. Such hurdles include heat, irradiation, low temperature, water activity, pH, redox potential, chemical preservatives etc. (Leistner and Gorris Leon 1995).

Chilling with Low Dose Irradiation

Food irradiation is one of the emerging technologies for the preservation and hygienic processing of perishable foods. It helps to extend the shelf life of fresh fish, by killing spoilage-causing microorganisms and eliminates pathogens and parasites of public health importance (Venugopal *et al.* 1999, Nickerson *et al.* 1983, WHO 1981). Irradiation treatment involves controlled exposure of the food to radiation sources such as isotopes of cobalt (^{60}Co) or caesium (^{137}Cs) which emit gamma rays, and also X-rays and electron beams (Lagunas-Solar 1995). The two principal microbiological advantages of radiation treatment of fish are extension of shelf life and improvement of hygiene and safety (WHO 1981). Processes that can be applied to fishery products include radurization (pasteurisation of chilled fish), radicidation (sanitization of fresh and frozen products including fish mince) by elimination of non-spore forming pathogenic bacteria (see section on freezing), and disinfestation (destruction of insect eggs and larvae in dried fish products) (Venugopal *et al.* 1999, Giddings 1984).

The treatment damages microbial DNA; gram-negative microorganisms are more sensitive to radiation than gram-positive bacteria, which are also responsible for the spoilage of muscle foods (Venugopal 1999). The D_{10} values (dose in kilo Gray, 'kGy', required for 90% of inactivation of initial population; one Gray is equivalent to 100 rads, one rad is equivalent to absorption of 100 ergs of energy per g) for several microorganisms when irradiated in fish media vary between 0.1 to 0.8 kGy depending upon the organism, medium in which it was suspended during irradiation and the temperature during the treatment (Venugopal *et al.* 1999). The treatment, however, is limited in its ability to eliminate viruses and spores of *Clostridium botulinum* because of their high radiation resistance. For example, the D_{10} values of *C. botulinum* type E spores, which can jeopardise the safety of seafoods under abuse of chilled storage, is as high as 1.37 kGy (Patterson 1990). It has been suggested that the storage temperature of irradiated fresh fish should not exceed 3.5°C in order to prevent germination of the spores (Hobbs 1977).

Radurization of fresh fishery products under ice at 1 to 3 kGy reduces initial microbial loads by 1 to 3 log cycle and extends their chilled storage life 2–3 fold. The treatment is effective for the extension of shelf life of most marine and freshwater fish species (Venugopal *et al.* 1999). Lean fish species show the least irradiation-induced rancidity, while fatty fish are treated only if adverse sensory changes as a result of irradiation are insignificant. Various aspects of radiation preservation of Indian mackerel (*Rastrelliger kanagurta*) have been studied which established that the shelf life of the fresh fish could be extended from 10 to 25 days when irradiated at a dose of 1.5 kGy. Combination of irradiation along with modified atmosphere storage has been found to enhance the shelf life of fresh fish (Licciardello *et al.* 1984). A code of practice for irradiation of seafoods has been prepared by Rodrick and Dixon (1998).

The wholesomeness of irradiated foods, including fish has been examined in detail in terms of four criteria, namely the presence of induced radioactivity, pathogens and their toxins, carcinogenic radiolytic products and loss of nutritive value. These studies have shown that irradiation of food at an overall average dose of 10 kGy produces no adverse effect and the treated foods are toxicologically safe for consumption (Rodrick and Dixon 1998, WHO 1981). Forty countries have given clearance to irradiation of more than 100 food items including fishery products and many are commercialising the technology (Venugopal *et al.* 1999, Lagunas-Solar 1995).

Chilling with Modified/Controlled Atmosphere Storage

The terms 'modified atmosphere (MA)' and 'controlled atmosphere (CA)' mean that the atmospheric composition surrounding a perishable product including fish, in a package has been maintained differently from that of normal air. In MA, only an initial air change is made, consistent with the expected requirement of the commodity and transportation duration. In CA, the selected atmospheric concentration of gases such as carbon dioxide, oxygen, nitrogen, etc. is maintained throughout the storage period (Flick *et al.* 1991). The extension in shelf life is due to the inhibitory effect of gases such as CO_2 on spoilage-causing microorganisms such as *Pseudomonas, Acinetobacter, Alteromonas putrefaciens, Flavobacterium, Staphylococcus,* etc. This is essentially due to the fall in pH as a result of the carbonic acid formed during dissolution of the gas in the intracellular fluid. Since lactic acid producing bacteria such as *Streptococcus* and *Lactobacillus* are less affected by CO_2, these organisms are predominant in seafoods during storage. The use of CO_2 may cause noticeable odours and affect the texture of packaged fishery products. The odours, however, decrease after opening the package. MA has been reported to extend the shelf life of both finfish and shellfish species (Ashie *et al.* 1996).

The shelf life of irradiated cod fillets, vacuum packed or packed in 60% CO_2 — 40% air in barrier bags and stored on ice, showed that the fish samples retained their quality attributes much longer than the vacuum packed product, which also had a longer shelf life than the air packed fillets (Licciardello *et al.* 1984). In another study, it was noted that cod fillets stored at 0°C in a 25% CO_2 — 75% N_2 atmosphere had a storage life of at least 8 days more than cod stored in air (Villemure *et al.* 1986). Flick *et al.* (1983) studied the advantages of MA packaging on the shelf life of chilled stored fishery products. They found a significant increase in stability, mainly due to an extension in the lag phase of psychrotrophic organisms and to their reduced growth rate in the logarithmic phase. Successful studies were also reported on brown shrimp, rockfish fillets, rock cod and crab, where MAP storage at least doubled the shelf life as compared to that obtained when these samples were stored in air (Ashie *et al.* 1996). While MA or CA storage can control *Salmonella* spp., there are concerns about the safety of products due to hazards resulting from growth of *Clostridium botulinum* type B, E, and F in the food in case of inadvertent abuse of storage temperature (Stammen *et al.* 1990). The risk of food-borne botulism may be reduced by the application of antimicrobial agents. For example, the beneficial effect of nitrite on chilled storage life of certain species stored under MA has been reported (Reddy *et al.* 1992). A combination of radiation treatment with modified atmosphere (60% CO_2) has been found beneficial to enhance the shelf life of cod fillets stored at temperature below 3.3°C (Licciardello *et al.* 1984) .

Novel methods of atmosphere modification in the packages are through the use of oxygen absorbents—CO_2 generators are ethanol vapour generators, by keeping sachets containing appropriate compounds such as 'ageless' (made by Mitsubishi Gas Chemical Co., Japan) and 'Ethicap' or 'Antimould 102' (manufactured by Freund Co., Japan) within the pouches (Ashie *et al.* 1996). . Vacuum packaging is also used to extend the shelf life and keeping quality of chilled fish. While oxygen is reduced to less than 1%, carbon dioxide produced from tissue and microbial respiration eventually increases to 10 to 20% within the package headspace, which inhibit the growth of aerobic spoilage microorganisms. Vacuum packaging helps control of fat oxidation along with shrinkage and colour deterioration. Vacuum packaging has been used to rainbow trout, cod fillets, red snapper, herrings, sardines and mackerel (Ashie *et al.* 1996). The safety issues associated with vacuum packaged products are similar to that of modified and controlled atmosphere packaged products.

Chilling with Chemicals

Some chemical additives have been recognised to extend the shelf life of fresh fish and fish fillets. A dip in 1 to 3% aqueous solution of potassium sorbate can extend the shelf life of fish fillets by several days (Sikorski and Pan 1994). The chemical is a GRAS (Generally Recognised as Safe) food preservative, which has proven effective for inhibiting spoilage causing bacteria, yeast, and moulds in chilled fish (Pedrosa-Menabrito and Regenstein 1990, Ashie *et al.* 1996). The compound can also be used as an alternative to the use of nitrite in cured meat products (Sofos *et al.* 1979). Extensive studies have shown that combinations of nitrite (40–80 ppm) and sorbate (0.2%) can be as effective as 120 ppm nitrite for the control of *C. botulinum* (Sofos *et al.* 1979). Sorbates may also extend shelf life when used in combination with polyphosphates and/or CO_2 (Statham *et al.* 1985). Several polyphosphates including sodium tripolyphosphate, disodium orthophosphate, tetrasodium pyrophosphae and sodium hexametaphosphate are known to enhance the keeping quality of fresh as well as frozen fishery products (Pedrosa-menabrito and Regenstein 1990). These compounds preserve fishery products by acting as metal ion chelators, altering pH, interacting with proteins to promote hydration and water holding capacity, preventing lipid oxidation, and inhibiting microbial enzyme secretion and growth (Ellinger 1972, Ashie *et al.* 1996, Venugopal *et al.* 1984).

The preservative action of organic acids such as lactic and acetic acids and their salts has been well documented. Sodium lactate is used as humectant and flavour enhancer in meat and poultry products, which has also antimicrobial properties. Potassium and calcium lactates are equally effective in controlling the growth of aerobes and anaerobes in meats. Antibotulinal and antilisterial activities of the lactate anion have also been established (Shelef 1994). Catfish fillets dipped in chilled dilute solutions of lactic acid followed by sterile plastic bag packaging could enhance their refrigerated shelf life, as assessed by microbiological and sensory methods (Ingram 1989). Treatment of Chilean mackerel with dilute acetic acid extended shelf life associated with a decrease in the formation of total basic volatiles (Borquez *et al.* 1994).

Several antioxidants, both natural and synthetic, may be used to control rancidity in food products including seafoods. The natural antioxidants include extracts of plants such as cocoa shells, tea, olive, garlic, red onion skins, apple cuticle, nutmeg, mustard seed, peanut seed coat, rice hull, and extracts of spices. The synthetic antioxidants include butyl hydroxyanisole (BHA), butyl hydroxytoluene (BHT) and *tert*-butylhydro-quinone (TBHQ). Addition levels of these antioxidants are generally limited to 100 – 200 ppm (Loliger 1991). These compounds also inhibit several microorganisms (Kabara 1981). Fig extracts which contain 4-substituted resorcinols have been shown to have great potential in the food industry as novel anti-browning agents. These compounds are water soluble, stable, nontoxic, and enjoy GRAS status (Ashie *et al.* 1996). The enzyme, glucose oxidase, can be used as a preservative, through its action giving hydrogen peroxide and gluconic acid from glucose, which have antibacterial properties against microorganisms such as *Pseudomonas* (Haardy 1986, Venugopal *et al.* 2000). Ascorbic acid and its isomer erythorbic acid have also been used in the food industry as antioxidants (Ashie *et al.* 1996). The antimicrobial potential of chitosan, a deacetylated product of chitin (abundantly available by-product from the shellfish industry) has also been reported (Chang *et al.* 1989). Benzoic acid is a permitted preservative in herring and mackerel, in many countries. In spite of availability of various preservatives, it should be noted that chemical treatment of foods and fish may not be popular (Hardy 1986).

The beneficial effects of chemical additives in combination with chilling and irradiation in extending the shelf life of fishery products have been pointed out earlier. The FAO/IAEA Advisory Group Meeting on Radiation Treatment of Fish and Fishery Products held in 1978 approved certain fungistatic agents such as sorbic acid combined with radiation at insect disinfestation doses to render the treated products free from both mould and insects.

Chilling Chemicals

Heat pasteurisation in combination with chilling is a popular method of shelf life extension up to several months of shrimp, crab and other crustacean products (Flick *et al.* 1991). Such products have a limited shelf life due to their sensitivity to microbial growth including pathogens. Combination of blanching (mild heating), chilling and irradiation has been found successful for shelf life extension of a number of fishery products including shrimp (Venugopal *et al.* 1999). Steaming of the fish fillets prior to processing reduces the urea content, leading to better odour and flavour retention. A combination treatment involving steaming for 3 to 5 min followed by irradiation at 1 kGy was found to extend the refrigerated shelf life of shark up to 30 days (Ghadi *et al.* 1978). The use of preservative such as sorbic acid (2,4-hexadienoic acid) in combination with irradiation has been reported to control off-flavour development during irradiation, besides providing safety against outgrowth of spores of *C. botulinum*. A combination of sorbic acid treatment and irradiation at 1 kGy had a synergistic effect on the keeping quality of iced cod fillets (Licciardello *et al.* 1984). A method of on-board processing of shrimp in the Netherlands involves boiling for 6 to 10 min in either plain seawater or seawater enriched with cooking salt. The blanching treatment inactivates the polyphenol oxidase enzyme involved in the black discolouration of the shellfish. The cooked shrimp containing 1 to 3% salt are cooled in seawater, which is prone to re-infection with marine microflora. In order to control this, 0.4% benzoic acid is incorporated in the material. The product has a shelf life of 8 days at 4°C. A dose of 1 kGy (100 Krad) could further extend the shelf life of the shrimp up to 16 days (Houwing *et al.* 1978). Blanching for 2 to 5 min followed by irradiation at 2 kGy extended the shelf life of Norway lobster tails and deep sea shrimp up to 7 weeks in ice, whereas the unblanched and unirradiated samples were acceptable only for a few days (Hannesson and Dagbjartsson 1971).

Sodium metabisulphiate is used in the shrimp industry in the form of dips or addition to seawater, to retard blackening of the shells and muscles during refrigerated storage. Different procedures for applying the bisulphite are used and different levels of residual sulphite are allowed in shrimp meat in different countries. Generally the sulphite content should not exceed 100 ppm, although this limit has been brought down by many countries. The treatment prevent spot blackening in shrimp and lobster for approximately 8 days when stored in ice (Finne *et al.* 1986).

Chilling with High Hydrostatic Pressure

Exposure of foods to high hydrostatic pressures (HHP) in the range of 300 to 600 MPa (0.1 MPa = 1 Bar = 1 kg/cm^2, or 1 Kbar = 100 MPa) causes inactivation of microorganisms, denaturation of proteins, association and gel formation of bypolymers resulting in modification of foods such as appearance, flavour and texture. The treatment can help hygienisation and shelf life extension of foods (Venugopal *et al.* 2000). The pressure applied to the food is transmitted in a uniform manner. Further, the pressurisation process is independent of sample volume, product size and geometry, in contrast to thermal processing. Processing by high hydrostatic pressure is usually carried out in a low compressibility liquid such as water. The effect of hydrostatic pressure in a food product is dependent upon the amount of pressure applied, duration of compression, depressurisation rate, temperature of treatment, product pH, water activity and salt concentration. The treatment promotes phenomena such as phase transition and chemical changes which are accompanied by a decrease in volume. As compared with thermal processing, HHP treatment consumes less energy (e.g. energy required for pressurisation at 400 MPa is equivalent to heating the same material to 30°C). The effect of HHP on proteins and carbohydrates of the food result in their conformational changes associated with modification of functional properties. A large volume of research has been performed on the effect of HHP on microorganisms. Most of the spoilage causing bacteria are sensitive to high hydrostatic pressure, the inactivating effect operating due to change in permeability of the bacterial cell membrane and associated leakage of intracellular material.

The sensitivity of microorganisms to HHP can be advantageously used for preservation of perishable foods including fishery products. For example, subjecting squid mantle flesh and tuna meat to 450 MPa at 25°C for 15 min, the initial counts could be reduced from 5200 per g to 300 per g. Such non-sterile products should be stored under refrigeration to control proliferation of surviving microorganisms and also to ensure safety from any contaminated spores of *Clostridium botulinum*. Combination of other processing techniques such as modified atmosphere storage, salting, heat pasteurisation, gamma irradiation etc. along with HHP can enhance chilled storage life of several fishery products because of the synergistic effect of the processing techniques on viability of the surviving microorganisms. At the pressures usually employed for food processing, sterilisation cannot be achieved.

Table 2 summarises the various techniques in combination with chilling that can be used for fish conservation.

Freezing of Fishery Products

Freezing, perhaps, is the major processing operation in industrial seafood processing. The topic has been discussed by several authors (Venugopal and Shahidi 1998, Gould 1996, FAO 1994, Pedrosa-Menabrito and Regenstein 1990, Hardy 1986). Once brought to the processing factories, the fish/shellfish is pre-processed by operations which include peeling, deveining, evisceration, beheading, filleting etc. after which the material is subjected to freezing. Quality of raw material is important since any delay between preparation and lowering of temperature will result in an irreversible loss in quality. The above pre-processing operations also need

Table 2. Combination of chilled/frozen storge with other processes for seafood preservation

Temperature conditions	Processes	Remark
Chilled storage (-1°to +7°C)	Active packaging	Change in package environment such as oxygen
	Aseptic packaging	After heat processing transferred to sterile and hermetically sealed containers
	Drying	Lowering of water activity
	Heat pasteurisation	Partial reduction of microorganisms
	High pressure	4000 to 7000 atmospheres
	Microwave heating	500 to 1000 MHz electrical fields heats the product
	Modified atmosphere	Elevated CO_2 and reduced O_2 levels
	Pickling	Salt and oil prevents microorganisms
	Radiation	A dose of 1 to 3 kGy enhances chilled storage life
	Sous vide	Prepared food is vacuum sealed, cooked and stored below 4°C.
Frozen storage (-10° to -30°C)	Radiation	Inactivation of pathogens at a dose of 4 to 6 kGy
	Blanching	Mild heat treatment
	Breading and battering	Seafoods and other muscle foods
	Surimi and imitation products	Restructured products prepared from washed fish meat
	Packaging	Glazing and packaging prevents surface desiccation

to be carried out under most hygienic conditions to prevent any contamination by pathogenic microorganisms. Depending upon the rate of cooling, freezing operation may be quick or slow. For *quick freezing*, the temperature of the product should decrease to -20°C within a 2 hr period, while in *slow freezing* the required time could be 3–72 hr. The best method of preserving fish is quick freezing, because the texture and flavour of the product are less affected. Quick freezing after rigor mortis is ideal for fish. Raw materials should be frozen in blocks of limited thickness which can be thawed in industrial plants within a few hours. Freezing time depends on freezer type, freezer operating temperature, air speed in a blast freezer and temperature, thickness, shape, packaging of the product. Figure 3 indicates comparison of the influence of temperature on water frozen out of some food items including fish. The quality of frozen fish depends, apart from its primary quality, on the design of the freezing process (which determines freezing time), storage temperature and duration, protection of the frozen material during storage, and thawing techniques. The ideal temperature for frozen storage of fishery products is around -30°C. Nevertheless, well-wrapped fishery products kept at -20°C undergo little deterioration when stored for up to one year.

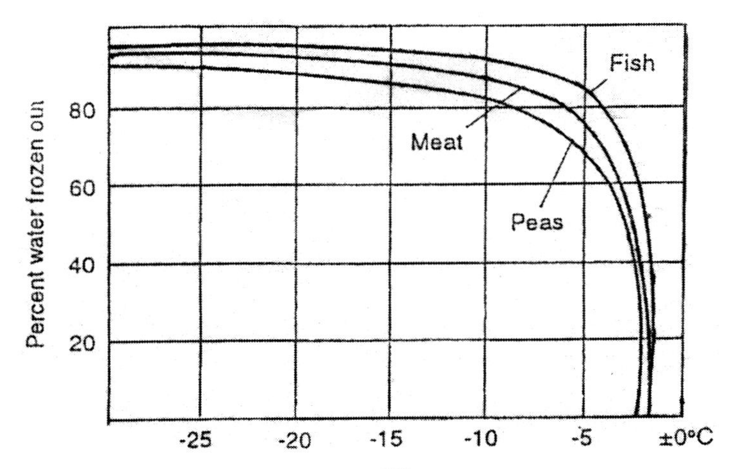

Figure 3. Frozen-out water from food items at different temperatures.

Fish freezers fall into three main categories: direct contact or cryogenic freezers and indirect mechanical freezers. In the former type, the fish is in direct contact with the refrigerant such as nitrogen, air, carbon dioxide or freon substitutes. Cryogenic freezers employ super-cold gases like liquid nitrogen or carbon dioxide as the freezing medium. These freezers may be of different designs, such as straight belt, multitier, spiral belt as well as immersion designs. In principle, the same basic equipment can be used for both gases, but with slight modifications. Liquid nitrogen at -196°C is sprayed into the freezer and circulated by small fans. The freezant partially evaporates immediately on leaving the spray nozzles and on contact with the products. The freezant consumption is in the range of 1.2 to 2.0 kg per kg of the product. The capacity can vary from 150 to 1000 kg/hr. In indirect, mechanical freezers, the fish is frozen by a secondary medium such as air or ammonia which is recirculated and re-cooled by a refrigeration coil. The equipment may employ one or more of the following systems: plate freezers or spiral freezers. Plate freezers are probably the most common contact freezers used in the fish industry. They may be vertical, horizontal, or rotary and may be manual or automatic. Freezing at sea is carried out in vertical plate freezers where round gutted fish are being frozen; horizontal plate freezers are used to freeze fillet packs. Their main advantage is in the production of regular and uniform thickness of the frozen blocks or cartons.

The requirements of a fish freezing plant include: whole fish and fish block weighing machines, washing, sorting and filleting machines, inspection and trimming tables, fillet packaging table, fish block weighing table, strapping machine, conveyor belts, tray and box washers, freezing trays and plastic boxes. The refrigeration equipment includes a blast freezer (5 tonnes per day), plate freezer (500 tonnes per load), chill room (0°C) for 20 tonnes of material, cold storage (-30°C) for 60 tonnes of finished product and ice making machine (300 tonnes per day). The design of cold storage for subsequent holding of frozen fish is important, and must be capable of maintaining a temperature of at least -20°C without serious temperature fluctuations.

Freezing is the most important process employed worldwide for preservation of highly perishable shrimp. The various pre-processing steps prior to freezing of the shellfish include cleaning, beheading, peeling, cooking, icing, dip treatment in preservative solutions such as bisulphite and dressing. Dressing include preparation of whole-head on, headless-shell-on (HL), peeled and undeveined (PUD), peeled and deveined (PD), fan-tail/butterfly, cooked, peeled and deveined, peeled deveined and cooked (PDC). Shrimp are graded into various sizes designated by the number of individuals required to weigh 450 g. These size grades include 0-10, 10-15, 16-20, 21-30, 31-50, 51-70, 71-90, 91-110 and 110 up pieces per 450 g. (Govindan 1985). The methods for pre-processing and freezing of shrimps have been summarised by Chandrasekaran (1994).

There are five main processes involved in deterioration during frozen storage. These are: (i) oxidation of the lipids present in the fish, resulting in rancidity, (ii) bacterial growth in the tissues, breaking down tissue structure and formation of trimethylamine and other compounds which affect flavour, (iii) enzymatic processes within fish muscle resulting in formation of hypoxanthine and bitter flavour, (iv) changes in protein in fish muscle affecting the texture causing it to become fibrous and chewy, due to the reaction between protein and oxidised lipids and (v) dehydration and weight loss (Hultin 1994, Mackie 1993).

Frozen storage life of fish and shellfish extends from a few months to one year depending upon the type of fish, its initial quality, oil content and frozen temperature, as mentioned earlier. Peeled and deveined shrimp in 2 kg blocks can be stored up to one year at -18° to -20°C without significant quality changes. As compared with non-fatty fish, oily fish such as sardines have limited shelf life, the limiting factor of shelf life being rancidity development and associated flavour changes. At -18°C, the storage life ranges from 20-40 weeks for fish having an oil content varying from 42 to 10%. Storing the fish in ice before freezing reduces the subsequent storage life. Water-glazed oil-sardines can be stored for up to 20 weeks at -12°C or -23°C, but this period falls progressively if the fish have been stored in ice for a few days before freezing. Generally, keeping fish in ice prior to freezing reduces its frozen storage life, determined by the duration of pre-freezing ice storage. Glazed blocks of Indian mackerel have a shelf life of 16 weeks at -20°C, before rancid flavour and a hard texture develop, whereas, low fat fish (fat, < 2%) become texturally unacceptable after six months at that temperature. Seer fish, whether frozen quickly or slowly, has a shelf life of 10 weeks at -23°C. If the fish is stored six days in ice prior to freezing, frozen storage life is only 6 weeks. Swordfish and tuna discolour during frozen storage. Swordfish fillets can be stored for 3 months at -8°C, nine months at -18°C and a few years at -26°C. Skipjack tuna stored at -4°, -17° and -25°C remains acceptable for a period of 1-2, 6 and 12 months, respectively. Discolouration of tuna is a problem; the fish stored at -5° and -10°C rapidly turns brown; at -20°C, the change is slower, taking up to two months and at -35°C the colour remains acceptable for at least 9–13 months.

Shrimp are one of the most important commercial seafoods. A large amount of information is available on the requirements and quality aspects of freezing shrimp. Shrimp are frozen into a variety of frozen products such as whole, headless, peeled undeveined (PUD), peeled and deveined (PD) etc., as discussed earlier. Individually quick frozen (IQF) shrimp is a popular item in international seafood trade. Figure 4 shows the various steps involved in shrimp processing and preservation.

Dehydration and weight losses are major problems of frozen fish and shrimp kept in commercial cold storage. In a good cold storage, the weight loss will be about 1 $g/m^2/day$. Reasons for higher losses in weight include higher temperature of the material during loading, frequent opening of the cold storage door, excessive defrosting and high air velocities (above 0.1 metre per second). Prevention of weight loss can be achieved by ensuring a lower product temperature during freezing and suitable packaging of the product prior to their transfer to cold storage. Frozen fish is generally wrapped in polyethylene or cellophane and packaged inside paraffin-coated paper board cartons. All packaging should be tightly fitted to the product to avoid free air space which may cause 'in-pack desiccation'. To control the loss in quality during frozen storage, the products are normally glazed with 5-10% of water immediately after freezing. Glazing is done by spraying cold water over the product or sometimes by dipping. For best quality

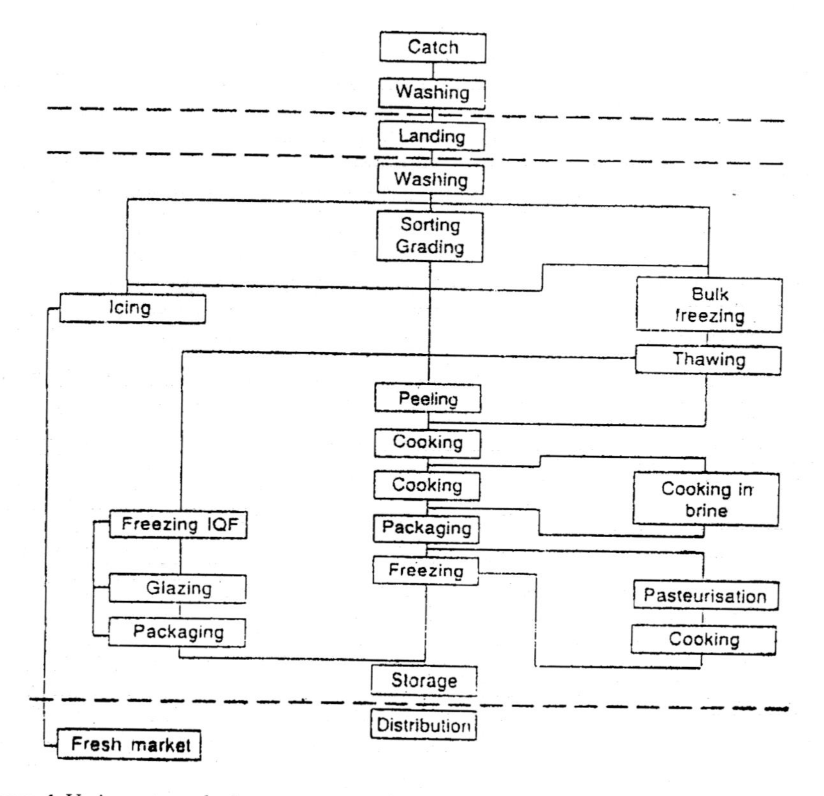

Figure 4. Various steps during processing of shrimp.

product, the glazed products should be refrozen to -20°C before packing and cold storage, ideally at -25°C. This helps in minimising weight loss during thawing.

Hygienisation of Frozen Fishery Products by Gamma Irradiation

As mentioned earlier, several public health issues are associated with contamination of seafood products such as shrimp, crab meat, oysters, mussels etc. by pathogenic bacteria. The incidence of pathogens including *Salmonella, Vibrio, Listeria* etc. in frozen fishery items has been a great concern in international trade of the commodities. In addition, a variety of parasites including flatworms, roundworms, and protozoa infest the gills, viscera and skin of marine, freshwater and farm-raised fish species. Fishery items dried under open sun are infested with eggs and larvae of insects such as flesh flies, beetles and mites. Many of the above mentioned problems can also be present with respect to aquacultured fish (WHO 1999).

Understanding of the profound influence of factors such as environment, process and distribution conditions on the fishery products has, therefore, necessitated a need for quality assurance and adoption of standards in the fish industry in order to safeguard the interests of the consumer. Standards for imported fishery products have been stipulated by regulatory authorities such as US FDA, EU etc.

Radicidation is sanitation of fresh and frozen products including fish minces by elimination of non-spore forming pathogenic bacteria such as *Salmonella, Vibrio* and other species at a dose of 4 to 6 kGy, as already discussed. It may be noted that in spite of good manufacturing practices and other cares taken, frozen fishery products may still harbour pathogens. Such processed items include not only frozen finfish items, but also shrimp products, individually quick frozen items and aquacultured fish and shellfish. Radicidation, however, is limited in its ability to eliminate viruses and *Clostridium botulinum* type E spores, which can jeopardise the safety of seafoods through production of lethal botulinum toxin. However, the botulism hazard can be avoided by storing the irradiated fresh fish below 3.5°C to prevent germination and outgrowth of the spores. Table 3 gives the different applications of radiation treatment for fishery products.

Value Addition of Fishery Products

Traditional Methods

The above discussion essentially related to techniques for shelf life extension of raw material. It is important that for better marketability of the fish catch, the raw material needs to be converted to products that can command consumer acceptability. A number of techniques are available for value addition of marine and freshwater resources. These include both traditional as well as novel techniques developed in the recent past. In general, these processes may be grouped into three categories, namely, flavourization techniques, texturization techniques and isolation of edible components such as protein concentrate, protein hydrolyzate, shark fin, gelatin, flavour extracts, among others. Flavourization involves applications of techniques that add flavour to the finished product such as fermentation, marination, pickling, and also by canning in special juices, among others. Texturization processes are intended to modify the textural profiles of fish muscle to products to suit the palatability of consumers. These techniques include development of fish/shellfish analogues through surimi (washed fish mince), extrusion cooked products and sausages (Venugopal 1992, Venugopal and Shahidi 1998, 1995). The traditional techniques include different methods of curing, fermentation, marination, as will be discussed below.

Table 3. Important radiation processes for conservation and hygienisation of seafoods under different storage conditions

Storage Conditions	Radiation process	Benefits
Chilled storage (-1°to +3°C)	**Radurization** (Radiation pasteurization) Dose: 1–3 kGy Reduction of initial microbial content by 1 to 2 log cycles. Specific reduction of spoilage causing organisms. Additional benefit includes elimination of non-spore forming pathogens Post-irradiation storage: under ice. While most of the laboratory studies were done using packaged fish, large quantity of fish can be stored under melting ice in insulated boxes at 10–13°C	Extends chilled shelf life of fishery products 2 to three times. Species examined: Different marine species such as Indian mackerel, white pomfret, Bombay duck, Indian salmon etc. Potential scope exists for shelf life extension of freshwater and aquacultured fishery items in ice.
Frozen storage (-10° to -20°C)	**Radicidation** Elimination of non-spore forming pathogens such as *Salmonella, Vibrio, Listeria* etc. Dose required: 4–6 kGy Packaged and frozen, ready-to-export fish can be treated before shipment	Ideal for hygiene improvement of export materials such as frozen shrimp, cuttle fish, squid, finfish, fillets etc. IQF products can also be treated.
Ambient storage (20° to 30°C)	**Radiation disinfestation** Elimination of eggs and larvae of insects (<1kGy) Inactivation of *Salmonella* sp. and other pathogens (4–6 kGY)	Dried fishery products Fish meal, feed for aquaculture

See Venugopal *et al.* 1999.

Curing

Underutilised fish provide a major portion of raw material for curing, most of which are landed as by-catch during fishing of targeted species. The inherent problems of these by-catch are their extreme heterogeneity of composition, bony structure, dark flesh, small size, unattractive appearance and texture, strong flavour and the possible presence of toxic species. These fish include shark, ray, sardine, sprat, Indian anchovy, anchovy, croaker, Bombay duck, catfish, jewfish, rock cod, leather jacket, jelly fish, herring, garfish, flying fish, barracuda, grey mullet, perch, horse mackerel, pony fish, jewfish, spotted bat, tilapia, mackerel, skipjack and Spanish mackerel, among others (Venugopal and Shahidi 1998, Horner 1992a, Infofish 1983).

Curing is a combination of processes such as salting, drying, pickling, smoking and marinading. In many developing countries, salted and sun-dried fish constitute a major portion of the staple diet (Infofish 1983). The basic principle involved in curing is rendering the medium unsuitable for the growth of microorganisms by the control of water activity (a_w) of the product. While, prior to drying, spoilage could be caused by psychrotrophic bacteria, during drying, the spoilage is mostly caused by the action of gram positive micrococci and bacilli (Liston, 1980). The infusion of soluble substances such as salt leads to a reduction of a_w and controls bacterial activity and also the autolytic spoilage. The a_w required for the control of spoilage causing bacteria is about 0.90, yeasts, 0.88, moulds, 0.88 and halophilic bacteria, 0.75 (Alur and Venugopal 2000). Reduction of a_w by combination of salting, dehydration and the addition of chemical preservatives helps retention of food stability and safety even without refrigeration, based on the 'hurdle concept', as described earlier. As a guide to storage stability of dried foods, the concept of 'alarm water' has also been suggested. The alarm water content is the water content that should not be exceeded if mould growth is to be avoided. Generally at 25% moisture level, bacterial activity of muscle foods is inhibited and at 15% moisture, mould growth is also arrested (Alur and Venugopal 2000). *Fusarium* spp. was present in samples of sun- or oven-dried rohu (*Labeo* spp), having a moisture content of 46%, while below this level, *Aspergillus* spp. and *Penicillium* spp. were present (Venugopal *et al.* 1999).

Salting

Salting of fish is generally practised prior to drying. High salt concentrations inhibit the activity of both microorganisms and enzymes. Penetration of salt in fish flesh is quite rapid and helps to lower water activity. Heavily salted fish, theoretically, have a water activity of 0.75, which is comparable to that of a saturated common salt solution. For dry curing, fish are stacked in salt and the brine formed is drained off. In wet curing, fish are immersed in a strong brine. Depending upon the fish type, it may be salted whole (e.g. anchovies and small herring), eviscerated and split open (e.g. mackerel) or in small pieces or as fish mince. Large fish are split dorsoventrally, gills and intestines removed, washed clean, and salt is scrubbed into it at a ratio of 1:10 to 1:3 (salt to fish) depending upon the size of fish followed by staking in a curing tank made of brick and cement for 24 hr. The salted fish are given a light rinse in fresh potable water to remove the adhering salt crystals and dried on mats in the sun for 2 – 3 days. In the case of 'mona curing' the fish (mackerel, lactarius etc.) are not split open, but the viscera and gills are pulled on through the mouth and cleaned before salting. In 'Colombo curing', pieces of tamarind are placed in the body cavities of salted fish, which are then placed in wooden barrels and filled with saturated brine (Govindan 1985). Brining procedures for fish (Kosak and Toledo 1981), theoretical aspects of salting (Horner 1992a), and sorption isotherms for predicting the course of drying and the selection of optimum storage conditions for dried and salted/dried seafoods (Curran and Poulter 1983) have been discussed. Sorption characteristics are generally independent of fish species, salt concentration or drying temperature. A comprehensive review of salting-curing techniques of fish has been provided by Mendelsohn (1974).

Small whole fish may be salted at sea. The fish are mixed with coarse salt and loaded into a container slowly adding concentrated brine from below without displacing the salt from the fish. By osmotic dehydration, the fish becomes firm and resistant to handling. Such fish are ideal for curing (Hansen 1996).

Drying

Drying of fish can be achieved by air (sun, solar) or contact (where heat is transferred to the fish from heated air or a heated surface), vacuum (by enhancement of evaporation rate by reduced pressure) or freeze drying (employing very low pressures to sublime ice from frozen fish). Very small-size lean fish species are generally dried without salting after preliminary washing in clean water. During glut, large quantities of such types of fish are spread on open sandy beaches for drying. The disadvantage of this method is that the fish may get contaminated with sand, microorganisms, and insects. In some cases, the fish are suspended from bamboo or wooden scaffolds.

Detailed aspects of the salting and drying behaviour of several fish species have been reported. Berhimpon *et al.* (1990) studied wet salting of low fat yellow tail using a 15–21% brine or a saturated salt solution and drying at three temperature and humidity conditions. The brine concentration influenced the drying rate. An improved process was developed to produce salted dried jellyfish (Huang 1988). The process consisted of a three-phase salting. In the first phase, freshly dressed jellyfish was brined in 7.5% salt and 2.5% alum for 2 days. In the second phase, 15% salt and 1% alum were used to soak the jellyfish for 2 days. In the third phase, salted jellyfish were drained and mixed with dried salt at 1:2 ratio and kept for 2 days. The treated product was dried at 30°C and 45% relative humidity to reduce the moisture content to 68 to 70%. The product containing 5% protein and 25% salt had a shelf life of 6 months at 10°C. A process to prepare dehydrated laminates of ribbon fish has been reported by Jeevanandam *et al.* (2001). It consisted of gutting and heading the fish, pressing in a screw press to reduce the moisture content, drying at 60°C, passing the semi-dried fish between two stainless steel rollers to flatten the fish, followed by further dehydration and packaging in polyethylene pouches.

The common defect of cured fish products is the presence of 'dun', 'rust' and formation of pink or red patches on the product. 'Dun' is the formation of brown or yellowish brown spots on the flesh and is caused by the growth of pigmented, salt tolerant microorganisms at ambient temperatures. 'Rust' is characterised by the colour of rusted iron and is formed through the reaction of oxidised lipids with fish proteins. The quality of dried fish depends on properties of the raw material, possible delays in the preparation of fish for curing, nature of the salt, procedure employed for curing/drying, sanitary conditions during drying, infestation during drying, barrier properties of the packaging material used and time and conditions of storage. The high fat content in some fish may result in development of rancid flavours, which is also enhanced by salt (Smith and Hole 1991, Hanson and Esser 1985, Maruf *et al.* 1990). For good quality product, raw materials should be as fresh as possible, wholesome and if necessary, gutted, headed, cleaned, and washed prior to curing. When drying naturally, protection against direct sunlight is often necessary in order to avoid partial cooking and the break-up of the flesh or case-hardening. If artificial dryers are employed, drying should be carried out as swiftly as possible to reduce the risk of spoilage. The air humidity needs to be low and the temperature as high as possible, to avoid cooking and early case-hardening. The storage life of the product is also influenced by the storage temperature (Srikar *et al.* 1993). In tropical countries, when the relative humidity of 90% and above, the dried fish can take up sufficient moisture to allow growth of halophilic bacteria which give rise to pink patches on the surface.

Nutritional evaluation of salted and sun-dried fish has been reported. Maruf *et al.* (1990) observed that during processing and storage of Indonesian mackerel, polyunsaturated fatty acids of the fish were oxidized to form thiobarbituric acid reactive substances and fluorescent compounds. TBA values increased with increasing salt content in the product. Further, positive correlations were found between odour of the stored products and both fluorescent pigment content and the degree of browning. Rat feeding trials showed massive decrease in net protein utilization of stored fish.

Another major problem of sun-dried fish stored in tropical countries is infestation by flesh flies (Sarcophagidae), beetles (*Dermestes, Cornestes* and *Necrobia* spp.) and mites (*Lardoglyphus* spp.) (Dollar 1989). Approximately 25% of cured fish are lost in Southeast Asia due to infestation. Several studies have shown the feasibility of low dose gamma irradiation at a dose of 1 kGy for the disinfestation of dried fish (Dollar 1989, Venugopal *et al.* 1999). FAO/WHO has suggested an approved insecticide, pirimiphos-methyl, as an effective and financially viable solution in resolving the infestation problem (Hansen and Esser 1985). Polyethylene film having a thickness of

0.1 mm and above is suitable for protecting dried fish from insect penetration, bacterial and water permeability, and weight loss of the product. In addition, use of solar dryers is increasingly becoming popular for quality products. These dryers help concentrate solar radiation to achieve a higher temperature, lower RH and higher drying rates (Curran and Trim 1983, Doe 1983, Mishkin *et al.* 1982).

Smoking

Smoked fish is a delicacy in several parts of the world. The preservative effect of smoking is due to a combination of four factors: (i) salting, which reduces water activity and inhibits the growth of many spoilage causing organisms including pathogens; (ii) sun-drying which provides a physical barrier to the passage of microorganisms, (iii) creation of a hostile environment for microbial proliferation by deposition of antimicrobial substances such as phenol; and (iv) deposition of phenolic antioxidant substances from smoke, which delays autooxidation and rancidity development.

Smoking is defined as the process of the penetration of food products by volatiles resulting from incomplete burning of wood. The essential steps in fish smoking are splitting, gutting, heading and washing, salting (usually in saturated brine to obtain a final salt concentration of 2–3%), hanging the brined fish on racks or in a kiln and smoking under careful control of temperature and air velocity. Fresh fish are ideal for smoking for good quality products, since prolonged frozen storage prior to the treatment can affect the final quality. Salt denatures the surface proteins, and with drying, an artificial, glossy layer forms on the cut surface. This finish helps to seal in the natural juices and flavours (Dillon *et al.* 1994). The wood used for smoke is usually in the form of sawdust or chips of wood such as maple, oak etc. Hardwood is preferred because it imparts a milder flavour, whereas softwood such as fir and pine impose a more resinous flavour.

The process of smoking may be 'cold' or 'hot' depending upon the temperature of the treatment. For cold smoking, the temperature is usually below 30°C, sometimes, the temperature is raised to 40°C towards the end. A relative humidity of 60 to 70% is most satisfactory for cold smoking for controlled dehydration. Cold smoked fish, containing about 5% salt and smoked for 7 hr can be kept for about 2 months at low temperature (Sikorski and Pan 1994). Such products are not necessarily cooked before consumption. Cold smoking has been employed for herring, haddock, cod, salmon, trout and eel. Hot smoking, on the other hand, is carried at temperature of 50° to 90°C for 4 to 12 hr. The treatment is done in three stages, a tempering stage where the fish is smoked at 30°C for about 1 hr (which allows toughening of the skin), a heating step where the temperature is raised to 50°C for 1 hr and finally a cooking stage at a temperature of 70° to 90°C for 1 to 2 hr. The hot smoked products are then cooled slowly to room temperature and chilled to 4°C prior to packaging (Dillon 1994, Bannerman 1980). A two-stage smoking process consisting of exposure to 70°C and RH of 60% for 35 min followed by at least 33 min smoking at 93°C has also been reported (Chang *et al.* 1975). Vacuum packaging, MA and CA packaging can increase the shelf life of the smoked products. Mackerel, herring, haddock, trout and eel have been studied for hot smoking (Horner 1992a). Smoked eel and trout fillets in vacuum packs have gained popularity in Europe.

Smoked seafoods may contain up to 0.5 g of smoke constituents per 100 g tissue. Some of the major components identified in smoke vapours include acids (formic, acetic, butyric etc.), alcohol (ethanol and methanol), carbonyl compounds (e.g. formaldehyde, prionic aldehyde, octyl aldehyde, acrolein, methyl ethyl ketone etc), hydrocarbons (naphthalene, stilbene, phenanthrene) ammonia, carbon dioxide, carbon monoxide, esters, furans, nitrogen oxides, phenols and sulphur compounds, among others (Maga 1988). However, along with these volatiles, several carcinogens are also produced. Of these, the prominent one is benzopyrene, which, however, may decrease during storage (Simko 1991). Characteristic colour formation in the treated products is due to the interaction of carbonyls with amino components on the flesh surface. Improvements in the smoking technologies that have been made relate to temperature control, electrostatic filtration, and treatment of the product with liquid smoke to control the formation of carcinogens and development of liquid smoke (Kyzling 1990). Mechanical and automated processing equipments are also available for smoking such as automatic briners and controlled mechanical kilns. Many types of dried and smoked fish with traditional seasonings have been used in Japanese foods. A process to make instant soup from smoked, dried and powdered sardine has been described (Oh *et al.* 1988).

Table 4 summarises various problems associated conventional processing of fish and possible solutions.

Innovative Techniques for Value Addition

The innovative methods for value addition of fishery products include individual quick freezing, *sous vide*, cook-chill, high pressure processing, breading and battering, collection of meat from low cost fish and development of products from the fish mince such as surimi and imitation foods and composite fillets and steaks etc.

Table 4. Some quality problems in processed fish

Processing	Cause	Effect	Solution
Chilling	Delay, poor ice quality, insufficient icing, improper mixing	Rapid spoilage, bacterial contamination	Avoid delay, improve handling practices, use good quality ice
Freezing	Poor initial quality	Reduced acceptable shelf life	Use fresh quality
	Filleting during pre-rigor	Shrinkage, loss of weight	Fillet after rigor
	Slow freezing	Poor texture	Adopt quick freezing
	Fluctuation in temperature	Loss of texture Reduced shelf life	Stringent temperature control during freezing and storage
	Dehydration	Freezer burn, inedible Product	Glazing
	Oxidation rancidity	Loss of flavour, texture	Glazing, antioxidants
	Protein denaturation	Loss of texture	Lower temperature, use of polyphosphates
Canning	Poor initial quality, frozen and thawed fish	Poor appearance, texture, and flavour	Use good initial quality, post-rigor fish
	Insufficient exhausting	Bulging of cans	Adequate exhausting
	Insufficient/over retorting, insufficint cooling, poor cooling water quality	Microbiological hazards including botulism, overprocessing, can bulging	Proper retorting, sufficient cooling in chlorinated water
	Defective cans, absence of lacquer in tin cans	Spoilage, internal corroosion in cans	Use good cans with proper lacquering. Use aluminium
Salting	Poor/low quality salt, high humidity, and low temperature of drying	Inferior products "putty," pink and dum, microbiological hazards	good quality salt (heat to inactivate microorganisms), drying conditions
Smoking	Fat content	Low (1%) fat: dry taste, succulent, glossy High (> 10%) fat: rancidity	Select fish having required fat contents
	Inadequate control of fire, high temperature	Cooking of fish flesh, poor product quality	Proper temperature control
	Excessive smoking	Case hardening	Process control
	Over-brining	Too salty product	Proper brining
	Poor gutting, beheading, filleting, etc.	Poor product appearance	
	Delay in cooling after smoking	Microbial spoilage	Cool rapidly after smoking
Marinading	Improper mixing in curing solution	Poor shelf life, bacterial spoilage	Proper pickling in first curing solution
	Lack of sorting after pickling	Mixed quality	Sorting after first pickling
	Lack of checking of final acid/salt concentration	Bacterial and autolytic spoilage, off-odours, discoloration, softening	Maintain required pH and salt in final product

Adapted from Connell (1995).

Individual Quick Frozen (IQF) Products

Traditional bulk freezing of raw material is slowly getting replaced by individually quick frozen products. The individual quick freezing method allows the processor to supply customer with frozen seafood in small, ready-to-cook quantities instead of large solid blocks which had to be cut or thawed prior to packaging or use. Currently, IQF shrimp and fish fillets are commercially available (Nickolson and Johnston 1993). Fluidised bed cryogenic freezers (straight line belt freezers) are normally used for small size shrimp. The shrimp is frozen in a tray with perforated bottom, through which cold air is circulated. The upward movement of the air carries the product through the freezer by the same blast that freezes them. The through fluidisation guarantees a full individually quick frozen product and a low weight loss. A shiny ice crust forms around each shrimp ensuring the best texture and flavour of the shrimp. It also helps in moisture loss as well as discolouration. The product is then transferred to long-term storage in a separate fish hold, usually equipped with a blast air unit and kept at -29°C. Normal handling capacities in the shrimp industry are 250–1000 kg per hr. The storage life of IQF frozen shrimp is about a few months when held at -18°C. Vacuum packing along with 8% water glazing can extend the shelf life up to 10 months at -18°C.

For larger shrimp and prawns, belt freezers are commonly used. Spiral belt freezers, with attached cleaning systems, are available with a variable speed of belt against a controlled counterflow air distribution system. For best hygiene, the preference is all for stainless steel construction with easily accessible coils. IQF fish fillets are made using spiral belt freezers, since the products take a longer time when subjected to conventional freezing. On board immersion freezing systems have also been developed for preparation of IQF products. The immersion system consists of an immersion freezing module and an insulated tank, made of fiberglass utilising a foam insulation. The tank is filled with seawater, and by adding common salt, the freezing temperature of the water can be kept at -10° to -20°C.

Cook-Chill Products

Supermarkets, nowadays, are turning to high quality 'heat-n-eat' refrigerated foods through 'cook-chill' technology, which imparts convenience, freshness and quality to the product. These foods, generated individually packed in a variety of retail packages, allow consumers to make their own favourite meals by mixing various side dishes, within a short time after purchase. In *sous vide* processing, the food is prepared, seasoned, vacuum sealed and subsequently cooked at controlled temperatures followed by immediately chilling to a temperature below 4°C and storage between 0° to 4°C. The treatment gives the product a shelf life of about 3 weeks. It is reheated before consumption. The technology has been used for a number of items including fish-based ready meals. Many processing plants throughout the European Union are at present using this technology.

A combination process for cook-chilled products from peeled shrimp (*Penaeus indicus)* and white pomfret (*Stromateus cinereus*) steaks consisted of dipping in 10% brine, 5% tripolyphosphate and steam treatments with storage at 3 ± 0.5°C. The products remained in acceptable microbiologial and sensory condition for 25 days (Venugopal 1993).

High Pressure Processing

High pressure treatment of foods can help modify the functional properties of proteins such as oil emulsification and foaming capacities, solubility, textural attributes etc. Whereas thermal processing causes covalent bond breakage in the proteins, resulting in formation of new compounds, HHP affects only non-covalent interactions in the proteins, and therefore does not cause the formation of new compounds. The non-covalent interactions in proteins at 400 to 600 MPa may lead to unfolding of the polypeptide chains causing denaturation, gel formation, association, aggregation, polymerisation and coagulation of proteins. Starch components undergo gelatinization and modification of its susceptibility to enzymatic hydrolysis. Since enzymes are not inactivated, food products may have to be subjected to an initial blanching before HHP treatment to inactivate the enzymes. HHP treated foods have glossy, smooth, soft texture, and enhanced gel hardness (Venugopal *et al.* 2000). For example, subjecting mackerel to HHP at refrigerated temperatures can render it opaque, as if it has been grilled or boiled. Similarly, the transparent nature of cod meat could be changed into opaque by HHP at 608 MPa for 15 min. The changes are due to modification of the myofibril structure. Development of fish analogues from surimi can be achieved by subjecting it to HHP instead of conventional heating, which causes gelation of the meat. Such pressure-induced surimi gels from marine species are smoother and more elastic than those produced by heat and are considered organoleptically superior. Excellent gels could be produced from pollock, sardine, skipjack and tuna meat at 400 MPa and from squid meat at 600 MPa. HHP is also effective in producing highly appealing kamaboko, a Japanese product made from surimi. However, if fatty fish are used for surimi making, HHP may lead to lipid oxidation and flavour changes (Ohshima *et al.* 1993).

Canned Products

Thermal treatment of fish in sealed cans eliminates bacterial as well as autolytic spoilage and gives products with a shelf life of 1 to 2 years at ambient temperature. In canning, fish are dressed, brined, cooked, or smoked prior to being put into a metal can in the presence of hot oil, sauce or brine. The filling medium accelerates heat transfer to the fish and avoids overcooking at points closest to the can walls. The main objective of brining is the enhancement of flavour of the final product. The cans are heated (exhausted) in steam to displace air, sealed and then heat sterilised at 110° to 121°C. The duration of heating depends on the size and dimensions of the can. The sterilised cans are then cooled rapidly to achieve a 'commercially sterile' product. The cans normally will have a 5 mm vacuum after processing. There are three essential rules for the safety of canned products, namely, container seal integrity, adequate thermal process to eliminate the most dangerous and heat resistant microorganisms including *C. botulinum*, and post-process hygiene (Horner 1992b).

A number of fish species such as tuna, mackerel, sardine etc. and molluscs like clams and oysters are ideal for canning. The skin is removed from many species, especially tuna and mackerel, for better product appearance. If a chemical skinning process is used, fish are briefly immersed in a hot (70-80°C) dilute sodium hydroxide solution at pH 11 to 14. The loosened skin is removed by water-jet sprays followed by immersion in dilute hydrochloric acid to neutralise the alkali. High-acid (pH < 4.5) products, such as fish marinades and pickles which contain acetic, citric or lactic acids will not support the growth of spore-forming, human pathogenic microorganisms. They also require lower heat treatment (e.g. at less than 100°C). On the other hand, medium acid (pH 4.5–5.3) products such as fish canned in tomato sauce and other low acid (>pH 5.3) products require full sterilisation (Horner 1992b).

Canning is an ideal method of processing small pelagic fish that do not have a ready market (Hardy 1986). Several marine species that are good candidates for canning include anchovy, herring, mackerel, sardine, scad, sprat and different species of tuna and pilchard apart from freshwater fish such as salmon and a variety of shellfish and mollusks (Govindan 1985, Pedraja 1987). Canned tuna products are highly popular in several countries including USA, France, Spain, Italy and Germany (Subasinghe 1996). Canned products (in fancy packs) such as tuna-based salads, marinated items and elaborated mackerel-based preparations are becoming popular in Europe. Salted and dried small fish such as anchovies is a popular dish in Japan, where dried fry and half-grown anchovies are referred to as *shirasu-boshi* and are delicate snack items. Catfish is a major freshwater item which has the potential for canning, apart from processing by other techniques such as smoking, filleting, or freezing (Fernandez 1985, Raksakulthai 1996). Canning is also employed to preserve fish pastes and spreads. The equipment required for canning has been described by Zugarramurdi et al. (1995).

Fish Meat Mince and Mince-Based Products

Recovery of flesh by mechanical means is perhaps the only viable means of utilising many under-utilised fish species, particularly pelagic fish. Mince separation techniques have also been applied for isolation of meat from fish frames generated during filleting of commercially important species of both marine and freshwater origin. A wide range of machines is available for this purpose. A common type is the belt and drum type. The headed and gutted fish are passed between a counter rotating belt and perforated drum. The fish meat, fat, and blood are squeezed through the drum having an orifice of 3 to 5 mm. Another machine used is the stamp type in which the fish are compressed against a perforated plate in a continuous operation. Mince yield ranges between 50 to 70% in the case of dressed fish. The mince may contain traces of bone. The quality of the mince is reflected by its colour, flavour, and functional properties, and depends on the type of fish as well as its initial quality. Mincing, however, enhances total bacterial count and lipid oxidation, and has a limited storage life even under frozen conditions.

The most important quality of fish mince is its gel-forming capacity. A gel is an intermediate between solid and liquid, in which strands or chains of proteins (myosin) are cross-linked to form a continuous network, in which flavouring compounds and other food additives such as starch could be entrapped to get the desired texture and flavour of the products. Several such products, including pastes, sauces, sausages, patties, sticks, balls and dried products, can be manufactured from the mince by incorporation of ingredients such as vegetable oil, milk proteins, gluten, alginates, carrageenan, xanthan gum or pectin, sugar, salts and preservatives. The fish mince technology has been discussed by several authors (Venugopal and Shahidi 1995, Venugopal 1992, Flick and Martin 1990, Regenstein 1986, FAO 1985).

One of the major uses of fish mince is for the preparation of surimi and surimi-based fish analogues. Surimi is myofibrillar protein concentrate produced by repeated washing of the fish mince in order to remove water soluble nitrogenous and flavour compounds and enhances the gel forming capacity of the proteins. About 2 to 4% sugar or sorbitol and 0.3% food-grade polyphosphates are added to the washed mince as cryoprotectants prior to freezing in order to retain water holding capacity and prevent protein denaturation. Surimi can be kept frozen at –20°C for up to 6 months without any deleterious consequences. Surimi generally contains 75% water, 16% protein and <1% fat. While Alaska pollock has been identified as the most suitable for surimi preparation, recently, several other fish species have also been successfully used for the purpose.

The characteristic property of surimi is its gel-forming capacity, which is dependent upon the temperature of treatment. Gelation is observed when surimi is incubated for setting at 0° to 4°C for 12 to 18 hr (low temperature setting), 25°C at 2 hr (medium temperature), or 40°C for 30 min (high temperature). The low temperature set gel is weak and has high elasticity and transparency. After the initial heat treatment, the surimi is heated at 90°C for 15 min. Heating of the gel at 90°C gives an opaque, rather than translucent and rigid gel. The setting of gel (known as 'suwari' in Japanese) is through cross-linking of the myofibrillar proteins, particularly, myosin (Lanier and Lee 1992). Surimi is used as the raw material for preparation of seafood analogues, such as shrimp, crab legs, scallop, and lobster tail, that possess the accepted texture, flavour and appearance of the natural products. For this, surimi is mixed with binding agents, such as salt, soy protein, starch, egg white and alginate in order to modify the texture and to fabricate the analogues. Chopping and comminution of surimi in presence of these additives helps to break up the muscle tissue, solubilisation of protein in salt solution, and gelation of the proteins. The comminuted mince in the form of sheets is extruded through a rectangular nozzle into thin sheets and the sheets are partially heat set at controlled temperatures. These sheets are further converted into ropes and then developed into imitation products (Lanier and Lee 1992).

Several traditional products can be manufactured from whole fish mince. These include sausages, patties, balls, wafers, loaves, burgers, fish fingers, fish fritters, and pickled products, many of which can be used in developing countries (Reiger and Raizin 1988, Grantham 1985). A process described for preparing restructured sardine meat involved washing the fish mince, mixing it with alginate, and then dialyzing it against calcium chloride solution. Pre-cooked frozen sardine burgers have been also developed. Processing of smoked, dried and powdered sardines into instant soups has also been described. Other products include cutlets, nuggets, chowders, scalloped fish, cakes and patties and fish balls and instant soup mixes (Venugopal and Shahidi 1995).

Extrusion Cooked Products

Thermoplastic extrusion is being commercially used for the development of a variety of snack foods from starch, particularly soy flour. The food extruders are high-temperature short-time (HTST) reactors that transform a variety of ingredients into modified intermediate and finished products. The advantages of this technology are its versatility, high productivity, quality, possibilities for product design and absence of effluents. The extrusion of proteins, including fish proteins, has been examined. For example, Noguchi (1989) described the texturisation of sardine meat. The fish meat and defatted spun fibre were mixed at a ratio of 7:3 at a moisture of 50% and were extruded in a twin-screw extruder equipped with a long cooling die. The product had the texture of animal meat and differed from fish products. Surimi is ideal for extrusion cooking, because of its low lipid content. Aoki *et al.* (1989) studied the extrusion of surimi using a twin-screw extruder. The texturized product compared well with the texture of lobster, crab and squid. An extruded crab analogue prepared from Alaska pollock surimi is already in commercial production in Japan. Similar products are also being produced in Newfoundland, Canada, by Terra Nova Fishery Company (Clareville, NF). The enormous potential of extrusion cooking to produce texturized protein from seafood is yet to be exploited.

Sausages

Sausages are complex mixtures of muscle tissue, solubilized proteins and binders, fat , spices, salt and water. The stabilisation of sausages is through an emulsion formation of protein films around fat globules. Mince from both freshwater and marine species has been shown to be useful for sausage preparation. In Japan, the principal raw material for fish sausage is frozen red flesh from fish such as tuna, marlin and shark as well as whale meat. Surimi, because of its high functionality, is ideal for sausage manufacture. Usually about 10% fat from either animal or vegetable origin is added to the mixture. The resulting product is stuffed into casings and is either steamed or smoked. At present polyvinylidene chloride (PVDC) casings, which are impermeable to oxygen and water, are used for sausages. Chemical preservatives used in formulation include nitrofurylamide, nitrofurazone and sorbic acid.

Fermented Products

Fermentation of fish products has been used intensively in Southeast Asia for preparation of flavoured products, which add variety in the diet and also contribute to general nutrition (Beddows 1985, Lee 1989, Leistner 1990). Fermented products from fish may be classified broadly into two types: (i) fish and salt formulations and (ii) fish, salt and carbohydrates (Owens and Mendoza, 1985). In the former category, fermentation results from autolytic enzymes present in tissue while high levels of salt (>20%) prevent microbial deterioration of the meat. Fish sauce is produced by this method. The long fermentation of fish sauce (5 to 12 months) may be reduced by addition of exogenous proteases. Lactic acid fermentation of fish/carbohydrate mixtures in the presence of small amounts of salt (6 to 10%) provides possibilities for developing a number of products, such as bonito from under-utilised fish species. The principal carbohydrate used in such fermentation is cooked rice. The bacteria used include *Lactobacillus, Streptococcus, Pediococcus* or *Leuconostoc.* Fermented sausages from under-utilised fish species can also be prepared. Fish pastes are another group of fermented products that are widely produced and consumed in many countries. Traditionally, pastes are fermented for a shorter period of time than sauces. Fish protein hydrolyzates and spray dried fish proteins can have significant nutritional importance (Venugopal 1992, 1998). Spray drying of fish meat was made possible by converting the meat into a thermostable dispersion in water, the process of which has been described recently (Venugopal 1998).

The various possibilities for product development from fish mince are summarized in Figure 5.

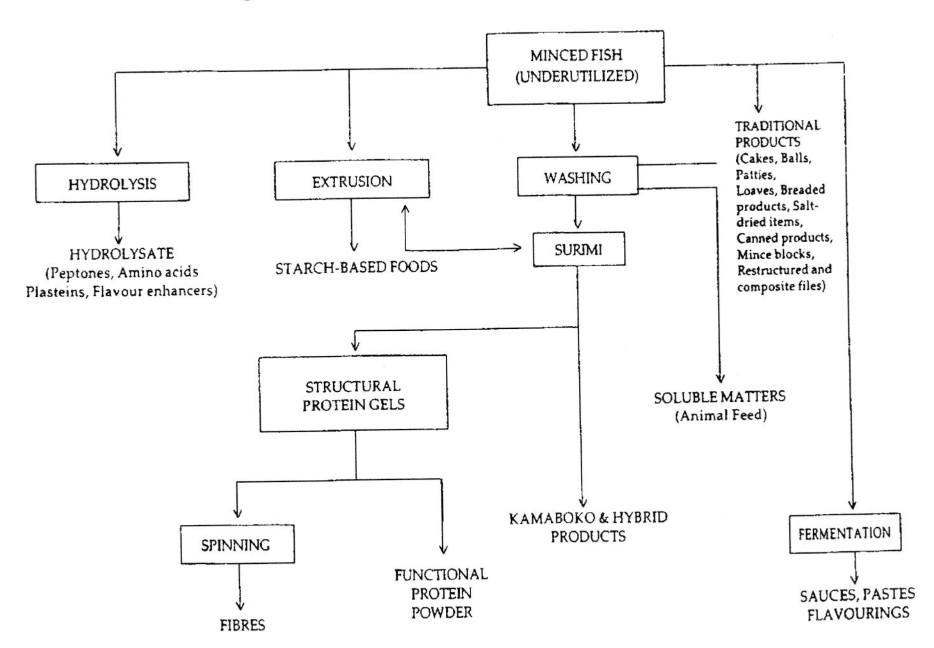

Figure 5. Product possibilities from fish mince.

Breading and Battering

From a commercial point of view, breaded items such as fish sticks and fingers are perhaps the most important. Breading and battering is a technique to prepare tasty products from diverse seafood items including crustaceans, finfish and cephalopods. The raw material is first pre-dusted with flour or dry batter mix, and subsequently conveyed through a special batter. The battered shrimp are usually frozen before or after frying. The technique of batter frying has allowed development of several breaded items that can be used as such or leavened (Roessink 1989). Machinery for automation of process is also available The fried products are heated in microwave oven before consumption. Fish mince can be also used as raw material for the products. Preparation of fish fingers from croaker (*Sciaenid* spp.) and pink perch (*Nemipterus* spp.) consisted of extruding the mince in the presence of additives, dipping the 6 cm-long pieces in batter, rolling in a breading mix, followed by frozen storage of the packaged products. The fried products has a moisture content of 20 to 22%.

Packaging Aspects

Different packaging materials are required for the transportation of fresh fish from ship to processing centres and also for storage of the processed items. The main requirements of the packaging of food items, including fishery products, are insulation properties, control of dehydration and oxidation, elimination of drip, prevention of odour-permeation protection from bacterial and chemical spoilage as well as mechanical damage, and enhancement of economic viability. The packaging requirements are also influenced by the nature of the product, viz., fresh, frozen, dried or otherwise value added (Gopakumar 1996, Narayanan 1997).

The boxes generally used for handling catches on board can be made of hard plastic, suitable for stapling and washing. Sometimes, the boxes are moulded from light, expanded insulating material that is very suitable for air transportation of chilled, valuable fish. The ice, needed to keep the product cold, may be added to the boxes in plastic bags, or else the box has a false bottom and is supplied with a water absorbing pad. Insulated containers have been developed for handling fish in chilled seawater. The developments in retail packages are affected by the demands of the supermarkets. A chilled food package should be compatible with the food, non-toxic, and be able to withstand packaging stress, prevent physical damage, possess appropriate gas permeability, prevent chemical or microbial contamination, protect odours and taints, communicate product information, have saleability as well being cost effective.

Since frozen fishery products form a major item of trade, packaging of such products has received much attention. A suitable frozen fish package must be able to meet the following requirements: (i) protection against moisture loss, oxidation and change of odour and flavour, (ii) retention of volatile flavours, (iii) flexibility to fit the contour of the food, (iv) resistance to puncture, brittleness and deterioration at low temperature, and (v) ease of filling. Paper board (duplex board) carton is the most important primary package used in the export of frozen seafood including shrimp. These cartons are given a coating to protect the contents from loss of moisture. Frozen blocks weighing usually 2 kg are packed in such cartons lined inside with 100/125 gauge low density polyethylene. The unit carton is made of solid white duplex board of chemical wood pulp and coated with wax on both sides. Low density polyethylene (20 GSM) can also be used for coating. There are no standards for duplex cartons for export of frozen shrimp.

However, the minimum requirements are: grammage, 300 GSM, caliper, 0.4 ± 0.02 mm, wax content, 10 GSM n each side, or, polyethylene coating, 20 GSM on each side, ring stiffness, 270 N, bursting strength, 400 Kpa and water proofness (Cobb 30° value) of 10. Ten such duplex cartons are packed in a 5 ply corrugated fibre board boxes which serve as shipping container. The master carton should have a minimum bursting strength of 12 kg/cm^2, water proofness of 120, wax coating, 20 GSM on each side, grammage of liners and corrugating medium, 140 GSM, and puncture resistance of 200 beach units and compression strength of 350 kg. The cartons used presently by the industry are nearly cubical in shape (355 ´ 330 ´ 285 mm). Ideally, a corrugated box can hold 12 liner cartons for better stacking.

For IQF products, the packaging systems are not standardised. Generally, exporters heat seal the product in plastic film pouches, which may be either provided with unit/intermediate cartons or directly placed in master cartons. The unit/intermediate cartons are made from duplex board or 3 ply corrugated fibre board. The most functional cost effective films is BOPP. The transit package should have a compression strength of 500 Kg for reasonable safety of the product. The Central Institute of Fisheries Technology, Cochin, India has developed flexible packaging material for IQF shrimp based on 12 micron plan polyester laminated with 250 gauge low density polyethylene laminate for consumer packs to hold 100 to 200 g for IQF shrimp. The Institute has also developed packaging systems for bulk frozen as well as dried seafood (Gopakumar 1996).

Apart from metallic cans, retortable flexible pouches are now available for ambient temperature storage of shelf stable products. Polyester laminated with LDPE-HDPE co-extruded film or metallised polyester laminated with LDPE-HDPE co-extruded film or nylon/surlyn polymers have been developed for fish pickles instead of conventional glass containers. Conventional packaging materials like flexible plastic films alone are not suitable for products such as fingers, cutlets or patties developed from fish mince, since they give little mechanical protection to the products during transportation. The thermoformed trays produced from food grade materials like polyvinyl chloride or high impact polystyrene are suitable for packaging of value added fishery products. Such trays are ideal for self service merchandising systems and constitute unit packs (Gopakumar 1996).

Crab meat packed in copolymer polyethylene/polypropylene cups, Saran over-wrapped or vacuum skin packaged polystyrene trays, and non-barrier pouches. Meat pasteurised in plastic and aluminium cans had better sensory and microbiological quality and longer shelf life than meat packed in steel cans. Oxygen-barrier pouches had the lowest quality and shortest shelf life, while non-barrier pouches gave extended shelf life. Vacuum skin packaging resulted in improved sensory quality of freshly cooked and picked meat (Gates *et al.* 1993). The best packaging material for dried fish are flexible plastic films, plastic containers or bitumen-lined brown paper. High density polyethylene or polypropylene pouches could extend the shelf life of intermediate moisture mackerel up to 12 weeks at ambient temperature (Che-Man *et al.* 1995).

Summary

The foregoing discussion summarises the major techniques for conservation and value addition of fishery products. As pointed out earlier, consumers are becoming increasingly aware of the health impact of foods and generally consider fish products as low calorie, low-cholesterol, and easily digestible proteins. Therefore, the need for conservation of highly perishable fish items cannot be over-emphasised. Integration of novel techniques with conventional processes such as chilling and freezing can help in better handling and preservation of the commodity. Innovations in packaging technology together with developments of processing technology can give positive advantages in order to make better use of fish items including currently under-utilised resources for human consumption.

Acknowledgement

The author thanks Dr. D.R. Bongirwar, Head, Food Technology Division, Bhabha Atomic Research Centre, Mumbai for his interest and his colleagues Dr. S.B. Warrier and Ms. Manisha Karani for critically going through the manuscript.

References

Alur, M.D., S.N. Doke, S.B. Warrier, and P.M. Nair. 1995. Biochemical methods for determination of spoilage of foods of animal origin. A critical evaluation. *J. Food Sci. Technol.(India)* 32: 181-189.

Alur, M.D. and V. Venugopal. 2000. Dried Foods. Pages 530–537 in *Encyclopedia in Food Microbiology*, Vol. 3, R.K. Robinson, C.A. Batt, and P.D. Patel, eds., Academic Press, New York.

Ando, M., H. Toyohara, Y. Simizu, and M. Sakaguchi. 1991. Post-mortem tenderisation of rainbow trout caused by gradual disintegration of the extracellular matrix structure. *J. Sci. Food Agric.* 55: 589-599.

Aoki, K., F. Hava, M. Ohmichi,, N. Nakatani, and H. Hosaka. 1989. Texturisation of surimi using a twin-screw extruder. *J. Jap. Soc. Food Sci. Technol.* 36: 748-751.

Ashie, I.N.A., J.P. Smith, and B.K. Simpson. 1996. Spoilage and shelf extension of fresh fish and shellfish, *Crit. Rev. Food Sci. Nutr.* 36: 87-121.

ASHRAE. 1978. Fishery Products. Chapter 30. ASHRAE Handbook and Product Directory: American Society of Heating, Refrigeration and Air Conditioning Engineers Inc., New York.

Bannerman, A. M. 1980. Hot smoking of fish. Torry Advisory Note No. 82, Torry Research Station, Ministry of Agriculture, Fisheries and Food, UK, pp. 3-12.

Bartholmey, D. S. 1994. New developments in ice production and storage. *Infofish Internat.* 4: 64-68.

Beddows, C.G. 1985. Fermented fish and fish products. Pages 1–45, in *Microbiology of Fermented Foods*, Vol. 2, B.J.B. Wood, ed., Elsevier, New York.

Berhimpon, S., R.A. Souness, K.A. Buckle, and R.A. Edwards. 1990. Salting and drying of yellow tail (*Trachurus meculloch*i Nichols). *Int. J.Food Sci. Technol.* 25: 409-419.

Bilinski, E.R., E.E. Jonas, and Peters, M.D. 1983. Factors controlling the deterioration of spiny dogfish (*Squalus acanthias*). *J. Food Sci.* 48: 808-812.

Boeri, R.L., L.A. Davidovich, D.H. Giannini, and H.M. Lupin. 1985. Method to estimate the consumption of ice during fish storage. *Int. J. Ref.* 8: 97-101.

Borquez, R., M. Espinoza, and R. Peters. 1994. The effect of storage time and chemical preservatives on the total volatile basic nitrogen content in Chilean mackerel (*Trachurus murphy*) prior to fish meal production. *J. Sci. Food Agric.* 66: 181-186.

Botta, J.R. 1994. Freshness quality of seafoods: a review. Pages 140–167 in *Seafoods: Chemistry, Processing Technology and Quality.* F. Shahidi and J.R. Botta, eds., Chapman & Hall, Glasgow.

Bremner, H.A. 1992. Fish flesh structure and the role of collagen–its post-mortem aspects and implications for fish processing. Pages 39–52 in *Developments in Food Science. Quality Assurance in the Fish Industry.* H.H. Huss, M. Jackobson and J. Liston, eds. Elsevier Science, Amsterdam.

CAC, Codex Alimentarius Commission. 1996. *Proposed Draft Guidelines for the Sensory Evaluation of Fish and Shellfish,* Codex Committee on Fish and Fishery Products, 22nd Session, Bergen, May CX FFP96/9, Food and Agricultural Organisation of the United Nations, Rome, Italy.

Chandrasekaran, M. 1994. Methods for preprocessing and freezing of shrimps: A critical evaluation. *J. Food Sci. Technol. (Mysore)* 31: 441-452.

Chang, D.S., H.R. Cho, H. -Y. Good, and W.K. Choe. 1989. A development of food preservative with the waste of crab processing. *Bull. Korean Fish Soc.* 22: 70.

Chang, W.S., R.T. Toledo, and J. Deng, 1975. Effect of smoke house temperature, humidity and air flow on smoke penetration into fish muscle. *J. Food Sci.* 40: 240-243.

Charm, S.E., R.E. Learson, L.J. Ronsivalli, and M. Schwartz. 1972. Organoleptic technique predicts refrigeration shelf life of fish. *Food Technol.* 26: 65-68.

Che-Man, Y.B., J. Bakar, and A.A.K. Morri. 1995. Effect of packaging films on storage stability of intermediate-moisture deep-fried mackerel fillets and mince. *J. Food Sci.* 58: 1208-1211.

Connell, J.J. 1995. Control of Fish Quality. Fishing News Books, London, U.K.

Curran, A.C. and R.G. Poulter. 1983. Isohalic sorption isotherms. III. Application to a dried salted tropical fish. *J. Food Technol.* 18: 739-746.

Curran, C.A. and D.S. Trim 1983. Comparative study of solar and sun-drying of fish. Pages 69-80 in *Production and Storage of Dried Fish,* James, D. ed., FAO Fisheries Report No. 279.

Cydesdale, F.M. 1978. Colourimetry: Methodology and Applications. *Critical Rev. Food Sci. Nutr.* 10: 243-301.

Dainty, R.H. 1996. Chemical/biochemical detection of spoilage. *Int. J. Food Microbiol.* 33: 19-33.

Dalgaard, P., L. Gram, and H.H. Huss. 1993. Spoilage and shelf life of cod fillets packed in vacuum or modified atmospheres. *Int. J. Food Microbiol.* 19: 283-294.

Daniels, J.A., R. Krishnamurthi and S.H. Rizvi. 1986. Effect of carbonic acid dips and packaging films on the shelf life of fresh fish fillets. *J. Food Sci.* 51: 929-931.

Davis, P. 1986. Fish. Pages 189–228, in *Principles and Applications of Modified Atmosphere Packaging of Foods,* T.R. Parry, ed, Blackie Academic, London.

Dillon, R., T.R. Patel, and A.M. Martin. 1994. Microbiological control of fish smoking operations. Pages 51–81, in *Fisheries Processing: Biotechnological Applications,* A.M. Martin, ed., Chapman & Hall, New York.

Doe, P.E. 1983. Spoilage of dried fish. The need for more data on water activity and temperature effects on spoilage organisms. Pages 209–244 in *Production and Storage of Dried Fish,* FAO Fisheries Report No. 279, Supplement.

Doke, S.N., S.B. K. Warrier, V. Ninjoor, and G.B. Nadkarni. 1979. Role of hydrolytic enzymes in the spoilage of fish. *J. Food Sci. Technol.*16: 223-226.

Dollar, A.M. 1989. Insect disinfestation of dried fish products by irradiation. Pages 19–27 in *The Prevention of Losses in Cured Fish,* FAO Fisheries Technical Paper, 219, International Atomic Energy Agency, Vienna.

EC, 1996. Council Regulations (EC No. 2406/96). 'Laying down common marketing standards for certain fishery products', Official J. European Community, No. L-334/2, 23.12.1996.

Ellinger, R.J. 1972. Phosphates in food processing. Pages 617-780 in *Handbook of Food Additives,* 2nd Ed. T. Furia, ed., CRC Press, Cleveland, Ohio.

FAO/IAEA. 1978. Advisory Group Meeting on Radiation Treatment of Fish and Fishery Products. Food and Agriculture Organization, Rome and International Atomic Energy Agency, Vienna, March 13–16, Manila, Philippines, International Atomic Energy Agency, Vienna.

FAO. 1973. Code of practice for fresh fish. FAO Fisheries Circular, C318, Food and Agricultural Organization, Rome, Italy.

FAO. 1985. *Storage of Tropical Fish and Product Development,* in *Fisheries Report.(Suppl.),* A. Reilly, ed., Food and Agriculture Organization, Rome.

FAO. 1993. *Ice in Fisheries,* FAO Fisheries Technical Paper No. 331, Food and Agriculture Organization, Rome.

FAO. 1994. *Freezing and Refrigerated Storage in Fisheries.* Fisheries Technical Paper 340, Food and Agriculture Organization, Rome.

FAO. 1995. Quality and quality changes in fresh fish. FAO Fisheries Technical Paper 348, H.H. Huss, ed., Food and Agriculture Organization, Rome.

Fernandez, M. 1985. Carp — a relatively easy to raise freshwater fish. *Infofish Marketing Digest* 2: 30-33.

Finne, G., T. Wagner, D. De-Witt, and R. Martin. 1986. Effect of treatment, ice storage and freezing on residual sulfite in shrimp. *J. Food Sci.* 51: 231-232.

Flick, G.J., G.P. Hong, and G.M. Knabi. 1991. Non-traditional methods of seafood preservation. MTS J. 25: 35-42.

Flick, G.J. and R.E. Martin, 1992. (eds.) *Advances in Seafood Biochemistry, Composition and Quality,* Technomic Publ. Co., Lancaster.

Garrett. E.S., M.L. Jahncke, and J.M. Tennyson. 1997. Microbiological hazards and emerging food safety issues associated with seafoods. *J. Food Prot.* 60: 1409-1415.

German, J.B. and J.E. Kinsella. 1985. Lipid oxidation in fish tissue. Enzymatic initiation via lipoxygenase. *J. Agri. Food Chem.* 33: 681-686.

Ghadi, S.V., V.N. Madhavan, U.S. Kumta, and N.F. Lewis. 1978. Effect of irradiation on storage stability of Elasmobranchs. *Fleischwirtschaft* 2: 1351-1355.

Gibbson, D.M. 1984. Predicting the shelf life of packaged fish from conductance measurements. *J. Appl. Bacteriol.* 58: 465-470.

Gill, T.A. 1990. Objective analysis of seafood quality. *Food Rev. Int.* 6: 681-714.

Gopakumar, K. (1990). Post-harvest technology for tropical fish — a review. Pages 77–86, in *Post Harvest Technology: Preservation and Quality of Fish in South Asia,* P.J.A. Reilly, R.W.H. Parry, and L.E. Burtle, eds., International Education for Science, Stockholm, Sweden.

Gopakumar, K. 1996. Processing for fresh and processed marine products, *Seafood Export J.,* February issue, 7-15.

Gould, G.W. 1996. The methods for preservation and extension of shelf life. *Int. J. Food Microbiol.* 33: 51-64.

Govindan, T.K. 1985. *Fish Processing Technology*, Oxford & IBH Publishing Co., New Delhi.

Gram, L. and H.H. Huss. 1996. Microbial spoilage of fish and fish products. *Int. J. Food Microbiol.* 33: 121-138.

Grantham, G.J. 1981. *Minced Fish Technology: A Review, FAO, Fish Tech. Paper* No. 216, Food and Agriculture Organization, Rome.

Gray, J.I. 1978. Measurement of lipid oxidation: A Review. *J. Am. Oil.Chem. Soc.* 55: 539-546.

Gray, R.J.H., D.G. Hoover, and A.M. Muir. 1983. Attenuation of microbial growth on modified atmosphere packaged fish. *J. Food Prot.* 46: 610-613.

Haard, N.F. 1992a. Technological aspects of extending prime quality of seafood, a review. *J. Aquatic Food Prod. Technol.* 1: 9 -15 .

Haard, N.F. 1992b. Biochemical reactions in fish muscle during frozen storage. Pages 1–176 in *Advances in Seafood Biochemistry: Composition and Quality*, G.J. Flick and R.E. Martin, eds., Technomic Publ. Co., New York.

Haard, N.F. 1992c. Biochemistry and chemistry of colour and colour change in seafoods. Pages 305–345 in *Advances in Seafood Biochemistry: Composition and Quality*, G.J. Flick and R.E. Martin, eds., Technomic Publ. Co., New York.

Haardy, R. 1986. Fish Processing. *Proc. Royal Soc. Edinburgh*, 87B: 201-220.

Hannesson, G. and B. Dogbjartsson. 1971. *Radurization of Scampi, Shrimp and Cod*, Technical Report Series No. 124, International Atomic Energy Agency, Vienna.

Hansen, P. 1996. Food uses of small pelagics. *Infofish Internat.* 4: 46-52.

Hansen, P. and J. Jensen. 1982. Bulk handling and chilling of large catches of small fish. Part I. Quality and storage life. *Infofish Marketing Dig.* 11: 26-28.

Hanson, S.W. and J. Esser. 1985. Investigation into post-harvest losses of cured fish in SE Asia (ODA Project Report), Overseas Development Administration, London, UK.

Himmelbloom, B.H., C.A. Crapo, E.K.Brown, and J.P. Doyle. 1994. Factors affecting quality of rock sole fillets. *J. Aquatic Food Prod. Technol.* 3: 45-56.

Hobbs, G.1977. *Clostridium botulinum* in irradiated fish. *Food Irrad. Inf.* 7: 39.

Horner, W.F.A. 1992a. Preservation of Fish by Curing (Drying, Salting and Smoking), Pages 31–72 in *Fish Processing Technology*, Blackie Academic Professional, London.

Horner, W.F.A. 1992b. Canning of fish an fish products. Pages 114-155 in *Fish Processing Technology*, Blackie Academic Professional, London.

Hoston, J. and J.W. Slavin. 1969. Technological problems in the preservation of fresh, iced fish, in *Technology of Fish Utilisation*, Kreuzer, R., ed., Fishing News Books, London.

Hseigh, R.J. and J.E. Kinsella. 1986. Lipoxygenase-catalyzed oxidation of N-6 and N-3 polyunsaturated fatty acids. Relevance to activity in fish tissue. *J. Food Sci.* 51: 940-945.

Houwing, H., J. Ohdam, and J.J. Oosterhuis. 1978. Irradiation of fishery products, especially shrimp and cod/plaice fillets. Pages 333-346 in *Food Preservation by Irradiation*, Vol. I, International Atomic Energy Agency, Vienna.

Hultin, H.O. 1994. Oxidation of lipids in seafoods. Pages 47-74 in *Seafoods: Chemistry, Processing Technology and Quality*, F. Shahidi and J.R. Botta, eds., Blackie Academic, London.

Hultin, H.O. 1992. Biochemical deterioration of fish muscle. Pages 125-138 in *Quality Assurance in the Fish Industry*, H.H. Huss, M. Jakobsen, and J. Liston, eds., Elsevier Science Publishers, Amsterdam..

Huang, Y.W. 1988. Cannonball jellyfish as a food source. *J. Food Sci.* 53: 341-343.

IAEA. 1996. International Atomic Energy Agency, *Food Irradiation Newsletter, Suppl.* 2.

Infofish. 1983. Dried fish, an Asian staple food. Pages 18–43 in *The Production and Storage of Dried Fish*, FAO Fisheries Report No. 279 Supplement, Food and Agricultural Organization, Rome.

Ingram, S. 1989. Lactic acid dipping for inhibiting microbial spoilage of refrigerated catfish fillet pieces. *J. Food Qual.* 12: 433-443.

Jeevanandam, K., A. Kakatkar, S. N. Doke, and V. Venugopal. 2001. Preparation and storage characteristics of dehydrated ribbon fish laminates. *J. Aquatic Food Prod. Technol.* (in press).

Jhaveri, S.N. and S.M. Constantinides. 1982. Atlantic mackerel (*Scomber scombrus*): shelf life in ice. *J. Food Sci.* 47: 188-192.

Johnson, P.B. 1989. Factors influencing the flavour quality of farm-raised catfish. *Food Technol.* 43: 94-97.

Kabara, J.J. 1981. Food grade chemicals for us in designing food preservative systems. *J. Food Prot.* 44: 633-647.

Kanner, J. and J.E. Kinsella. 1983. Lipid deterioration initiated by phagocytic cells in muscle foods: beta carotene destruction by hydrogen peroxide-halide system. *J. Agri. Food Chem.* 31: 370-376.

Kawai, T. 1996. Fish flavour, *Crit. Rev. Food Sci. Nutr.* 36: 257-298.

Kosak, P.H. and R.T. Toledo. 1981. Brining procedures to produce uniform salt content in fish. *J. Food Sci.* 46: 874-876.

Kraft, A.A. 1992. Spoilage of eggs and fish. Pages 99–111 in *Psychrotrophic Bacteria in Foods, Disease and Spoilage*. CRC Press, Boca Raton, Florida.

Labuza, T.P. and B. Fu. 1995. Use of time/temperature integrators, predictive microbiology and related technologies for assessing the extent and impact of temperature abuse on meat and poultry products. *J. Food Safety* 15: 201-227.

Lanier, T.C. and C.M. Lee. 1992. *Surimi Technology*, Marcel-Dekker, New York.

Lee, C.M. 1993. Surimi processing from lean fish. Pages 263-287 in *Seafoods: Chemistry, Processing Technology and Quality*, F. Shahidi and J.R. Botta, eds., Blackie Academic Publ. Co., London.

Leistner, L. and G.M. Gorris Leon. 1995. Food preservation by hurdle technology *Trends Food Sci. Technol.* 6: 41-46.

Leistner, L. 1990. Mould-fermented foods: recent developments. *Food Biotechnol.* 4: 433-440.

Licciardello, J.J., E.M. Ravesi, B.E. Tukkunen, and L.D. Racicot. 1984. Effect of some potentially synergistic treatments in combination with 100 Krad irradiation on iced shelf-life of cod fillets. *J. Food Sci.* 49: 1341-1344.

Lima dos Santos, C. A. M. 1981. The storage of tropical fish in ice, a review. *Trop. Sci.* 23: 97-131.

Lindsay, R.C. 1991. Chemical basis of the quality of seafood flavours and aromas. *Mar. Technol. Soc. J.* 25: 16-22.

Liston, J. 1980. Microbiology in fishery science. Pages 139–157 in *Advances in Fish Science and Technology*, J.J. Connell, ed., Fishing News Books Ltd., Surrey, UK.

Lindner, P., S. Angel, Z.G. Weinberg, and R. Granit. 1988. Factors inducing mushiness in stored prawns. *Food Chem.* 29: 119-123.

Loliger, J. 1991. The use of antioxidants in foods. Pages 121-150 in *Free Radicals and Food Additives*, O.I. Aruoma and B. Halliwell, eds., Taylor and Francis Ltd., London.

Lupin, H.M. 1985. Measuring the effectiveness of insulated fish containers. *FAO Fisheries Rep.* 329, Suppl., pp. 47-55.

Mackie, I.M. 1993. The effect of freezing on flesh proteins. *Food Rev. Int.* 9: 575-610.

Maga, J.A. 1988. Pages 61–68 in *Smoke in Food Processing*, CRC Press, Boca Raton, Florida.

Marrakchi, El A., M. Bennour, N. Bouchriti, A. Hamama, and H. Tagafait. 1990. Sensory, chemical and microbiologial assessments of Moroccan sardines stored in ice. *J. Food Prot.* 53: 600-605.

Martinez, A. and A. Gildberg. 1988. Autolytic degradation of belly tissue in anchovy. *Int. J. Food Sci. Technol.* 23: 185-194.

Martinez, I., R.J. Olsen, H. Nilsen, and N.K. Sorensen. 1997. Seafood: Fulfilling market demands. *Outlook on Agriculture* 26: 107-114.

Maruf, F.W., D.A. Ledward, R.J. Neale, and R.G. Poulter. 1990. Chemical and nutritional quality of Indonesian dried salted mackerel (*Rastrelliger kanagurta*). *Int. J. Food Sci. Technol.* 25: 66-77.

McDonald, R.E. and H.O. Hultin. 1987. Some characteristics of the enzymic lipid peroxidation system in the microsomal fraction of flounder skeletal muscle. *J. Food Sci.* 52: 15-19.

Medina, I., S. Saeed, and N. Howell. 1999. Enzymatic oxidative activity in sardine and herring during chilling and correlation with quality. *Eur. Food Res. Technol.* 210: 34-38.

Mendelsohn, J.H. 1974. Rapid techniques for salt curing fish: a review. *J. Food Sci.* 39: 125-131.

Mishkin, M., M. Karel, and I. Saguy. 1982. Optimization of dehydration. *Food Technol.* 36: 101-103.

Narayanan, P.V. 1997. Packaging of seafoods. *Seafood Export J. (India)* 28: 23-26.

Neilson, J. 1993. Quality management of the raw material in the food fish sector. European Economic Community Contract No. UP-2-452.

Nickerson, J.T.R., J.J. Licciardello, and L.J. Ronsivalli. 1983. Preservation of food by ionizing radiation. Pages 13–82 in *Radurization and Radicidation of Fish and Shellfish*, E.S. Josephson and M.S. Paterson, eds., CRC Press, Boca Raton.

Nickolson, F.J. and W.A. Johnston. 1993. Individual quick freezing (IQF). *Infofish Int.* 3: 51-54.

Noguchi, A. 1989. Extrusion cooking of high moisture protein foods. Pages 143–171 in *Extrusion Cooking*, C. Mercier, P. Linko, and J. Harper, eds., Am. Assoc. Cereal Chem.. St. Paul, Minnesota.

Oh, K.S., B.K. Chung, M.C. Kim, N.J. Sug,. and E.H. Lee. 1988. Processing of smoked, dried and powdered sardines into instant soups. *J. Korean Soc. Food Nitr.* 17: 149.

Ohshima, T., H. Ushio, and C. Koizumi. 1993. High pressure processing of fish and fish products. *Trends Food Sci. Technol.* 4: 370-375.

Olafsdottir, G., E. Martinsdottir, J. Oehlenschlager, P. Dalgaard, B. Jensen, I. Underland, I. M. Mackie, G. Henehan, J. Neilsen, and H. Nilsen, 1997. Methods to evaluate fish freshness in research and industry. *Trends Food Sci. Technol.* 8: 258-265.

Olley, J. and D.A. Ratkowsky. 1973. Temperature function integration and its importance in the storage and distribution of flesh foods above the freezing point. *Food Technol. (Aust.)* 25: 66-73.

Owens, J.D. and L.S. Mendoza, 1985. Enzymatically hydrolyzed and bacterially fermented fishery products. *J. Food Technol.* 20: 273-286.

Owen, D. and M. Nesbitt. 1984. A versatile time temperature function integrator. *Lab. Practice,* 33: 70-75.

Papadakis, S.E., S. Abdul-Malek, R.E. Kamdem, and K.I. Yam. 2000. A versatile and inexpensive technique for measuring colour of foods. *Food Technol.* 54: 48-51.

Pedrosa-Menabrito, A. and J.M. Regenstein, 1990. Shelf life extension of fresh fish — A review. Part II. Preservation of fish. *J. Food Qual.* 13: 129-146.

Pedraja, P. 1987. Opportunities for development of processed (value added) fishery products in India. Report. FAO Project TCP/RAS/6653, Cochin, India.

Poulter, N.H. and L. Nicolaides. 1985. Studies of the iced storage characteristics and composition of a variety of Bolivian freshwater fish. 2. Parana and Amazon basis fish. *J. Food Technol.* 20: 451-465.

Raksakulathai, N. 1996. Processing of hybrid *Clarias* catfish. *Infofish International* 3: 33-36.

Ratkowsky, D.A., J. Olley, T.A. McMeekin, and A. Ball. 1982. Relationship between temperatures and growth rate of bacterial cultures. *J. Bacteriol.* 149: 1-5.

Raveshi, E.M., J.J. Licciardello, B.E. Tuhkunen, and C. Lundstrom. 1985. The effect of handling or processing treatments on storage characteristics of fresh spiny dogfish. *Mar. Fish Rev.* 47: 48-67.

Reddy, N.R., D.J. Armstrong, E.J. Rhodehamel, and D.A. Kautter. 1992. Shelf life extension an safety concerns about fresh fishery products packaged under modified atmosphere. A review. *J. Food Safety* 12: 87-118.

Regier, W.L. and M.A. Raizin. 1988. Fish mince — its potential for less developed countries and others. Pages 202-206 in *Post Harvest Fishery Losses*, M.T. Morrisay, ed., University of Rhode Island, Kingston.

Regenstein, J.M. 1986. The potential for minced fish. *Food Technol.* 40: 101-105.

Reineccius, G. 1991. Off-flavours in foods. *Crit. Rev. Food Sci. Nutr.* 29: 381-412.

Rodriguez, C.J., I. Besteiro, I. and C. Pascual, 1999. Biochemical changes in freshwater rainbow trout (*Oncorhynchus mykiss*) during chilled storage. *J. Sci. Food Agri.* 79: 1473-1480.

Rodrick, G.E. and D. Dixon.1999. Code of practice for the irradiation of seafoods. International Atomic Energy Agency, Vienna.

Roessink, G.L. 1989. Battered and breaded products. *Infofish International* 4: 17-21.

Rohr, von R. 1995. Ice in fisheries. *Infofish Int.* 1: 47-50.

Ronsivalli, L.J., R.J. Learson, and S.E. Charm. 1973. *Mar. Fish. Review* 35: 34-36.

Saito, T., K. Ari and M. Matsuoyshi. 1959. A new method for estimating the freshness of fish. *Bull. Jap. Soc. Sci. Fish.* 24: 749-753.

Savagaon, K.A. and A. Sreenivasan. 1978. Activation mechanism of pre-phenoloxidase in lobster and shrimp. *Fish Technol.* 15: 49-53.

Shelef, L.A. 1994. Antimicrobial effects of lactates. A review. *J. Food Prot.* 57: 445-450.

Sherekar, S.V., S.N. Doke, M.S. Gore, and V. Ninjoor. 1986.. Proteinases of tropical fish and their role in autolysis. *Indian J. Exp. Biol.* 4: 440-446.

Sikorski, Z.E. and B.S. Pan. 1994. Preservation of seafood quality, Pages 165–195 in *Seafoods: Chemistry, Processing Technology and Quality,* F. Shahidi and J.R. Botta, eds., Chapman and Hall, London.

Simpson, K.L. 1985. Chemical changes in natural food pigments, Pages 409-418 in *Chemical Changes in Food During Processing,* T. Richardson, and J.W. Fineley, eds., Van Nostrand Reinhold Co., New York.

Sinnhuber, R.O. and T.C. Yu. 1958. 2-Thiobarbituric acid method for measurement of rancidity in fishery products. II. Quantitative determination of malonaldehyde. *Food Technol.* 12: 9-12.

Smith, G. and M. Hole, 1991. Browning of salted sun-dried fish. *J. Sci. Food Agri.* 55: 291-301.

Sofos, J.N. 1984. Antimicrobial effects of sodium and other ions in foods. A review. *J. Food Safety* 6: 45-78.

Sorensen, N.K., I.G. Helgasson, and R. Brataas. 1997. *Pre-rigor* fish: handling and processing. Pages 135-140 in A. Bremner, C. Davis, and B. Austin, eds., Making the most of the catch, Proc. of Symposium, Brisbane, July 1996.

Spencer, R. and C.R. Baines. 1964. The effect of temperature on the spoilage of wet white fish. I. Storage at constant temperatures between –1°C and 25°C. *Food Technol.* 18: 769-773.

Srikar, L.N., B.K. Khunta, G.V.S. Reddy, and B.R. Srinivasan, 1993. Influence of storage temperature on the quality of salted mackerel and pink perch. *J. Sci. Food Agri.* 63: 319-322.

Stammen, K., D. Gerdes, and F. Caporaso. 1990. Modified atmosphere packaging of seafood. *Crit. Rev. Food i. Nutr.* 29: 301-331.

Stanier, T.C. 1981. Effect of early evisceration on keeping quality of croaker and trout. *J. Food Sci.* 46: 863-868.

Statham, J.A., H.A. Bremner, and A.R. Quarmby. 1985. Storage of morwong in combination of polyphosphate, potassium sorbate and carbon dioxide at 4°C. *J. Food Sci.* 50: 1580-1585.

Subasinghe, S. 1996. Innovative and value added tuna products and markets. *Infofish International* 1: 43-50.

Suzuki, T. 1981. *Fish and Krill Protein: Processing Technology,* Elsevier, London.

Tarladgis, B.G., B.M. Watts, M.T. Younathan, and L. Dugan 1960. A distillation method for the quantitative determination of malonadehyde in rancid foods. *J Am. Oil Chemists' Soc.* 37: 44-48.

Tsukuda, N. 1970. Studies on the discolouration of red fishes. VI. *Bul. Jap. Soc. Sci. Fish.* 36: 725-328.

Venugopal, V. 1990. Extracellular proteases of contaminant bacteria in fish spoilage: A review. *J. Food Prot.* 53: 341-350.

Venugopal, V. 1992. Mince from low cost fish. *Trends Food Sci. Technol.* 3: 2-5.

Venugopal, V. 1993. Cook-chill process to extend refrigerated shelf life of peeled and deveined shrimp and pomfret. *Int. J.Food Sci. Technol.* 28: 273-278.

Venugopal, V. 1994. Preparation of fish protein hydrolyzates by microorganisms. Pages 223-239 in *Seafood Processing: Biotechnological Applications*, Martin, A.M., ed., Martin, Chapman and Hall, London.

Venugopal, V. 1998. Functionality and potential applications of thermostable water dispersions of fish muscle. . *Trends Food Sci. Technol.* 8: 271-276.

Venugopal, V., S.N. Doke, and P. Thomas . 1999. Radiation processing to improve the quality of fishery products. *Crit. Rev. Food Sci. Nutr.* 39: 341- 400.

Venugopal, V., R. Lakshmanan, S.N. Doke and D.R. Bongirwar. 2000. Enzymes in fish processing, biosensors and quality control. A review. *Food Biotechnol.* 14: 21-77.

Venugopal, V., A.S. Kamat, and D.R. Bongirwar. 2001. Processing of foods using high hydrostatic pressure. *Indian Food Industry* 20: 65-69.

Venugopal, V., N.F. Lewis, and G.B. Nadkarni. 1981. Volatile acid content as a quality index for Indian mackerel. *Lebensm. Wiss. U. Technol.* 14: 35-42.

Venugopal, V. and F. Shahidi. 1995. Value added products from under-utilised fish species. *Crit. Rev. Food Sci. Nutr.* 35: 431- 453.

Venugopal, V. and F. Shahidi. 1996. Structure and composition of fish muscle. *Food Rev. Int.* 12: 175-197.

Venugopal, V. and F. Shahidi. 1998. Conventional techniques to process under-utilised fish species. *Food Rev. Int.* 14: 35-97.

Venugopal, V., A.C. Pansare, and N.F. Lewis. 1984. Inhibitory effect of some preservatives on protease secretion of *Aeromonas hydrophila J. Food Sci.* 49: 1078-1081.

Villemure, G., R.E. Simard, and G. Picard. 1986. Bulk storage of cod fillets and gutted cod under CO_2 atmosphere. *J. Food Sci.* 51: 317-320.

Ward, D.R. 1994. Microbiological quality of fishery products. Pages 1-17 in *Fisheries Processing: Biotechnological Applications,* A.M. Martin, ed., Chapman and Hall, London.

WHO, 1981. Wholesomeness of Irradiated Foods, Report of Joint FAO/WHO/IAEA Expert Committee. Technical Report Series No. 604, World Health Organization, Geneva.

WHO. 1999. Food safety issues associated with products from aquaculture. Report of a Joint FAO/NACA/WHO Study Group, WHO Technical Report Series # 883, World Health Organization, Geneva.

Woyewoda, A.D. and P.J. Ke. 1980. Laboratory quality assessment of Canadian Atlantic squid. Fisheries and Marine Service Technical Report No. 902, Dept. of Fisheries and Oceans, Halifax, Nova Scotia.

Yan, X., K.D.A. Taylor, and S.W. Hanson. 1989. Studies on the mechanism of blackspot development in Norway lobster. *Food Chem.* 34: 273-278.

Yen, G.C. and C.L. Hsieh. 1991. Simultaneous analysis of biogenic amines in canned fish by HPLC. *J. Food Sci.* 56: 158-160.

Zugarramurdi, A., M.A. Parin and H.M. Lupin. 1995. Economic engineering applied to fishery industry. FAO Technical paper No. 351, Food and Agriculture Organization, Rome.

Recent Advances in Surimi Technology

Herbert O. Hultin

Department of Food Science, University of Massachusetts/Amherst
Gloucester Marine Station, Gloucester, Massachusetts 01930, U.S.A.

Introduction

Surimi is washed, minced fish muscle tissue to which has been added stabilizing agents called cryoprotectants and which is then frozen. As such, it is an intermediate in the production of several types of finished products, all of which are characterized as gels formed during the application of heat. The cooked gels take a wide variety of forms in Japan, where they have been a traditional food for centuries. Typical of these is kamaboko, which is served cold, but other products such as chikuwa, which is broiled, are also popular. The source of the fish muscle protein and the processing procedures determine the characteristics and thus the quality of the finished product.

Surimi

An outline of the traditional surimi process is given in Figure 1. The fish must first be pretreated to prepare the muscle tissue for further processing. The material used for the mincing process could be skinless fillets, fillets with the skin on, or even headed and eviscerated (gutted) fish. Tissues associated with the muscle such as skin, bones and scales, have to be removed from the product. Some components of these tissues, such as the heme protein hemoglobin, may not be so easily removed during the processing, and some may remain behind as a contaminating material. Heme proteins darken the color of the surimi and could initiate lipid oxidation, producing rancid, fishy odors (Richards and Hultin 2000). With small fish which are difficult to fillet, heading and gutting will give higher yields of product but generally at the price of lowering the quality. It is useful to head and gut fish before preparing fillets since this reduces contamination of the muscle tissue with the head and abdominal contents.

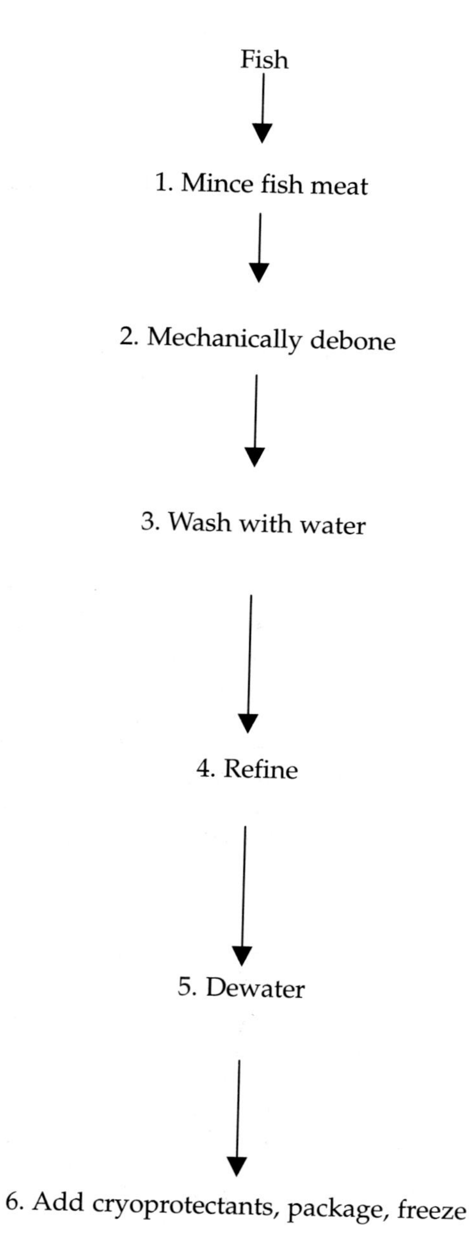

Figure 1. Traditional method of preparing surimi.

The material containing the fish muscle tissue is then minced (Step 1, Fig. 1). After mincing, the material is passed through a mechanical deboning machine which removes bone from the product (Step 2, Fig. 1). The machine used for deboning often includes a mincer so that both processes are done sequentially. A deboner works by forcing the tissue through small holes in a screen. Soft tissue passes through whereas particles of bone, skin and scales do not. The orifices in the screen can be very small to effectively remove most all the bone; under these conditions, however, the yield will be low since it is difficult to pass the muscle tissue through the small holes. Conversely, increasing the size of the holes produces better yields but allows some small bone fragments and other impurities to pass through.

The deboned, minced muscle tissue is then washed with water (Step 3, Fig. 1). The purpose of washing with water is to remove components that would adversely affect the ability of the proteins to form a gel when heated, as well as those components that would affect other quality attributes such as flavor and odor, stability to frozen storage, and color. Removal of the heme proteins will produce a whiter product, and removal of oil may delay the development of oxidative rancidity. The process of washing is also believed to concentrate the myofibrillar proteins which are thought to be primarily responsible for the ability of the muscle proteins to gel (Akahane *et al.* 1984, Lee 2000). It has been traditionally believed that removal of the soluble (sarcoplasmic) proteins prevents their interference with the gelation process although recent work has suggested that the soluble proteins improve gel characteristics (Morioka and Shimizu 1990, Ko and Huang 1995, Morioka *et al.* 1997).

In terms of muscle tissue used for food, the proteins are generally placed in three categories. First, there are the myofibrillar or contractile proteins which are responsible for the ability of muscle tissue to contract in the living animal. These proteins are soluble in concentrated salt solutions, i.e., greater than an ionic strength of 0.3 and usually around 0.6. There are also cytoskeletal proteins which are responsible for maintaining the overall structure in the muscle cell and are usually closely associated with the contractile elements. Some of these proteins are soluble in high concentrations of salt; some of them are not. The second group of proteins are those which are soluble in dilute salt and are extractable when the muscle tissue is treated with water or with salt solutions of a low ionic strength. These proteins are termed the sarcoplasmic proteins. The third group of proteins are called the stromal proteins and are not soluble in either high or low ionic strength salt solutions. The principal proteins associated with this group are the connective tissue proteins, but the proteins of the cellular membranes, denatured myofibrillar proteins, and some cytoskeletal proteins which are not extractable under the above conditions also contribute to the stromal proteins.

A major loss in protein occurs during the washing step where a considerable portion of the sarcoplasmic proteins are removed. The sarcoplasmic proteins account for as much as 25–30% of the total protein of the muscle tissue (Foegeding *et al.* 1996). There may be some removal of myofibrillar proteins with extended washing in water (Lin *et al.* 1995) since the myofibrillar and cytoskeletal proteins are soluble in water if the ionic strength is reduced sufficiently (Stefansson and Hultin 1994). The efficiency of the washing process in removing water-soluble components depends on a number of factors. These include the number of washes, the volume of water used per weight of minced fish in each wash, the time of contact between the wash water and the minced muscle tissue, the amount and type of agitation, and the temperature of the water which may affect denaturation of the proteins. A wide range of wash volumes and times are used in industry, ranging from one wash at a two-to-one water to mince ratio to five washes at a five-to-one water to mince ratio. Operations carried out on factory ships generally use small amounts of water due to the scarcity of water resources on board ship. Obviously, the greater the volume of water used and most especially the number of washes used will lead to removal of greater percentages of water-soluble materials. The washing process is inherently slow and represents a bottleneck in the process. The reason for this is that it takes time for the water-soluble components to diffuse out of the broken muscle cells. Excessive rupture of the tissue to remove this constraint on the process leads to excessive swelling of the washed tissue, making it extremely difficult to lower the water content to an acceptable level. The washing procedure is usually carried out in separate tanks which are filled with water and gently stirred to keep the minced tissue distributed throughout the water.

After a specified time, which can be anywhere from 9 to15 min, stirring is stopped, the particles are allowed to settle, and the water is decanted. This introduces another loss of proteins since small, insoluble particles will not settle out and are removed with the wash water. If these insoluble proteins are not recovered, there is a decrease in product yield. The temperature of the water should be low enough to prevent denaturation which varies according to species (Ohshima *et al.* 1993).

The washed minced muscle is then taken to a refiner (Step 4, Fig. 1). A refiner is a machine in which the washed mince is forced through small pores in a screen to further remove any bone pieces, skin, scales or connective tissue that may remain. This is usually done before the dewatering step since the high moisture content of the mince after the washes makes it easier to separate the undesirable material. The moisture content may be 90%, compared to the original moisture content of about 80%. This is a gain of water greater than 100% of the original. The swelling occurs because at the pH of the washed mince (about 6.4–7.0), the proteins have a net negative charge and repel each other. This repulsion allows for expansion and water is absorbed to minimize this repulsion by separating the proteins. The addition of a small amount of salt (0.1 to 0.3%) shields the negative charges, reduces the repulsive forces and allows the proteins to more easily approach one another, thus expelling water and reducing the tendency of the tissue to absorb water. Under pressure, excess water can be removed. In a factory, this dehydration is usually accomplished with a screw press and the moisture is lowered from 90% or more to around 82% (Step 5, Fig. 1).

Surimi production involves the use of a large volume of water. The organic contaminants in this water are relatively low in concentration, so it is critical to be able to efficiently remove these materials to allow discharge of the water. Ideally in future, processes will be developed that recycle more of the water by removal of organic and inorganic impurities by membrane processes.

The washed, refined, dehydrated minced fish muscle is usually frozen prior to its use at some future date. Fish proteins are susceptible to denaturation during frozen storage which will reduce their ability to form gels. To prevent this, various compounds may be added which protect the fish proteins from denaturation during frozen storage. These compounds are termed cryoprotectants. The most commonly used ones in the industry are a combination of sugars and polyphosphate salts. A typical cryoprotectant mixture consists of 4% sucrose, 4% sorbitol (a sugar alcohol), and 0.3% sodium tripolyphosphate (STPP). After thorough mixing the product is put in packages and frozen in blocks until used (Step 6, Fig. 1). This product is termed "surimi."

Gelation

The purpose of producing a washed, minced, stabilized, frozen fish protein product is to use it in the manufacture of gels. These gels take the form of a large variety of traditional products in Japan, or the relatively recent use of seafood analogs, such as imitation crab legs, in the Western world. In all of these cases, the ability to form strong elastic gels is paramount for their use. Although the original meaning of surimi was for the uncooked product, in the U.S. the cooked products are referred to as "surimi products." Whatever the definition of surimi, it would be inappropriate to talk of the production of the intermediate raw material without a consideration of the process of gelation which is typically achieved by heating.

To make a protein gel from the frozen, stabilized surimi, the material is first thawed after which salt is added, usually at the rate of 2–3%. It is then thoroughly chopped. It has been suggested that the salt is needed to solubilize the proteins to form the gel. There is not sufficient water at this stage to truly solubilize the proteins, but they form rather a sol, indicating that with higher amounts of water the protein would be solubilized. At this point, other ingredients can be added. In addition to materials that are used for flavor or color, components may be added that directly affect the physical character of the gels. Proteins such as egg white may be added to increase gel strength, and starches and hydrocolloids to improve water binding. Milk proteins, soy proteins and wheat gluten are sometimes added to modify functional characteristics and/or improve the economics of the process.

Two of the major characteristics of gels prepared from surimi are the gel strength and the water-holding capacity of the protein matrix. Gel strength has two components: a hardness or stress, and an elastic component or strain. Both torsion and gel penetration tests are carried out to evaluate gel quality. In the former, the gel is twisted until it breaks. The resistance to the twisting at breakage is measured (stress) as well as the distance the gel is turned before breakage (strain). In gel penetration, a plunger, often 5 mm in diameter, is forced into a gel and the force required for breakage and the distance the plunger moves before breakage are measured as indicators of gel hardness and gel elasticity, respectively. Gel strength is a product of the two.

The characteristics of the gel that is formed are highly dependent on the heating procedure. Different species respond in different ways to the same heat treatment. Some gels may be improved by holding the surimi at a temperature below 50°C for a period of time before the final heat treatment, often at around 90°C. With some species, however, this holding period does not improve the gelation characteristics of the material. Surimi prepared from some fish species will even gel at refrigerated temperatures, and it is a good idea to form the surimi into the shape desired as soon as possible after mixing with the salt and the other ingredients since the surimi gels are not reversible.

Loss of gel strength

In some species, there is a loss in gel quality if the temperature remains at 55–60 °C for too long a time during the heating process. The softening of the gel under these conditions is termed "modori" in Japanese and is probably caused by neutral proteases acting on the muscle proteins. In particular, the heavy chain of myosin is broken down into fragments of lower molecular mass. In some cases, the proteolytic activity is associated with a parasite, although it is not entirely clear whether the parasite produces the enzyme or whether it is the response of the fish tissue to invasion by the parasite. Proteolysis which causes gel softening can be reduced or prevented by the addition of inhibitors of these fish proteases (Kang and Lanier 2000). Among the more successful compounds that have been shown to act as inhibitors of proteolysis are beef plasma protein, egg white protein, whey proteins, and extracts of white potatoes. The various protease-inhibitors may perform differently depending on the nature of the specific enzyme causing the problem in a given surimi (Kinoshita *et al.* 1990). Protease inhibitors occur widely in nature to control proteolytic activities in living organisms, being found even in the sarcoplasmic protein fraction of fish muscle (Nomura *et al.* 1998).

Increasing Gel Strength

Chemical or enzymic processes may be used to improve the gelation characteristics of heat-gelled surimi. Oxidation of sulfhydryl groups to the corresponding disulfides may be initiated with sodium ascorbate and has been suggested to increase gel strength (Nishimura *et al.* 1992). Likewise, the enzyme transglutaminsase can increase gel strength by cross linking myofibrillar proteins. This is a widely occurring enzyme that catalyzes post translational cross linking by transferring the acyl group of a glutamine residue to a primary amino group of a lysyl residue. The commercial use of this enzyme was dependent on obtaining a microbial source which did not require calcium ion for activity (Ando *et al.* 1989). This made it economically feasible and relatively straightforward to incorporate into the process although the specificities of the tissue and microbial enzymes may be different (Nakahara *et al.* 1999).

Gelation by Non-Thermal Treatment

The muscle proteins of some fish species will gel at room temperature or below. This phenomenon may be related in part to the action of enzymes such as transglutaminase normally present in the muscle tissue (Gilleland *et al.* 1997). Gelation may also be induced with the use of high hydrostatic pressures (Ohshima *et al.* 1993). It is believed that the high pressure affects the conformation of the muscle proteins, exposing more groups that are capable of interacting with other groups, e.g., sulfhydryl groups and/or hydrophobic regions. In general, use of high hydrostatic pressure at low temperatures produces gels that are softer and more elastic than gels obtained by heating at atmospheric pressure (Fernandez-Martin *et al.* 1998).

Improved Recovery

One of the causes for unrecovered protein in normal surimi processing is the loss of insoluble proteins when the material is dewatered by straining and/or pressing. Increases in recovery have been achieved by the use of decanter centrifuges to remove the insoluble proteins from the wash water. In an extension of that idea, the washing is done in the centrifuge itself (Babbitt and Stevens 1998). The minced fish and wash water travel the length of the decanter centrifuge which gives time for the wash water to extract components from the broken muscle cells. The insoluble proteins are then efficiently separated from the aqueous phase by centrifugal force.

New Process Based on Solubilization of Myofibrillar/Cytoskeletal Proteins

The standard process for producing surimi involves the separation of soluble impurities by washing, whereas insoluble impurities like skin and bone have to be removed by techniques that separate one insoluble fraction from another insoluble fraction, e.g., by forcing the soft tissue through small orifices while retaining the more solid impurities. This is done in both the deboning and refining steps.

A new procedure has been developed based on selective solubilization and precipitation of the various proteins of the muscle tissue (Hultin and Kelleher 1999, 2000a,b). In this process, the muscle is first minced (Step 1, Fig. 2), homogenized (Step 2, Fig. 2), and then the myofibrillar and cytoskeletal proteins are dissolved (Step 3, Fig. 2) in either acid (pH 2.5-3.5) or alkali (pH 10.8-11.5). The sarcoplasmic proteins remain soluble under these conditions so that the only proteins which are not solubilized are a majority of the connective tissue proteins, membrane proteins, and perhaps some severely denatured myofibrillar/cytoskeletal proteins. With fresh muscle tissue which has not been exposed to a low pH for a long period of time, solubilities of greater than 95% of the total proteins can be routinely achieved. The proteins become soluble at pH values somewhat higher on the acid side and somewhat lower on the basic side, but the pH values cited decrease the viscosity of the protein solutions. When the viscosity is lowered to <50 mPa · s, insoluble materials like skin and bones, and, under the proper conditions even the cellular membranes can be separated by centrifugation from the soluble proteins (Step 4, Fig. 2). This is important since the cellular membranes are the site of much of the lipid oxidation reactions in the muscle tissue; thus, stability of the product can be expected to be improved if these membranes are removed.

If a fatty fish is used, the oil or fat will accumulate on the surface of the aqueous phase and can be removed either by centrifugation or by skimming. The pH of the aqueous supernatant phase is then adjusted to its isoelectric point (between pH 5.2–6.0) (Step 5, Fig. 2). This causes the precipitation of the proteins which can be collected by centrifugation in a decanter centrifuge or by other means such as filtration (Step 6, Fig. 2). The total yield of the collected proteins can be high, greater than 90%. This means that much of the soluble or sarcoplasmic proteins are also recovered, thus increasing the yield greatly over what is achieved in normal surimi processing. It had been earlier demonstrated that the proteins in surimi wash water could be recovered by selective adjustment of pH (Ninomiya *et al.* 1990). In this new

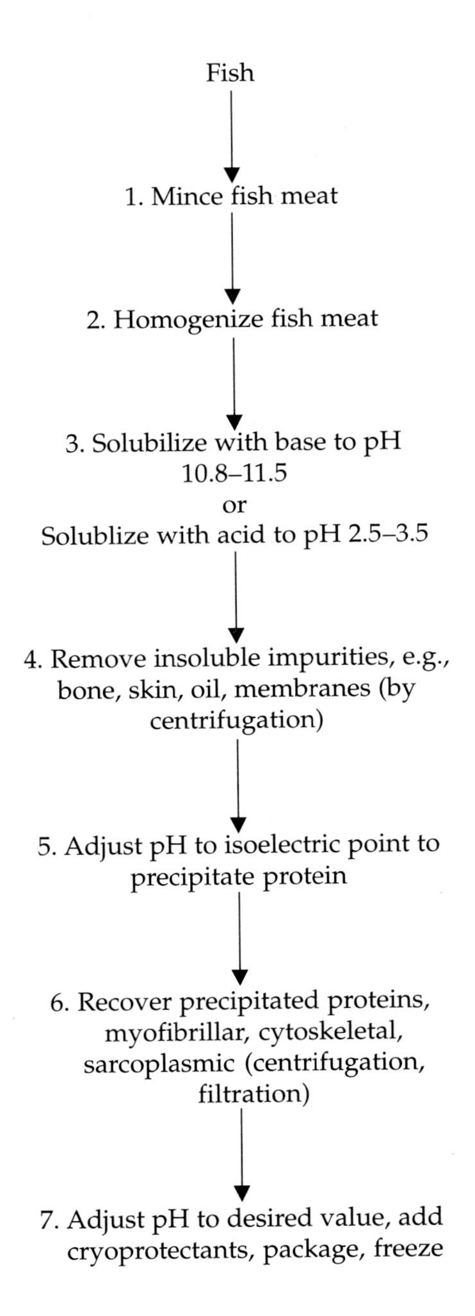

Figure 2. Solubilization process for preparing surimi.

process, it is not known how much of the sarcoplasmic proteins actually become insoluble because of the pH adjustments and how much co-precipitate with the insoluble myofibrillar/cytoskeletal protein fraction at that pH.

The supernatant fraction from this step contains water-soluble impurities such as low molecular weight compounds like glycolytic intermediates, TMAO, nucleotides, salts, and some proteins. The protein content is low. Thus, waste water treatment is minimized and the possibility for re-use and recycling of the water improved. Since the water in the protein isolate contains the same concentration of impurities as the wash water, the protein isolate can be washed further to remove most of the remaining impurities.

The pH of the isolated protein can now be adjusted to any desirable pH, e.g., 7.0-7.4, mixed with cryoprotectants, and frozen (Step 7, Fig. 2) the same as with surimi produced by the traditional process. This new process has several advantages over the standard process. The high yields obtained have already been mentioned. The high solubility of the proteins in either the acid or alkaline condition, along with their almost complete precipitation at the isolelectric point, leads to yields much higher than are achieved by the standard process. In addition, since the process liquefies the material of interest, it can be run on a continuous basis and the moving of product around the processing plant is easier than when dealing with a solid material. The lipid content can be greatly lowered and can include removal of a good part of the membrane lipids. At the same time, many lipid-soluble toxins can also be removed. The chemicals needed in the process are all GRAS (generally recognized as safe by the U.S. FDA). They include hydrochloric acid and/or sodium hydroxide or sodium carbonate; the net result of these when the final sample is brought to neutral pH is that there is some small amount of sodium chloride or sodium carbonate in the product.

The ability to selectively solubilize or precipitate the desirable proteins and remove the undesirable components as insoluble or soluble fractions opens the possibility of using the process in situations where it was not feasible to do so in the past, i.e., with small fish with a lot of skin and small bones. Undoubtedly our ability to process these underutilized resources will continue to improve, and they will become an important raw material for surimi fish protein isolates.

References

Akahane, T., S. Chihara, Y. Yoshida, T. Tsuchiya, S. Noguchi, H. Ookami, H., and J.J. Matsumoto. 1984. Roles of constituent proteins in gel properties of cooked meat gels. *Bull. Jap. Soc. Sci. Fish.* 50: 1029-1033.

Ando, H., M. Adachi, K. Umeda, A. Matsura, M. Noraka, R. Uchio, H.Tanaka, and M. Motoki. 1989. Purification and characteristics of a novel transglutaminase derived from microorganisms. *Agric. Biol. Chem.* 53: 2613-2617.

Babbitt, J.K and W.A. Stevens. 1998. Demonstration of a new decanter surimi process. *Final Report of ASTF, Grant No. 96-1-015,* Alaska Science & Technology Foundation, Anchorage, AK.

Fernandez-Martin, F., M. Perez-Mateos, and P. Montero. 1998. Effect of pressure/heat combinations on blue whiting (*Micromesistius poutassou*) washed mince: thermal and mechanical properties. *J. Agric. Food Chem.* 46: 3257-3264.

Foegeding, E.A., T.C. Lanier, and H.O. Hultin. 1996. Characteristics of edible muscle tissue. Pages 879–931 in *Food Chemistry,* O.R. Fennema., ed., Marcel Dekker, Inc., New York.

Gilleland, G.M., T.C. Lanier, and D.D. Hamann. 1997. Covalent bonding in pressure-induced fish protein gels. *J. Food Sci.* 62: 713-716.

Hultin, H.O. and S.D. Kelleher. Protein composition isolated from a muscle source. U.S. Patent No. 6,005,073, December 21, 1999.

Hultin, H.O. and S.D. Kelleher. High efficiency alkaline protein extraction. U.S. Patent No. 6136959, 24 October 2000a.

Hultin, H.O. and S.D. Kelleher. 2000b. Surimi processing from dark muscle fish. Pages 59–77 in *Surimi and Surimi Seafood,* J.W. Park, ed., Marcel Dekker, New York.

Kang, I.-S. and T.C. Lanier. 2000. Heat-induced softening of surimi gels by proteinases. Pages 445–474 in *Surimi and Surimi Seafood,* J.W. Park, ed., Marcel Dekker, New York.

Kinoshita, M., H. Toyohara, and Y. Shimizu. 1990. Diverse distribution of four distinct types of Modori (gel degradation)-inducing proteinases among fish species. *Nippon Suisan Gakkaishi* 56: 1485-1492.

Ko, W.-C. and M.-S. Hwang. 1995. Contribution of milkfish sarcoplasmic protein to the thermal gelation of myofibrillar protein. *Fisheries Sci.* 61: 75-78.

Lee, C.M. 2000. Surimi: science and technology. Pages 2229-2239 in *Wiley Encyclopedia of Food Science and Technology,* F.J. Francis, ed., John Wiley and Sons, Inc., New York.

Lin, T.M., J.W Park, and M.T. Morrissey. 1995. Recovered protein and reconditioned water from surimi processing waste. *J. Food Sci.* 60: 4–9.

Morioka, K., T. Nishimura, A. Obatake, and Y. Shimizu. 1997. Relationship between the myofibrillar protein gel strengthening effect and the composition of sarcoplasmic proteins from Pacific mackerel. *Fisheries Sci.* 63: 111-114.

Morioka, K. and Y. Shimizu. 1990. Contribution of sarcoplasmic proteins to gel formation of fish meat. *Nippon Suisan Gakkaishi* 56: 929-933.

Nakahara, C., H. Nozawa, and N. Seki. 1999. A comparison of cross-linking of fish myofibrillar proteins by endogenous and microbial transglutaminases. *Fisheries Sci.* 65: 138-144.

Ninomiya, K., T. Ookawa, T. Tsuchiya, and J. Matsumoto. 1990. Concentration of fish water soluble protein and its gelation properties. *Nippon Suisan Gakkaishi* 56: 1641-1645.

Nishimura, K., N. Ohishi, Y. Tanaka, M. Ohgita, Y. Takeuchi, H. Watanabe, A. Gejima, and E. Samejima. 1992. Effects of ascorbic acid on the formation process for a heat-induced gel of fish meat (Kamaboko). *Biosci. Biotech. Biochem.* 56: 1737-1743.

Nomura, A., Y. Itoh, Y. Osaka, Y. Kitamura, H. Miyazaki, and A. Obatake. 1998. A factor inhibiting myosin heavy chain degradation of washed meat paste around 40°C, in the sarcoplasmic protein of fish meat. *Nippon Suisan Gakkaishi* 64: 878-884.

Ohshima, T., H. Ushio, and C. Koizumi. 1993. High-pressure processing of fish and fish products. *Trends Food Sci. Technol.* 4: 370-375.

Richards, M.P. and H.O. Hultin. 2000. Effect of pH on lipid oxidation using trout hemolysate as a catalyst: a possible role for deoxyhemoglobin. *J. Agric. Food Chem.* 48: 3141-3147.

Stefansson, G. and H.O. Hultin. 1994. On the solubility of cod muscle proteins in water. *J. Agric. Food Chem.* 42: 2656-2664.

DNA-based Diagnostics in Sea Farming

Carlos R. Osorio and Alicia E. Toranzo

Departamento de Microbiología y Parasitología, Facultad de Biología, Universidad de Santiago de Compostela, 15782 Santiago de Compostela, Spain.

Introduction

Aquaculture has developed considerably during the last decades, and nowadays represents nearly 20% of the total fishing captures worldwide. In this scenario, fresh water aquaculture still amounts to more than 50% of the fish-farming production. In marine aquaculture, fish production represents a minor percentage compared with shellfish, but its commercial value and the expensive investments demanded by fish sea-farming clearly enhance scientific attention for diseases affecting marine fish production.

Microbial pathogens are of the greatest importance in terms of their impact on fish and shellfish populations, and in their economical consequences. They include viruses, bacteria, fungi, protozoans, and parasites, and they can kill or debilitate the host, as well as reduce their commercial value as food for humans.

A large body of literature exists on marine fish and shellfish diseases, with many books and reviews available to the reader. It would be too pretentious for this review to cover all topics on sea farming diseases, and for that reason, we will focus our attention on bacteria. With no doubt, the economical importance of many bacterial fish pathogens has enhanced research in this area. Though it was developed much later than human clinical microbiology, the study of the bacterial diseases of marine animals is nowadays an important field of study, and literature on this subject has increased enormously in the past three decades.

Moreover, certain bacterial species that occur in fish and seafood in general, are not only pathogenic for marine animals but some of them may also be of concern for human health. This is a particularly important problem in ready-to-eat seafood products that are ingested with no previous cooking steps.

One of the most critical steps in the study and control of bacterial fish pathogens is the correct identification of the infectious agent. In certain cases, bacterial infections in fish may take place as subclinical infections of asymptomatic pathogen carriers. Therefore, diagnosis of fish disease lies in the specific detection of the pathogens, using laboratory methods.

Traditional isolation and characterization schemes for analyzing bacterial pathogens of fish include direct sampling of fish tissues with a sterile loop and streaking onto general, selective, and/or differential agar media. It may also include homogenization of the sample in buffer and inoculation of a liquid medium, attempting to achieve an enrichment of the species of interest, and then streak a broth sample onto agar media. After several hours of incubation (overnight or longer), isolated colonies are subjected to further analysis. These may include a battery of conventional tube and plate tests, commercial multigallery systems (API), and serological agglutination tests. In fact, a fast and commonly used identification system consists in testing a collection of polyclonal antisera on bacterial colonies.

Although phenotypic methods are adequate for the identification of many bacteria associated with problems in sea-farming, they can also be of limited utility in certain cases. Both the biochemical and serological approaches rely compulsorily on the possibility that the bacterial species under consideration grows optimally on a synthetic culture medium. And, even when this is feasible, it takes additional time for bacteria to grow once samples arrive in the laboratory. This means that, by the time the pathogen has been identified, it may be too late for the fishfarmer to bring the outbreak under control. Similarly, international trade in aquatic animals and their products is increasing and intensifying in importance, which means that control measures have to be developed in order to avoid spreading pathogenic bacteria between different countries, areas, or even the propagation of a bacterial infection between fishfarms. Appropriate and fast responses to human public health problems arising after consumption of seafood contaminated with some pathogenic marine bacteria are often hampered by the long periods of time required to identify the causative agents. Moreover, if seafood samples are not processed quickly, bacteria may be inactivated and so they could not be recovered following standard culture-dependent microbiological methods.

During recent years, molecular techniques have been found to be a powerful tool for the diagnosis of bacterial pathogens, including those that infect fish. In fact, protocols for the DNA-based detection of many bacterial fish pathogens have been published during the last two decades. The availability of fast, sensitive and specific diagnosis methods for bacteria is of crucial importance for the sea-farming industry. Confirmation of the etiology of a bacterial disease provides information to apply control measures. In addition, correct identification of any bacterial isolate, as well as the possibility of detecting the presence of a bacterial pathogen in a sample, even without prior culturing of the bacteria, is crucial for epidemiological studies. In this sense, the development of DNA fingerprinting techniques that allow the molecular discrimination of strains within a species, may help to trace an outbreak back to the source of infection.

Molecular diagnosis protocols based on the polymerase chain reaction and DNA probes, when properly designed, provide those tools needed for effective bacterial diagnosis in sea-farming.

The main aim of this chapter is to review different DNA probe- and PCR-based methods that have been developed by different authors to test the presence of pathogenic bacteria affecting farmed fish. In addition, a few marine bacterial species which may affect humans after consumption of sea-farming products, such as *Vibrio cholerae* and *Vibrio parahaemolyticus*, will also be considered in this review, though they are rarely, if ever, isolated as primary fish pathogens. Special attention will be placed on those bacterial pathogens that cause high economic losses and for which more attention has been paid in the literature (see Tables 1 and 2), including gram-negative bacilli, and *Renibacterium salmoninarum* a gram-positive bacillus. Gram-positive cocci species with importance in sea farming have been reviewed by Romalde and Toranzo (2002).

Table 1. Main bacterial diseases affecting sea-farming, reviewed in this chapter

Disease	Aetiological agent	Host
Vibriosis	*Vibrio anguillarum*	Salmonids Turbot Sea bass Eels Ayu Cod
	Photobacterium damselae subsp. *damselae*	Turbot Sea bream Sea bass Yellowtail
	V. vulnificus	Eels
	V. salmonicida	Salmonids
Pasteurellosis	*Photobacterium damselae* subsp. *piscicida*	Sea bream Sea bass Yellowtail
Marine flexibacteriosis	*Flexibacter maritimus*	Turbot Salmonids
Rickettsial septicemia of salmonids	*Piscirickettsia salmonis*	Salmonids
Furunculosis	*Aeromonas salmonicida* subsp. *salmonicida*	Salmonids Turbot
Bacterial kidney disease (BKD)	*Renibacterium salmoninarum*	Salmonids

Table 2. Reports of the application of PCR-based methods for detection of different bacterial pathogens of importance for the sea-farming industry

Species	Target gene	Sample	Reference
Vibrio anguillarum	*rpoN*	Turbot tissues	González *et al.* 2000
Vibrio vulnificus	Cytolysin	Oyster meat	Hill *et al.* 1991
	Cytolysin	Culturable and non-culturable cells	Brauns *et al.* 1991
	23S rDNA	Eel tissue, tank water and sediment	Arias *et al.* 1995
	Cytolysin	Oyster and eel tissue	Coleman *et al.* 1996
	Cytolysin	Mixed cultures	Aono *et al.* 1997
	Cytolysin	Clinical samples	Lee *et al.* 1998
	Cytolysin	Seafood samples	Lee *et al.* 1999
Photobacterium damselae subsp. *piscicida*	DNA	Yellowtail tissues	Aoki *et al.* 1995
	Plasmid sequences	Yellowtail kidney	Aoki *et al.* 1997
	16S rDNA	Pure and mixed cultures Spiked trout kidney	Osorio *et al.* 1999
	16S rDNA and *ureC*	Pure culture	Osorio *et al.* 2000a
Photobacterium damselae subsp. *damselae*	16S rDNA	Pure and mixed cultures Spiked trout kidney	Osorio *et al.* 1999
	16S rDNA and *ureC*	Pure culture	Osorio *et al.* 2000a
	dly	Pure culture	Osorio *et al.* 2000b
Aeromonas salmonicida	Plasmid sequences	Pure culture	Hiney *et al.* 1992
	vapA	Salmon tissue, water	Gustafson *et al.* 1992
	Plasmid sequences	Hatchery effluent, feces	O'Brien *et al.* 1994
	Plasmid sequences	Salmon blood	Mooney *et al.* 1995
	Library clone	Salmon kidney, pure cultures	Miyata *et al.* 1996
	16S rDNA and plasmid sequences	Salmon kidney and gills	Høie *et al.* 1997
Flexibacter maritimus	16S rDNA	Pure culture	Toyama *et al.* 1996
	16S rDNA	Pure cultures	Bader *et al.* 1998
	16S rDNA	Fish tissues	Cepeda *et al.* 2001
Renibacterium salmoninarum	DNA	Salmon kidney	León *et al.* 1994
	p57	Salmonid eggs	Brown *et al.* 1994
	p57	Pure culture	Brown *et al.* 1995
	16S rRNA	Salmon ovarian fluid	Magnússon *et al.* 1994
	p57	Trout lymphocytes	McIntosh *et al.* 1996
	p57	Salmon ovarian fluid and kidney	Miriam *et al.* 1997
	p57	Salmonid kidney	Chase and Pascho 1998
	p57	Salmon ovarian fluid	Cook and Lynch 1999
Piscirickettsia salmonis	16S rDNA	Fish tissues	Mauel *et al.* 1996
	ITSs and 23S rDNA	Fish serum and tissue	Marshall *et al.* 1998
Vibrio cholerae	*ctxA*	Pure culture	Fields *et al.* 1992
	ctxAB	Food samples	Koch *et al.* 1993
	rfb	Pure culture	Falklind *et al.* 1996
	rfb	Stool samples	Albert *et al.* 1997

(Cont.)

	ctx	Seeded shellfish	Brasher *et al.* 1998
	ISR (Intergenic Spacers)	Pure culture	Chun *et al.* 1999
Vibrio parahaemolyticus	Cloned DNA fragment	Pure cultures and seeded oysters	Lee *et al.* 1995
	gyrB	Shrimp	Venkateswaran *et al.* 1998
	tl	Seeded shellfish	Brasher *et al.* 1998
	toxR	Pure cultures	Kim *et al.* 1999
	tl, tdh, trh	Pure cultures and seeded oysters	Bej *et al.* 1999

How different authors chose different genes as amplification and/or hybridization targets, the methods employed for DNA preparation, as well as the application of the procedures to field samples and the sensitivities of the different approaches will be described. A summary of the PCR primers described in this review is presented in Table 3.

A Summary of the Application of Molecular Tools

The field of microbial genomics, though still in its first half-decade of life, has elucidated so far the complete genome sequence of as many as thirty species of prokaryotes, and another 100 sequencing projects are under way. One of the landmarks in this regard, and particularly in marine microbiology, is the recent determination of the complete nucleotide sequence of *Vibrio cholerae* (Heidelberg *et al.* 2000).

The introduction of automated fluorescent sequencing methods in 1986 (Ansorge *et al.* 1986, Smith *et al.* 1986) has had a major impact on molecular diagnosis in microbiology, since the use of DNA probes and the development of PCR-based detection protocols are intimately associated with the progress in DNA sequencing. An increasing knowledge of the nucleotide sequence of many genes in a high number of bacterial species, and particularly the determination of the consensus 16S rRNA gene sequences of virtually all bacterial species described to date, has allowed the design of a large number of gene probes, and the selection of countless PCR-primer combinations that allow the characterization of bacterial isolates at the level of specific genera, species, subspecies, and even strains. Determination of the nucleotide sequence of virulence genes has also made possible the direct molecular detection of particularly virulent isolates without the need of carrying out animal bioassays.

In this scenario, it plays a crucial role in the availability of public DNA sequence databases, as EMBL database (European Molecular Biology Laboratory), GENBANK database, or RDP database (Ribosomal Database Project), among others. These public databases offer the possibility of obtaining virtually any known DNA or protein sequence that has been previously deposited by laboratories from all around the world.

Table3 : Primer pairs for PCR-based detection of bacteria reviewed in this chapter.

Organism	Target sequence	Primer name	Nucleotide sequence (5'-3')	Amplicon size (bp)	Reference
Vibrio anguillarum	*rpoN*	RponAng-5' RponAng-3'	5'-GTTCATAGCATCAATGAGGAG-3' 5'-GAGCAGATCAATATGTTGGATC-3'	519	González *et al.* 2000
Vibrio vulnificus	*vvhA*	VVp1 VVp2	5'-CCGGCGGTACAGGTTGGCGC-3' 5'-CGCCACCCACTTTCGGGCC-3'	519	Hill *et al.* 1991
	vvhA	Vv oligo 1 Vv oligo 3	5'-CGCCGCTCACTGGGGCAGTGGCTG-3' 5'-GTTGGTTGACGAGCCCGCAGAGCCG-3'	388	Brauns *et al.* 1991
	23S rDNA	Dvu9V Dvu45R	5'-CCACTGGCATAAGCCAG-3' 5'-CTACCCAATGTTCATAGAA-3'	978	Arias *et al.* 1995
	vvhA	primer 1 primer 2	5'-CGCCGCTCACTGGGGCAGTGGCTG-3' 5'-GCGGGTGGTTCGGTTAACGGCTGG-3'	344	Coleman *et al.* 1996
	vvhA	P1 P2	5'-GACTATCGCATCAACAACCG-3' 5'-AGGTAGCGAGTATTACTGCC-3'	704	Lee *et al.* 1998
	vvhA	P3 P4	5'-GCTATTTCACCGCCGCTCAC-3' 5'-CCGCAGAGCCGTAAACCGAA-3'	222	Lee *et al.* 1998
Photobacterium damselae subsp. *piscicida*	library clone	forward primer reverse primer	5′-GTAGCTCTTGTGGAGTAATGCT-3′ 5′-CATTCGTAGTGCTTACTGCCCA-3′	629	Aoki *et al.* 1995
	pZP1 plasmid	PZP1-1a PZP1-1b	5′-GCCCCCATTCCAGTCACACA-3′ 5′-TCCCTAAGCACACCGACAGG-3′	484	Aoki *et al.* 1997
	pZP1 plasmid	PZP1-4a PZP1-4b	5′-CTACGTAGCAAAAGGTGTTCC-3′ 5′-AAGGAGAGAAAGACGGGTTG-3′	321	Aoki *et al.* 1997
	16S rDNA	Car-1 Car-2	5′-GCTTGAAGAGATTCGAGT-3′ 5′CACCTCGCGGTCTTGCTG-3′	267	Osorio *et al.* 1999

(Cont.)

Photobacterium damselae subsp. *piscicida*	16S rDNA	Car-1 Nestcar-1	5′-GCTTAGAAGAGATTCGAGT-3′ 5′-GGTCTTGCTGCCCTCTG-3′	259	Osorio *et al.* 1999
Photobacterium damselae subsp. *damselae*	16S rDNA	Car-1 Car-2	5′-GCTTGAAGAGATTCGAGT-3′ 5′-CACCTCGCGGTCTTGCTG-3′	267	Osorio *et al. 1999*
	16S rDNA	Car-1 Nestcar-1	5′-GCTTGAAGAGATTCGAGT-3′ 5′-GGTCTTGCTGCCCTCTG-3′	259	Osorio *et al.* 1999
	UreC	ure-5′ ure-3′	5′-TCCGGAATAGGTAAGCGGG-3′ 5′-CTTGAATATCCATCTCATCTGC-3′	448	Osorio *et al.* 2000a
	dly	dly-5′ dly-3′	5′-CCTATGGGACATGAATGG-3′ 5′-GCTCTAGGCTAAATGAATC-3′	567	Osorio *et al.* 2000b
Aeromonas salmonicida	6.4 Kb plasmid	PAAS 1 PASS 2	5′-CGTTGGATATGGCTCTTCCT-3′ 5′-CTCAAAACGGCTGCGTACCA-3′	423	Hiney *et al.* 1992
	VapA	AP-1 AP-2	5′-GGCTGATCTCTTCATCCTCACCC-3′ 5′-CAGAGTGAAATCTACCAGCGGTGC-3′	421	Gustafson *et al.* 1992
	6.4 Kb plasmid	PASS 4 pASS 5	5′-AGGTAAGTCTATTAGGTTCG-3′ 5′-GTTACACTTTTTCCTTCCGC-3′	278	Mooney *et al.* 1992
	Asal-3 clone	primer1 primer2	5′-AGCCTCCACGCGCTCACAGC-3′ 5′-AAGAGGCCCCATAGTGTGGG-3′	512	Miyata *et al.* 1996
	16 rDNA	forward primer reverse primer	5′-GGCCTTTCGCGATTGGATGA-3′ 5′-TCACAGTTGACACGTATTAGGCGC-3′	271	Hoie *et al.* 1997
Flexibacter maritimus	16r DNA	MAR1 MAR2	5′-AATGGCATCGTTTTAAA-3′ 5′-CGCTCTCTGTTGCCAGA-3′	1088	Toyama *et al.* 1996
	16r DNA	Mar1 Mar2	5′-TGTAGCCTTGCTACAGATGA-3′ 5′-AAATACCTACTCGTAGGTACG-3′	400	Bader *et al.* 1998

(Cont.)

Renibacterium salmoninarum	Cloned DNA fragment	forward primer reverse primer	5′-GATCGTGAAATACATCAAGG-3′ 3′-GGATCGTGTTTTATCCACCC-3′	149	León *et al.* 1994
	p57 gene	RS1 RS2	5′-CAAGGTGAAGGGAATTCTTCCACT-3′ 5′-GACGGCAATGTCCGTTCCCGGTTT-3′	501	Brown *et al.* 1994
	16S rRNA	F176 R492	5′-TGGATACGACCTATCACCGCA-3′ 5′-GCAAGTACCCTCAACAACCACA-3′	316	Magnússon *et al.* 1994
	p57 gene	G6480 G6481	5′-GCGCGGATCCTTGGCAGGACCATC TTTGT-3′ 5′-GCGCGGATCCAAAATAAAAAAAA TTTTAGCGCTG-3′	376	McIntosh *et al.* 1996
	p57 gene	FL7 RL11	5′-CGCAGGAGGACCAGTTGCAG-3′ 5′-GGAGACTTGCGATGCGCCGA-3′	349	Miriam *et al.* 1997
	p57 gene	FL7 RL5	5′-CGCAGGAGGACCAGTTGCAG-3′ 5′-TCCGTTCCCGGTTTGTCTCC-3′	372	Miriam *et al.* 1997
	P57 gene	FL10 RL11	5′-GGTGTAACGATAATGCGCCA-3′ 5′-GGAGACTTGCGATGCGCCGA-3′	149	Miriam *et al.* 1997
	P57 gene	LP3 UP1	5′-TTACCCGATCCAGTTCCC-3′ 5′-ATGTCGCAAGGTGAAGGG-3′	1356	Cook and Lynch 1999
Piscirickettsia salmonis	16 rDNA	PS2S PS2A2	5′-CTAGGAGATGAGCCCGCGTTG-3′ 5′-GCTACACCTGCGAAACCACTT-3′	467	Mauel *et al.* 1996
	3S rDNA and ITS	RTS1 RTS2	5′-TGATTTTATTGTTTAGTGAGAATGA-3′ 5′-AAATAACCCTAAATTAATCAAGGA-3′	91	Marshall *et al.* 1998
	23S rDNA and ITS	RTS1 RTS4	5′-TGATTTTATTGTTTAGTGAGAATGA-3′ 5′-ATGCACTTATTCACTTGATCATA-3′	283	Marshall *et al.* 1998

(Cont.)

Vibrio cholerae	*ctxA*	CTX2 CTX3	5′-CGGGCAGATTCTAGACCTCCTG-3′ 5′-CGATGATCTTGGAGCATTCCCAC-3′	564	Fields *et al.* 1992
	ctxAB	P1 P3	5′-TGAAATAAAGCAGTCAGGTG-3′ 5′-GGTATTCTGCACACAAATCAG-3′	777	Koch *et al.* 1993
	putative *rfb* genes	O139-1 O139-2	5′-GCGTTATAGGTATCATCAAGAGA-3′ 5′-GTCATTATTAAAACTGCTCCATT-3′	419	Falklind *et al.* 1996
	putative *rfb* genes	O139-3 O139-4	5′-TCGAATTTTCAAAATATACACTT-3′ 5′-CAAACATCTTACAATAGAGTAGT-3′	528	Falklind *et al.* 1996
	ISRs	prVC-F PrVCM-R	5′-TTAAGCSTTTTCRCTGAGAATG-3′ 5′-AGTCACTTAACCATACAACCCG-3′	295-310	Chun *et al.* 1999
Vibrio parahaemolyticus	pR72H clone	VP33 VP32	5′-TGCGAATTCGATAGGGTGTTAACC-3′ 5′-CGAACCTTGAACATACGCAGC-3′	387	Lee *et al.* 1995
	gyrB	VP-1 VP-2r	5′-CGGCGTGGGTGTTTCGGTAGT-3′ 5′-TCCGCTTCGCGCTCATCAATA-3′	285	Venkateswaran *et al.* 1998
	tl	L-TL R-TL	5′-AAAGCGGATTATGCAGAAGCACTG-3′ 5′-GCTACTTTCTAGCATTTTCTCTGC-3′	450	Brasher *et al.* 1998
	toxR	forward primer reverse primer	5′-GTCTTCTGACGCAATCGTTG-3′ 5′-ATACGAGTGGTTGCTGTCATG-3′	349	Kim *et al.* 1999
	tl	L-t1 R-tl	5′-AAAGCGGATTATGCAGAAGCACTG-3′ 5′-GCTACTTTCTAGCATTTTCTCTGC-3′	450	Bej *et al.* 1999
	tdh	L-tdh R-tdh	5′-GTAAAGGTCTCTGACTTTTGGAC-3′ 5′-TGGAATAGAACCTTCATCTTCACC-3′	269	Bej *et al.* 1999
	trh	L-trh R-trh	5′-TTGGCTTCGATATTTTCAGTATCT-3′ 5′-CATAACAAACATATGCCCATTTCCG-3′	500	Bej *et al.* 1999

General DNA fingerprinting techniques

When we talk about molecular tools applied to the diagnosis of bacterial pathogens, we generally associate this idea with nucleic acid hybridization (DNA probe technology), and the polymerase chain reaction (PCR). Of course, these two are the most important and generally used techniques that allow the detection of DNA in a sample. However, additional techniques are employed, together with or apart from PCR and/or DNA hybridization. Some of these molecular procedures are currently employed for the analysis of the genetic variability among strains of a given bacterial species, as well as for typing, rather than for detection or diagnosis as, for example, the study of the restriction fragment length polymorphisms (RFLP) of the genome (Giovannetti and Ventura 1995). However, total restriction patterns of bacterial genomic DNA are often too difficult to interpret due to the complexity of the band pattern that is generated. This problem can be overcome with the use of restriction enzymes that cut the DNA very rarely, and thus generate a small number of fragments, that can be then resolved by pulse field gel electrophoresis (PFGE) (Hillier and Davidson 1995). Another solution is to carry out the RFLP analysis on a small piece of DNA that has been previously amplified by PCR. There is also another possibility, consisting of reducing the band pattern complexity using probe hybridization. Though the source of probes can be very varied, the approach that has become more popular is the use of rRNA gene sequences as probes (Grimont and Grimont 1986, 1995). Since rRNA genes contain some highly conserved regions that can hybridize to rRNA genes of any bacterial species, it is possible to type bacteria by studying their rRNA gene restriction patterns. This technique is known as ribotyping, and it has been applied to the typing and characterization of many bacteria, including bacterial fish pathogens (Magariños *et al.* 1997, Arias *et al.* 1997).

Polymerase Chain Reaction (PCR) and its variants: multiplex PCR and nested PCR

The Polymerase Chain Reaction (PCR) is an *in vitro* technique that makes possible the amplification of a specific DNA segment by using a pair of specific oligonucleotides that serve as primers, and which delimitate a section of DNA to be exponentially copied. It is a very sensitive technique, being possible to begin with only one copy of the DNA target, and produce as many as one billion copies within 3 h.

A number of books and reviews have been written which describe many aspects of this technique, as well as its application to bacteria and whose reading is recommended if more information on PCR applications is required (Erlich 1989, Innis *et al.* 1990, Hill 1996, Persing 1996).

One variant of PCR is the so-called "multiplex PCR". A multiplex PCR reaction provides the possibility of detecting simultaneously the presence of several genes and/or organisms, using multiple primer sets. One possibility is to employ three primers targeted to the same gene (two forward primers and one reverse, or two reverse and one forward), so that two different products that share one of their ends are amplified, thus providing confirmation of the reaction specificity.

For example, Trost *et al.* (1993) designed three sets of primers that amplify DNA stretches of the enterotoxin, the thermostable direct hemolysin and the cytolysin genes of *Vibrio cholerae*, *V. parahaemolyticus* and *V. vulnificus* respectively. Such primer sets have to meet the requirements for simultaneous detection of the three species, i.e., primers have to function under the same PCR conditions (similar Tm values, similar Cl_2Mg optimal concentration), and of course they have to generate PCR products of sufficiently different lengths, so that they can be well resolved by agarose gel electrophoresis.

Amplification products generated in a PCR reaction can be used as templates for a second amplification in which an internal sequence from the products of the first amplification is then amplified. This variant is known as "nested PCR". Normally, after the first amplification (carried out with the outer primer set) is completed, a small volume of the reaction (0.01 to 1 µl) is pipeted into a second tube that contains a second set of primers (inner primer set) that generate a shorter product. When one of the primers of the first reaction is maintained for the second reaction, while the other primer is designed *de novo*, the term "hemi-nested PCR" is used in many publications.

Nested PCR reduces the risk of false positives, since the inner primer set will yield amplification product only if the outer primer set has amplified the specific DNA fragment. A nested PCR approach greatly increases the sensitivity of detection, and can be really useful for the detection of very low copy number templates, as is the case, for example, of apparently healthy fish that contain bacterial cells in very low numbers, as carriers of the disease. In fact, nested PCR has been extensively applied to the detection of many bacterial fish pathogens (Arias *et al.* 1995, Chase and Pascho 1998, Osorio *et al.* 1999, Cook and Lynch 1999).

However, nested PCR has the disadvantage that tubes from the first amplification contain a high concentration of PCR products that enhances the possibility of cross-contaminations while manipulating them to set up the second PCR round. To avoid this problem, a method for nested PCR reactions with single closed reaction tubes has been proposed by some authors (Yourno 1992, Wolff *et al.* 1995). Reagents for the first PCR are added to the tube, while PCR reagents for the second PCR remain sequestered at the tube top, embedded in a gel matrix (a trehalose matrix). Once the first reaction has finished, reagents for the second round are introduced into the reaction tube by centrifugation. However, this imposes the restriction that mineral oil has to be used in the reaction instead of the thermal lid that forms part of most of the modern PCR thermocyclers. In addition, the primer concentrations for both the first and second rounds have to be well optimized, since an excess of first round primers would lead to amplification of bands resulting from combinations of first- and second-round primers.

Randomly Amplified Polymorphic DNA analysis (RAPD)

Also known by many authors as AP-PCR (arbitrarily-primed-PCR), RAPD analysis is a PCR-based methodology that uses short random primers for rapidly detecting polymorphisms under low-stringency conditions (Welsh and McClelland 1990, Williams *et al.* 1990). It has the advantage that no previous knowledge of DNA sequences from the genetic system under study is needed. Thus, it is very useful as a starting strategy to identify DNA stretches that may further be used as a source for DNA probes and PCR primers (Miyata *et al.* 1996, Falklind *et al.* 1996, Argenton *et al.* 1996). RAPD-generated bands can be sequenced after plasmid cloning. This is becoming increasingly important, as DNA-based diagnostic assays depend to a greater extent on sequence analysis.

Another generalized application of RAPD is the differentiation of strains by comparing polymorphisms (DNA-band patterns in agarose or acrylamide gels) generated with this technique (Mileham 1995). It has been applied to the study of intraspecies genetic diversity of many fish pathogens, as *Photobacterium damselae* subsp. *piscicida* (Magariños *et al.* 2000), *Vibrio vulnificus* (Vickery *et al.* 2000), and *Renibacterium salmoninarum* (Grayson *et al.* 1999, 2000).

16S rDNA amplification and sequencing

This has become a very powerful strategy, and can be used to assign a species name to a bacterial strain, to identify new pathogens, as well as to characterize previously recognized uncultured infectious agents. The 16S rRNA molecule is a crucial tool in molecular phylogeny studies (Woese 1987), and the determination and comparative analysis of its nucleotide sequence make possible the revealing of the evolutionary relationships between species.

Different pairs of universal primers that target conserved positions at both 5' and 3' ends of the 16S rRNA bacterial genes are available in the literature (Lane 1991, Alm *et al.* 1996, Marchesi *et al.* 1998). For pure cultures, this ribosomal gene can be amplified by PCR and subjected to direct sequencing.

In addition, a very useful approach of the 16S rRNA amplification and sequencing consists in amplification of all the 16S rRNA gene targets that are present in the sample, after a previous DNA extraction step. This amplification is followed by cloning of individual 16S rDNA molecules into suitable plasmids and further DNA sequencing. Then, database sequence comparisons allow identification of the different 16S molecules that have been amplified and cloned. This is possible when the obtained sequence matches another sequence previously deposited in public databases and for which a species epithet has been coined. Otherwise, the obtained sequence helps to increase the list of 16S sequences corresponding to putative new bacterial species. Thus, by this approach, it is possible to get a collection of 16S rDNA sequences that do not match any previously described species, and which have never been detected in culture media, indicating that an important percentage of the microbiota found in virtually any environment corresponds to non-culturable bacteria. This approach may be very useful in helping to determine the nature of bacterial diseases for which the causative agent remains unknown, due to the inability to culture the organism.

However, this strategy has been shown in certain cases to be source of a series of molecular artifacts that may mask the final results (Speksnijder *et al.* 2001). When a mixture of homologous genes, i.e., all the different rRNA genes present in a natural sample, are used as PCR templates, it is possible that a series of PCR-generated artifacts, as heteroduplexes and chimeras, can form (Qiu *et al.* 2001). This limitation, which leads to obtaining amplified and cloned DNA sequences that do not exist *in vivo*, has to be taken into account when using a 16S rRNA gene-based cloning approach to identify bacterial DNA sequences (Suau *et al.* 1999).

DNA probes

Nucleic acid probes are segments of DNA or RNA that can bind with high specificity to complementary sequences of nucleic acid. Probes have the characteristic that they are labelled in such a way that it is possible to detect if a probe is bound to its complementary sequence. Different methods are used for the labelling, which may include radioisotopes, enzymes, antigenic substrates (to be further recognized by antibodies), and chemiluminescent moieties. Probes can be directed to either DNA or RNA targets, and its size may vary from as few as 18–20 to thousands of bases long.

To design a DNA probe-based test for a given bacterial pathogen, a certain degree of knowledge of the nucleotide sequence of the target organism is required. In addition, since these DNA probes are going to be tested in some cases with DNA extracted from a complex sample, it is necessary that some genetic information about phylogenetically related organisms, and other organisms that might be present in the same sample, is available. Otherwise, there are no guarantees as to the reliability of the species-specificity of the designed molecular tool unless the possibility of undesirable cross-reactivities is ruled out.

However, more often than not, DNA probes intended to be species-specific have been designed to detect bacterial species for which the 16S rRNA gene sequence is almost, if not all, the only genetic information available. 16S rRNA gene sequencing is currently the more accurate method to assign a given bacterial isolate to a species epithet. This means that for nearly all the bacterial species nowadays considered as such, their respective partial or complete 16S rRNA gene nucleotide sequence is available for any researcher that may be interested. Thus, it is quite tempting to employ the 16S gene as a target for nucleotide probes. Once a bacterial species is chosen as the subject of study, one can withdraw its 16S rRNA gene sequence from public databases, get the 16S sequence of the closest relatives, compare all of them, and select for those DNA stretches along the gene that clearly differ between species, and which can be good candidates for species-specific DNA probes.

Several approaches are followed to apply DNA probe protocols for the detection of bacteria. Once the DNA is extracted, it must be denatured (for example, boiling samples for a few minutes) and loaded onto a nylon membrane, after which it has to be covalently fixed. This last step can be easily accomplished by treatment of the membrane with U.V. light for a few minutes. In other cases, confirmative probes are tested against PCR products that have been previously separated on agarose gels and transferred to nylon membranes by southern-blotting (Magnússon *et al.* 1994, Mooney *et al.* 1995, Osorio *et al.* 1999), or PCR products that are simply denatured and directly loaded onto the membrane.

Another critical step is the choice of probe-labelling. For many years, isotopic labelling of DNA probes has been the most used option for the detection of fish pathogens (Rehnstam *et al.* 1989, Aoki *et al.* 1989, Hiney *et al.* 1992). Use of isotopic labelling demands the existence of a laboratory area exclusive for manipulation of radioactive products, and it also needs qualified staff to work with it. Moreover, detection protocols include long exposure times to obtain an autoradiography film.

The direct labelling of oligonucleotide probes with enzymes, such as alkaline phosphatase or horseradish peroxidase was introduced in the mid-1980s (Jablonski *et al.* 1986). The design of non-isotopic labelling systems has contributed to simplification of the detection protocols of DNA hybridization. Commercial systems for non-isotopic labelling are currently available, as is the case of digoxigenin-labelling systems (Dig-DNA labelling and detection kit, Boehringer Mannheim), biotin-labelling kits (Rad-Free kit, Schleicher and Schuell), or the new generation of chemiluminescence-based labelling kits (ECL kits, Amersham-Pharmacia-Biotech). Use of digoxigenin-labelled probes has been reported for detection of several bacterial fish pathogens (Osorio *et al.* 1999, McCarthy *et al.* 1999). However, digoxigenin labelling still has the disadvantage of involving numerous steps of membrane blocking and washing, as well as the use of antibodies. These inconveniences are minimized in protocols that employ luminescent-labelled probes (ECL kits).

Dealing with the samples

The employment of DNA-based diagnosis methods for bacteria, needs first to overcome the inconveniences associated with DNA extraction from the sample.

Depending on the nature of the sample to be analyzed, it is of crucial importance to set up a convenient DNA purification method. Bacterial DNA can be purified directly from the sample, but some investigators prefer to harvest the microorganisms from the sample, or dilute the sample and then inoculate a culture medium so that the microorganisms are amplified *in vivo*.

In vitro amplification reactions conducted by thermostable polymerases, as it is the case of PCR reactions, can be extremely sensitive to the presence of inhibitor molecules or elements in the reaction tube, a consequence of its existence in the sample to be analyzed (Rossen *et al.* 1992). This means that the application of this molecular technique to the detection of bacteria in fish or shellfish may be problematic due to the complex nature of the sample.

Common inhibitors include various substances, as organic and phenolic compounds, detergents, antibiotics, buffers, enzymes, glycogen, fats, proteins, Ca^{2+}, non target DNA, and constituents of the bacterial cell itself (Wilson 1997). Inhibition of the PCR reaction can be total or partial, thus manifesting either complete reaction failure or reduced sensitivity in detection. Reduction of sensitivity in the detection of microorganisms by PCR in food samples has been extensively reported by many authors (Wernars *et al.* 1991, Wilson *et al.* 1991, Jaykus *et al.* 1996, Simon *et al.* 1996). It has also been reported that the sensitivity of the PCR may vary depending on the microbial species under analysis, even when identical DNA extraction protocols are used. This can be explained by the fact that rates of DNA recovery may be different between species (Steffan and Atlas 1988, Tebbe and Vahjen 1993).

When molecular protocols are to be assayed with pure or mixed bacterial cultures, this does not constitute a problem in most of the cases. DNA extraction from bacteria can be achieved by traditional treatments, as cell lysis with detergents (sodium dodecyl-sulfate, "SDS"; Lauril-Sarcosyl) and enzymes as lysozyme and proteinase K, followed by phenol-chloroform extraction and further precipitation of DNA with ethanol. This may provide a good quality DNA that can be directly used for PCR amplification. Other authors use a simple step of boiling the sample for a few minutes. Some authors also report the successful amplification of DNA targets by PCR, directly from whole bacterial cells added to the PCR tube without prior DNA extraction (Brauns and Oliver 1995, Coleman and Oliver 1996).

In recent years, all of these traditional extraction procedures are being replaced by a variety of commercial kits and reagents, which guarantee purification of good quality DNA in a matter of minutes. For example, simple boiling of samples can now be carried out in the presence of a chelating matrix, as for example Instagene matrix (Bio-Rad), or Chelex 100 (Sigma) (Walsh *et al.* 1991). These chelating resins, among other functions, serve to chelate metal ions and thus prevent the catalytic breakage of DNA at temperatures of 100°C. The successful employment of these chelating matrix-based commercial products has been reported for the PCR-based detection of fish pathogens (Brasher *et al.* 1998, Marshall *et al.* 1998). Other commercial kits are recommended for purification of high quality bacterial DNA (Easy-DNA kit, Invitrogen).

However, things are not so easy when it comes to purifying bacterial DNA mixed with a biological or environmental sample, since the complex nature of its composition may influence downstream enzymatic reactions. Though many authors have reported a successful PCR-based detection of bacterial cells in different fish tissues and processed seafood following simple DNA extraction protocols such as direct boiling of the sample (Lee *et al.* 1998), in other cases DNA purification remains the most fastidious and limiting step in the molecular procedures. Fish tissue samples, as well as food and environmental samples are very different in their characteristics. This includes their consistency, physical and chemical composition, bacterial load, and presence of proven PCR inhibitors. For many cases, DNA extraction must involve a series of purification steps including lytic enzymes, or chaotropic salts as guanidine isothiocyanate (Lippke *et al.* 1987). There is a variety of commercial kits which serve for the DNA extraction from virtually any biological source, including mammalian tissues, yeast, bacteria, or plant tissues (QIAamp DNA kits of Qiagen). Other commercial kits, for example the "EZNA Mollusc DNA Kit" (Omega Biotek), are specially designed to deal with very viscous samples as oyster meat and related samples. Another example is the "DNAzol Reagent" (Gibco BRL), based on the use of a novel guanidine-detergent lysing solution that permits selective precipitation of DNA from a cell lysate.

Another important aspect in molecular diagnosis of bacteria affecting sea-farming is the management of sediment samples. In certain cases, environmental conditions in the fish farm are not optimal for the development of a bacterial outbreak. Nevertheless, this is far from meaning that the fish farm in question is not in contact with the pathogenic agent. More often than not, certain bacterial pathogens may undergo a viable-but-non-culturable state, thus surviving in water and sediments, being impossible to detect by traditional culture-based methods. It is in these cases that PCR-based detection methods become really worthy. Applications of PCR to aquatic environmental samples have been extensively treated in the literature during the last decade (Steffan and Atlas 1991, Bej and Mahbubani 1992). However, it is important to consider the possibility that environmental samples may need processing steps that differ from those applied to fish tissue samples. Recovering bacterial DNA from sediment is a critical step in any method for the detection of specific microbial genotypes in environmental samples. In those cases where sediment samples are going to be analyzed without any previous step of culturing the microorganisms present in the sample, then efficient recovery of DNA must be achieved.

Due to the complex nature of such samples, it must be ensured that potential inhibitors of the PCR reaction and of the DNA hybridization are eliminated. Many articles have been published on DNA purification from sediment samples, and thus further reading is recommended in order to select the one that fits best with each particular case (Steffan *et al.* 1988, Somerville *et al.* 1989, Tsai and Olson 1992, Porteous *et al.* 1994, Volossiouk *et al.* 1995, Miller *et al.* 1999).

Applications of magnetic separations

For bacterial diagnosis in fish tissue samples, specific bacteria must be detectable from a complex matrix including fish cells and DNA, as well as other background microbiota. However, an approach that allows achieving a selective enrichment stage, which does not depend on previous culture, is the use of specific magnetic separation of the organism of interest, directly from the complex sample. Once separated, bacterial cells can be subjected to DNA extraction and further PCR or DNA probe protocols.

For magnetic separations, many magnetic carriers can be used. In most of the cases these particles are superparamagnetic, i.e., they only exhibit magnetic properties when a magnetic field is applied. Thus, these particles can be removed from a suspension with the help of a magnetic separator. Once the magnetic field is removed, particles can be resuspended again, if elution of the bound sample is desired.

One of the applications of magnetic separation is the so-called "immunomagnetic separation". This consists of the use of antibodies immobilized on magnetic particles for capturing the target cells. These particles are suspended in the sample suspension for an appropriate incubation time and then separated with a magnetic separator. The amount of immunomagnetic particles required for capturing target cells can vary. As an example, 2 ´ 10^{6}–10^{7} particles per ml have been reported for *Salmonella* and *E. coli* (Vermunt *et al.* 1992, Fratamico *et al.* 1992). General applications of immunomagnetic separation in both clinical and food microbiology have been published (Olsvik *et al.* 1994, Safarík *et al.* 1995). In the case of the bacterial species reviewed in this chapter, there is a report on the application of this technique for detection of *Vibrio parahaemolyticus* in food samples (Tomoyasu 1992).

An additional possibility of immuno-magnetic separations is the coupling with further DNA hybridization or PCR. The so-called Magnetic Immuno PCR Assay (MIPA), combines the polymerase chain reaction after immunomagnetic separation of the target cells (Widjojoatmodjo *et al.* 1991), thus removing PCR inhibitory compounds from the sample. Though this strategy has been reported for the detection of *Salmonella* and *Listeria* (Widjojoatmodjo *et al.* 1991, Fluit *et al.* 1993), to the best of authors' knowledge no reports are available for fish pathogenic bacteria.

Unfortunately, the use of immuno-magnetic separation implies that specific monoclonal or polyclonal antiserum for the species of interest will be obtained and bound to magnetic carriers, which makes it a non-straigthforward approach for many laboratories. However, magnetic separation systems have been designed for the purification of any kind of DNA from complex samples, thus helping to obtain inhibitor-free DNA samples for PCR. One of the magnetic carriers currently available for DNA-binding and purification is Dynabeads DNA DIRECT™ Universal, produced by Dynal (Oslo, Norway). Dynabeads are uniform, superparamagnetic polymer beads. This commercial kit provides a lysis buffer including the magnetic beads, which permit the adsorption of the released DNA after lysis. Then, by applying a magnetic field with a magnetic separator (Dynal MPC), the DNA-Dynabeads complex is pulled to the side wall of the tube. This allows removal of supernatant, as well as further washing steps to remove any possible contaminants and PCR inhibitors. Finally, the DNA-Dynabeads complex is resuspended for direct use in PCR reactions. This procedure can be completed in as little as 10 min, and has been successfully applied to DNA purification from many different tissues, as varied as bacteria, algae, urine, bile, and blood, among others. The successful application of Dynabeads to the detection of fish pathogenic bacteria in fish tissues has been recently reported (González *et al.* 2000, Cepeda and Santos 2000, Cepeda *et al.* 2002).

Species-Specific Protocols for DNA-Based Detection

Vibrio anguillarum

Vibrio anguillarum is a halophilic Gram-negative, polarly flagellated curved rod. Vibriosis caused by *Vibrio anguillarum* is one of the most prevalent fish diseases caused by a member of the genus *Vibrio* and constitutes a serious problem in the marine culture of different species of fish, being particularly devastating in cultures of salmonids (Trust 1986). *V. anguillarum* was first described in 1909 as the causative agent of the "red pest of eels" in the Baltic Sea, and since then this bacterium has been implicated in numerous outbreaks of vibriosis worldwide (Toranzo and Barja 1990).

Sørensen and Larsen (1986) recommended an international harmonization of the serotype classification, and presented an antigenic typing scheme in which 10 distinct O serotypes of *V. anguillarum* were recognized, setting the basis of the current European serotyping system for this species. Following this division, nearly all the isolates implicated in fish mortalities belong to serotypes O1 and O2 and, to a lesser extent, to serotype O3 (Sørensen and Larsen 1986, Toranzo and Barja 1990, Toranzo *et al.* 1997, Pedersen *et al.* 1999).

Knowledge about the *V. anguillarum* genome is increasing as new genes are sequenced and made available to the scientific community. By April

2001, as many as 89 entries of DNA sequences from this species were available in the EMBL database, including a variety of virulence factor genes which are of great interest for the molecular characterization of isolates from this species.

Rehnstam *et al.* (1989) obtained partial 16S rRNA gene sequences from seven *V. anguillarum* strains. A 25-mer oligonucleotide was selected and radioactively labelled to be used as a probe in RNA-DNA colony hybridization. The authors proved that no cross-reactivities occurred with several other non-*V. anguillarum* bacterial species tested. A detection level of 5×10^3 bacteria per ml was achieved with this probe in slot-blot hybridization. Homogenized kidney samples from both healthy and experimentally infected fish were dotted on nitrocellulose filters and subjected to RNA-DNA hybridization. It was found that at least 6×10^5 bacteria per ml of kidney homogenate were detected. However, since more 16S rRNA gene sequences were subsequently reported for other *Vibrio* species (Dorsch *et al.* 1992), comparative sequence analysis shows that the oligonucleotide probe designed by Rehnstam *et al.* (1989) has a single nucleotide difference with the respective sequences of other *Vibrio* species. This implies that the probe may not be suitable for the species-specific detection of *V. anguillarum*, since other *Vibrio* species present in the sample may lead to a false-positive result.

A procedure for colony-hybridization-based detection of *V. anguillarum* was also proposed by Aoki *et al.* (1989). These authors selected a *V. anguillarum* 562-bp DNA fragment cloned in pUC9 vector, to be radiolabelled and used as a probe in colony-hybridization. At least 100 ng of chromosomal DNA spotted on nitrocellulose filters could be detected with this probe. When ten-fold dilutions of a bacterial cell suspension were also spotted on filters, positive hybridization was observed with 2×10^5 *V. anguillarum* cells.

According to Aoki *et al.* (1989), this 562-bp DNA fragment is specific for *V. anguillarum* serotypes A and H (following the serotyping system described by Kitao *et al.* 1983). But since *V. anguillarum* is not a homogeneous group of strains, additional research should be conducted in order to make sure that this probe can effectively detect the presence of any *V. anguillarum* isolate pathogenic to fish. In addition, no information is available on field studies employing this probe. The detection limits reported above are acceptable for colony hybridization, but may not be sensitive enough to allow the detection of low levels of *V. anguillarum* cells in infected fish tissues.

Another DNA probe-based approach was described by Powell and Loutit (1994a). In this case, a 310-bp *V. anguillarum*-specific DNA fragment was isolated by differential hybridization for its use as a probe. The nucleotide sequence of this 310-bp DNA fragment showed no homology to sequences in EMBL database. The authors proved its specificity in hybridization assays conducted with more than 200 marine bacterial isolates. Though this probe failed to hybridize with *V. anguillarum* serotypes O7 and O9, this is not considered an important limitation, since O7 and O9 isolates are environmental, not usually associated with fish disease. Similar to other *V. anguillarum*-specific described probes, these authors report a detection limit of 100 ng of purified chromosomal DNA. Later, Powell and Loutit (1994b) developed a method which combines membrane filtration and DNA probing. The 310 bp *V. anguillarum* specific-DNA fragment described previously (Powell and Loutit 1994a) was used as a probe for the detection of culturable *V. anguillarum* cells in water and chinook salmon (*Oncorhynchus tshawytscha*) tissue samples.

The 16S rRNA of *V. anguillarum* was choosed by Martínez-Picado *et al.* (1994) as the target for a DNA probe-based detection protocol. A 24-base DNA oligonucleotide, named VaV3, was selected as specific for *V. anguillarum* after partial sequencing and comparative analysis of 16S genes from *Vibrio* species. This probe was radiolabeled and tested with chromosomal DNA blotted on nylon membranes. It also was used in RNA-DNA colony hybridization, on bacteria previously patched on nylon membranes. Both experiments confirmed that the VaV3 probe recognized *V. anguillarum* chromosomal DNA, and no cross-reaction was detected with other *Vibrio* species. The sensitivity of this VaV3 probe was evaluated by slot blot hybridization, confirming that as little as 150 pg of purified *V. anguillarum* chromosomal DNA can be detected by this procedure.

Later, Martínez-Picado *et al.* (1996) reported an approach for the detection of *V. anguillarum* which included the combined use of culture on a selective medium (Alsina *et al.* 1994) and further DNA hybridization using the VaV3 probe described previously (Martínez-Picado *et al.* 1994). The authors evaluated more than 400 bacterial cultures, and when both the selective medium and the specific probe gave positive results, the cultures were always identified as *V. anguillarum*.

Surprisingly, despite the economic importance of *V. anguillarum* in fish farming, it was not until the year 2000 that a PCR-based approach was described for detection of this pathogen (González *et al.* 2000). The target gene of choice was *rpoN* (also named *ntrA* or *glnF*), a gene that codes for the sigma factor σ^{54}. The nucleotide sequence of the *V. anguillarum rpoN* gene was retrieved from EMBL database and compared to *rpoN* gene sequences from other bacterial species, including *Vibrio cholerae*, *V. alginolyticus*, and *Escherichia coli* among others. Two regions of high sequence variability were selected for the design of a pair of PCR primers, RponAng-5' and RponAng-3' respectively (see Table 3), flanking a 519-bp internal stretch of the *rpoN* gene. Extensive database comparisons showed that no significant homology exists between the sequence of the designed primers and the rest of DNA sequences deposited in EMBL database, except for *rpoN* gene of *V. anguillarum* (EMBL Accession number U86585). Specificity of the described primers was tested with more than 20 *V. anguillarum* strains belonging to serotypes O1 to O10 (Sørensen and Larsen 1986), as well as with a collection of other 60 marine bacterial isolates, including *Vibrio*, *Aeromonas*, and *Photobacterium* strains. Since *Vibrio ordalii* is considered to be one of the closest relatives of *V. anguillarum*, PCR parameters were adjusted so that no cross-reactivity occurred with that species. The result was that an annealing temperature of 62°C during the PCR program completely avoided any false-positive amplification, whereas all the strains of the 10 *V. anguillarum* serotypes tested gave a clear positive amplification band. When primers targeted to the *rpoN* gene were tested for its suitability to detect *V. anguillarum* cells in experimentally infected fish tissue, a detection limit of ca. 11 cells per PCR tube was achieved.

Vibrio vulnificus

Vibrio vulnificus is a gram-negative, estuarine halophilic bacterium that may cause disease in cultured eels (*Anguilla japonica*). This bacterium is also pathogenic for humans (Blake *et al.* 1980), causing gastroenteritis, fatal primary septicemia and necrotizing wound infections. Primary septicemia caused by *V. vulnificus* occurs following ingestion of raw seafood or exposure of wounds to seawater or shellfish where the bacterium is present. This septicemia progresses robustly, with high fatality rates (more than 50% within a day or two), most readily among patients who usually have a pre-existing chronic illness, hepatic diseases, alcohol drinking habit, diabetes mellitus, immunosuppression from corticosteroid therapy and AIDS, among others.

Many studies have reported physiological, serological and genetic comparisons among clinical, environmental and eel isolates of *V. vulnificus*. One of the conclusions of these studies is that eel isolates have distinct phenotypical and serological characteristics. Only eel isolates are pathogenic for eels, as well as to mice, whereas typical *V. vulnificus* isolates are only pathogenic for mice (Tison *et al.* 1982). Due to these differences, eel isolates were all included in a new group, as *V. vulnificus* biogroup 2, keeping the designation of *V. vulnificus* biogroup 1 for clinical and environmental strains. However, further serological studies indicated that biogroup 2 is just a serotype of biogroup 1 that is adapted to infect eels, and thus it is now designated as the *V. vulnificus* serotype E (Biosca *et al.* 1996).

Biogroup 1 is responsible for 95% of shellfish-related deaths in the USA (Oliver *et al.* 1991). On the other hand, serotype E is highly virulent for juvenile European and Japanese eels, having been isolated in numerous outbreaks in eel farms that resulted in high mortalities (Muroga *et al.* 1976, Biosca *et al.* 1991).

Knowledge about the *V. vulnificus* genome comprises about 40 entries of DNA sequences available in the EMBL database. Among these, the gene coding for the cytolysin-hemolysin (*vvhA*), cloned and sequenced by

Yamamoto *et al.* (1990), has been the most widely used target sequence for molecular protocols designed for the detection of this bacterium, as it will be summarized below.

Other genetic techniques apart from DNA sequencing have been extensively applied to study the intraspecies variability of this species. Arias *et al.* (1997) found that ribotyping is a good fingerprinting choice to discriminate eel-pathogenic strains from clinical and environmental isolates. Similarly, Høi *et al.* (1997) applied both ribotyping and RAPD analysis to a collection of *V. vulnificus* strains, corroborating that this species shows a high heterogeneity. Later, Warner and Oliver (1999) carried out a RAPD analysis on different *V. vulnificus* strains, as well as with several other *Vibrio* species. The resulting RAPD profiles obtained, showed that each *V. vulnificus* strain produced a unique band pattern, and some differences were also found between clinical and environmental strains. These findings thus indicate that this species has an important intraspecies heterogeneity. Similar conclusions can be drawn from the results reported by Vickery *et al.* (2000), who applied arbitrarily-primed-PCR (AP-PCR) to a collection of *V. vulnificus* isolates from patients fatally infected after consumption of raw oysters. The DNA fingerprints obtained also showed significant genetic heterogeneity among strains.

This genetic variability has to be taken into consideration when selecting gene targets for either PCR or DNA-probe detection protocols. The target sequence of choice must be present, and its sequence conserved, in all the *V. vulnificus* genetic variants whose diagnosis may be of interest, otherwise there is a risk of obtaining false negatives.

Molecular tools for the detection of this species have been extensively published by many authors, both for applications to the fish-farming industry and for human health purposes.

Morris *et al.* (1987) selected a 3.2 Kb DNA fragment encoding the *V. vulnificus* cytotoxin-hemolysin gene that was radiolabelled and used as a probe in colony hybridization. It hybridized with a collection of *V. vulnificus* environmental isolates, as well as 20 *V. vulnificus* reference strains, including both biotype 1 and 2. The probe did not hybridize with other lactose-positive marine vibrios nor with many other *Vibrio* species tested. The authors also screened spiked oyster homogenates to determine if the probe could be useful to detect the bacterium in natural samples. Serial 10-fold dilutions of a *V. vulnificus* cell suspension with concentrations ranging from 10^7 to 10^{10} CFU/ml were used to seed oyster homogenates. Ten ml samples of the spiked homogenates were spotted onto nitrocellulose filters and incubated overnight on L agar. After this, filters were processed and hybridized with the cytotoxin probe. Positive hybridization was confirmed for spotted 10 μl homogenate samples containing 6 ´ 10^3 CFU/ml, which represents detection of as few as 60 cells.

The nucleotide sequence of the *V. vulnificus* cytotoxin-hemolysin gene was used by Hill *et al.* (1991) to select a pair of oligonucleotides (VVp1 and VVp2) flanking a 519-bp gene fragment, to be employed in a PCR protocol for the detection of *V. vulnificus* in oyster homogenates. Since oyster meat represents a very complex matrix for the extraction of DNA for PCR, the authors examined different DNA purification procedures. Aliquots of 1 ml oyster homogenates in alkaline peptone water seeded with serial 10-fold dilutions of a *V. vulnificus* cell suspension were pelleted and washed twice with saline solution (0.85% NaCl), and then subjected to three different DNA extraction protocols. The one that proved to yield the best quality DNA for PCR amplification is described as follows. Cells were disrupted with 25 µl of 5.9 M guanidine isothiocyanate and the suspension was incubated for up to 90 min at 60°C. The suspension was then diluted with sterile water until the concentration of guanidine isothiocyanate was lowered to 0.3 M. Then, the lysate was extracted twice with phenol-chloroform and precipitated with 95% ethanol. This protocol yielded a DNA template with fewer inhibitors than did the other methods tested.

The sensitivity of PCR was determined by seeding oyster homogenates with 10^2 to 10^7 CFU per g of oyster meat. After an enrichment step of 24 h in alkaline peptone water at 35ºC, DNA was extracted from 1 ml of homogenate by the method described previously, and PCR amplification was carried out. The 519-bp band was amplified in all the samples. Though this method proved to be very useful for detection of *V. vulnificus*, a prior enrichment step is necessary, since the authors report that incubation periods of less than 24 h in alkaline peptone water give variable results.

A methodology that relies on a prior enrichment step before PCR amplification would be suitable as long as culturable *V. vulnificus* cells are present in the sample. It has been reported that *V. vulnificus* may enter a viable but nonculturable state (Linder and Oliver 1989, Warner and Oliver 1998), and thus it cannot be detected by conventional culture-dependent methods. Here is where PCR-based detection becomes very useful, eliminating the problem of nonculturability. Brauns *et al.* (1991) also chose cytotoxin-hemolysin gene for the design of a PCR primer pair, flanking a 388-bp gene fragment, and developed a protocol for the detection of both culturable and nonculturable *V. vulnificus* cells. Culturable cells were obtained by growth in standard media, and non-culturable cells were produced following a variation of the method described by Linder and Oliver (1989). DNA was extracted by lysis with lysozyme, chloroform-isoamyl alcohol separation, and ethanol precipitation. For DNA extracted from culturable cells, 40 amplification cycles were selected in the thermocycler program, and 50 cycles for the DNA from non-culturable cells. Detection limits were 72 pg of DNA from culturable cells, and 31 ng from nonculturable cells.

The cytolysin gene was also employed by Wright *et al.* (1993), for the identification of *V. vulnificus*. In this case, an oligonucleotide DNA probe (VVAP) was constructed and labelled with alkaline phosphatase. With this probe, naturally occurring *V. vulnificus* cells were detected in both seawater and unseeded oyster homogenates. Thus, the work reported by these authors represents a successful use of a non-isotopic probe, as well as the detection of the bacterium in field conditions. Samples of water and oyster homogenates were serially diluted in PBS, and 200 µl were spread on different agar media which permitted differentiation of *V. vulnificus* colonies due to its cellobiose- and lactose-fermenting activities. Presumptive *V. vulnificus* colonies were then picked and transferred to L agar for further hybridization with the VVAP probe. This probe specifically detected those isolates previously identified as *V. vulnificus*, and did not cross-react with the other *Vibrio* species tested. Since the process of picking single colonies from selective media is time-consuming, Wright *et al.* (1993) also tested the VVAP probe directly on colony blots prepared from L agar plates on which seawater and oyster homogenates had been plated. The VVAP probe was useful for enumeration of the *V. vulnificus* colonies.

Using seeded oyster homogenates, this probe detected between 10 and 100 bacteria per ml. However, since it relies on a previous plate culture step, it is not valid for the detection of viable but non-culturable cells.

An additional improvement in *V. vulnificus* DNA-based detection methodology was that described by Arias *et al.* (1995). These authors were the first to apply a nested-PCR approach for the detection of this bacterium in fish, sediment and water samples. The selected target for the design of PCR primers was 23S rRNA gene. The first PCR round was carried out with a pair of universal primers complementary to highly conserved regions of eubacterial 23S genes, flanking a 1828-bp fragment. A sample of 1 µl from this first PCR reaction was used as a template for a second PCR reaction which was conducted with specific primers for the *V. vulnificus* 23S rRNA gene, flanking a 978-bp fragment. To test the sensitivity, DNA was purified from three types of samples: 1) *V. vulnificus* pure cultures, 2) eel tissue homogenates free of *V. vulnificus*, 3) eel tissue homogenates free of *V. vulnificus* and spiked with 10^9 to 10^2 CFU per g. Then, DNA was extracted following the procedure described by O'Brien *et al.* (1994). With the nested approach, after the second PCR round as little as 10 fg of *V. vulnificus* DNA were detected by amplification of the specific 978-pb fragment, even when mixed with variable amounts of eel DNA. In the artificially seeded eel-homogenates, between 12 and 120 *V. vulnificus* cells were detected after the two rounds of nested-PCR. These detection limits obtained by the methodology reported by Arias *et al.* (1995) not only are very satisfactory, but also do not require neither a culture step nor a time-consuming use of DNA probe technology.

PCR amplification was also developed by Coleman *et al.* (1996) and applied to the detection of *V. vulnificus* biotypes 1 and 2 in eels and oysters. Primers used by these authors flanked a 344-bp portion of the *V. vulnificus* cytolysin-hemolysin gene (see Table 3). After DNA extraction without prior enrichment steps, positive PCR amplification was possible when starting with as few as 41 cells per g of oyster homogenate. Using the same primers in samples of unseeded oysters that presumably contained non-culturable *V. vulnificus* cells (as no colonies were developed after plating on agar), positive results were observed when the number of PCR cycles was increased to 50. PCR was also employed with DNA isolated from artificially infected eel tissues. Coleman *et al.* (1996) also point out that eel tissues can be preserved in 4% formalin without further affecting amplification of *V. vulnificus* DNA.

Aono *et al.* (1997) also employed the cytotoxin-hemolysin gene as a target for PCR detection of *V. vulnificus* isolates from both the marine environment and fish, using the same primers described by Hill *et al.* (1991). PCR amplifications were carried out with pure cultures, and positive ones were further confirmed by DNA hybridization. These results can be considered additional confirmation of the usefulness of the work previously reported by Hill *et al.* (1991).

This gene was also used by Lee *et al.* (1998) in a nested-PCR protocol to detect *V. vulnificus* in clinical specimens. Two primer sets were designed, the first flanking a 704-bp fragment, and the second flanking a 222-bp fragment of the hemolysin gene (see Table 3). A direct DNA extraction method consisted of boiling the sample pellet in a 1mM EDTA-0.5% Triton X-100 solution. As little as 1 fg of chromosomal DNA (ca. 1 *V. vulnificus* CFU) was detected with this nested-PCR approach. This protocol was applied to clinical samples from patients suffering from septicemia, finding that 94% of the cultures positive for *V. vulnificus* were also positive by nested-PCR.

Later, this same nested-PCR protocol allowed Lee *et al.* (1999) to detect *V. vulnificus* from a variety of seafood samples, including octopus, oysters, and eel among others. As in previous reports of nested-PCR applications to *V. vulnificus* detection, in this case a second PCR round made it possible to identify specific amplification bands that otherwise, i.e., with a single PCR round, would not have been detected.

Recently, Cerdá-Cuéllar *et al.* (2000) reported the successful combined use of a selective medium and a DNA probe for the detection of *V. vulnificus*. This medium (VVM) contains D(+)-cellobiose as the main carbon source, as well as polymyxin B, colistin, electrolytes ($MgCl_2$-$6H_2O$ and KCl) and moderate alkalinity and salinity. In this medium, *V. vulnificus* grows forming flat, bright yellow colonies, whereas most *Vibrio* species either do not grow, or grow as green-bluish colonies. An oligonucleotide probe of 24 nucleotides based on 16S rRNA gene was chosen for dot blot and colony hybridization. The combined use of the VVM and of this DNA probe, was applied to the differential detection of *V. vulnificus* in mixed bacterial suspensions and spiked mussels.

Vibrio salmonicida

Vibrio salmonicida is the causative agent of "Hitra disease", or "cold-water vibriosis". This fish pathogen appeared for the first time in Norwegian salmonid farms around the island of Hitra in 1979 (Egidius *et al.* 1986), and since then, it has been the cause of serious losses in fish farms of Atlantic salmon, particularly in Norway and Scotland. However, this disease is currently under control by the use of effective vaccination programmes (Toranzo *et al.* 1997).

This bacterium shows a very slow growth in laboratory media, requiring a temperature of incubation below 15°C, and the addition of blood to the culture medium. Knowledge of gene sequences from this bacterium is very limited, since as few as 5 entries of DNA sequences are currenty deposited in the EMBL database. Some molecular studies have been focused on the analysis of plasmid profiles of *V. salmonicida* (Wiik *et al.* 1989; Sørum *et al.* 1990). However, despite the economical consequences of the problems caused by this species in salmon farms, to the best of our knowledge no DNA-based methods for the detection of *V. salmonicida* have been designed so far.

Photobacterium damselae subsp. *piscicida (Pasteurella piscicida)*

Photobacterium damselae subsp. *piscicida* (*Pasteurella piscicida*) is the causative agent of pasteurellosis or pseudotuberculosis, one of the most important fish diseases in marine aquaculture. It causes substantial economic losses especially in yellowtail (*Seriola quinqueradiata*), gilthead sea bream (*Sparus aurata*), and sea bass (*Dicentrarchus labrax*) cultures worldwide (Magariños *et al.* 1996). Since its initial isolation from natural populations of white perch (*Morone americanus*) and striped bass (*Morone saxatilis*) in 1963, studies on this pathogen have shown that the different strains are biochemically and serologically homogeneous (Magariños *et al.* 1992, 1996). However, some variability has been detected by DNA fingerprinting methods as Ribotyping (Magariños *et al.* 1997), AFLP (Thyssen *et al.* 2000) and RAPD (see Figure 1) (Magariños *et al.* 2000).

Our knowledge of the genetic baggage of *P. damselae* subsp. *piscicida* was nonexistent until 1989. The first nucleotide sequence determined for this bacterial subspecies consisted of that obtained by Zhao and Aoki (1989) to be used as a probe. Later, Kim and Aoki (1993, 1994, 1996a,b), sequenced a series of different antibiotic resistance genes contained in plasmids of certain strains of this bacterium. However, no chromosomal gene was sequenced for this pathogen until 1995 when the nearly complete 16S rRNA gene

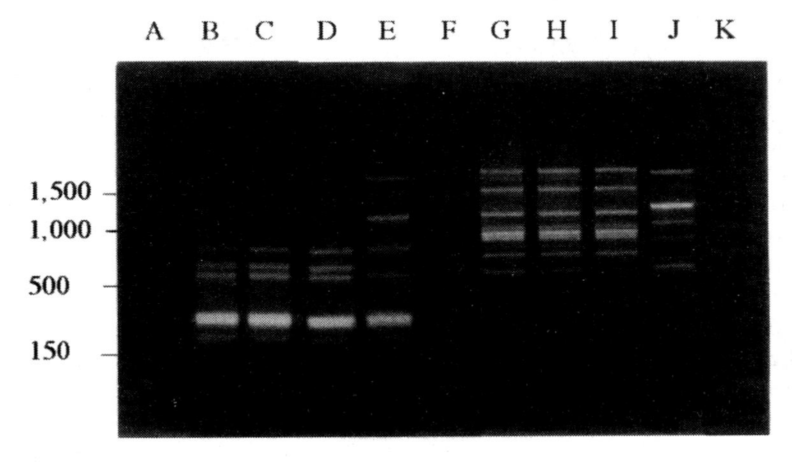

Figure 1. RAPD fingerprints obtained for *Photobacterium damselae* subsp. *piscicida*. Lanes: A, F and K, Amplisize Molecular Ruler (50-2000 bp ladder; Bio-Rad); B, DI-21; C, IT-1; D, B3; E, ATCC 29690; G, 666.1; H, 10831; I, 069A; J, MZS-8001. Molecular sizes (in bp) are indicated on the left. Courtesy of Dr. Jesus L. Romalde.

sequence was available (Gauthier *et al.* 1995). It made possible a comparative phylogenetic analysis with other related bacteria which, in fact, led to the reassignment of the formerly named *Pasteurella piscicida* to the genus *Photobacterium*, as *Photobacterium damselae* subsp. *piscicida*, and sharing species epithet with the formerly named *Vibrio damsela*, which in turn changed its name to *Photobacterium damselae* subsp. *damselae*.

Since 1997 the determination of the nucleotide sequence of a variety of gene loci and non-coding genomic regions of *Ph. damselae* subsp. *piscicida*, as well as of the subsp. *damselae*, provided a new insight into the inter-subspecies and intra-subspecies variability of these taxa. There are currently deposited in the EMBL database about 20 entries of DNA sequences from *Ph. damselae* subsp. *piscicida*. It also settled the methodological basis for the design of a variety of DNA-based protocols for detection of the causative agent of pasteurellosis in fish. It is one of the few fish pathogenic bacteria for which the nearly complete sequence of all the ribosomal operon, including the three rRNA genes and their respective internal transcribed spacer regions have been determined (Osorio 2000). There is also a high amount of genetic information obtained from this bacterium, after a great effort of partial sequencing of a *Ph. damselae* subsp. *piscicida* genomic library, that remains to be published and/or made available in public databases (Naka *et al.* 1999).

The need for fast, specific and sensitive molecular tools for the detection of *Ph. damselae* subsp. *piscicida* is particularly crucial in the context of the sea-farming. There are many reasons for this: First of all, pasteurellosis is one of the most devastating bacterial diseases affecting sea-farming worldwide. Second, adult fish may remain covertly infected, being asymptomatic "carriers" of the pathogen, and helping to spread the disease. Moreover, this bacterium grows very slowly on synthetic media, and is normally overgrown by most of the species coexisting in the sample, which makes very difficult its visualization and isolation in a mixed culture. In addition this species has been reported to undergo a viable-but-non-culturable state (Magariños *et al.* 1994), which makes its detection even more difficult. The striking genetic similarity of this pathogen with *Ph. damselae* subsp. *damselae* also represents a challenge for any molecular tool that intends to be subspecies-specific.

The first reported DNA-based tool for the detection of *Ph. damselae* subsp. *piscicida* was that described by Zhao and Aoki (1989). A 692-bp DNA fragment from a pUC9-cloned library of this bacterium was radiolabelled and used as a hybridization probe for the detection of the causative agent of fish pasteurellosis. With this probe, it was possible to detect 3.9 ng of purified DNA in dot-blot hybridization. Also, it served to detect *Ph. damselae* subsp. *piscicida* on nitrocellulose filters smeared with infected kidney and spleen tissue of yellowtail (*Seriola quinqueradiata*). This 692-bp DNA fragment was used by Aoki *et al.* (1995) to design a pair of PCR primers (see Table 3) flanking a 629-bp fragment, for the detection of *Ph. damselae* subsp. *piscicida*, and which constituted the first report of the application of PCR to the diagnosis of this pathogen. These primers were tested with pure cultures of *Ph. damselae* subsp. *piscicida* and other bacterial species, as well as with fish that had been naturally and artificially infected with the causative agent of fish pasteurellosis. However, since this work reported by Aoki *et al.* (1995) was carried out before the reassignation of *Pasteurella piscicida* to the species *Photobacterium damselae*, unfortunately they did not test these primers with strains of *Ph. damselae* subsp. *damselae*. In our laboratory, we have evaluated the specificity of these primers, and we have found the existence of cross-reactivities with strains of the subsp. *damselae*.

Further attempts to detect *Ph. damselae* subsp. *piscicida* chose different DNA targets for both PCR- and DNA probe-based diagnosis protocols. Thus, Aoki *et al.* (1997) designed two primer sets for the amplification of two sequences from the pZP1 plasmid of this pathogen (see Table 3). This approach was useful for detecting the bacterium in kidney homogenates of naturally infected fish, and showed no cross-reactivity with other closely-related marine bacterial species. However, since the plasmid profile of *Ph. damselae* subsp. *piscicida* strains has proved to be different depending on the geographical origin of the isolates (Magariños *et al.* 1992), this plasmid-based PCR protocol only guarantees detection of Japanese isolates of the pathogen, as well as some American isolates, whereas it remains to be demonstrated whether this protocol would allow diagnosis of European isolates of *Ph. damselae* subsp. *piscicida*.

Gauthier *et al.* (1995) obtained the first sequence of the 16S rRNA gene from one *Ph. damselae* subsp. *piscicida* strain. Later, Osorio *et al.* (1999) sequenced 16S rRNA gene of 18 *Ph. damselae* subsp. *piscicida*, and 8 *Ph. damselae* subsp. *damselae* strains from many geographical origins and isolation sources. Results showed that the consensus sequence of this gene is 100% similar in all the strains studied of the two subspecies. Osorio *et al.* (1999) used this information to select variable regions within the 16S rRNA gene of *Ph. damselae*, by comparison with 16S rRNA genes from many *Vibrio* and *Photobacterium* species. Three different primers were designed for a nested-PCR approach (see Table 3). For the first PCR round, a set of primers (car1/car2) flanking a 267-bp 16S gene fragment were chosen. For the second PCR round, primer car1 was maintained as the 5'-primer, whereas a new 3'-primer (nestcar-1) was selected. Car1 and nestcar-1 primers flanked a 259-bp fragment. In addition, a confirming short oligonucleotide probe that hybridizes to both the 267- and 259-bp PCR fragments was digoxigenin-labelled and used to confirm the PCR results by southern-blot. Specificity of primers was tested with a collection of *Ph. damselae* strains and other *Vibrio* and *Photobacterium* species.

Nested PCR was applied to the detection of the causative agent of fish pasteurellosis in DNA extracted from mixed plate cultures obtained after sampling of asymptomatic carrier gilthead sea bream (*Sparus aurata*) (Osorio *et al.* 1999). For this case, the nested-PCR approach proved to be very useful, since after the second PCR round specific amplification bands were seen in all the samples, whereas a single PCR round showed many nonspecific bands.

In addition, 50-mg pieces of kidney from rainbow trout were spiked with 10-fold dilutions of a cell suspension of *Ph. damselae* subsp. *piscicida*, and total DNA was extracted using a commercial EZNA Mollusc DNA kit (Omega Biotek). The detection limit in tissues was 10^3 bacterial cells per 100 mg of tissue homogenate with a first PCR round. However, in this case, a second PCR-round did not show improvement in the sensitivity. When purified chromosomal DNA was used as template for the PCR, amplification of the expected DNA fragment was evidenced with 10 fg of DNA. This good detection level obtained with purified DNA, compared with the detection level in fish tissues, tells us that further improvements are necessary in order to avoid PCR inhibition and get better results for fish tissue samples.

Since 16S sequence is identical in the two subspecies, the designed primers will invariably detect *Ph. damselae* subsp. *damselae* and subsp. *piscicida*. Though these two subspecies seem to have different ecological habitats and distinct host specificities, it is pertinent that subspecies-specific diagnosis methods are desirable in order to avoid possible cross-reactivities.

This has been partially overcome by the use of a multiplex PCR approach including the use of primers for the 16S rRNA gene of *Ph. damselae*, as described above, and primers for a gene that is uniquely encountered in one of the subspecies. The work reported by Osorio *et al.* (2000a) makes use of primers (ureC-5'/ureC-3'; see Table 3) targeted to the *ureC* gene coding for a subunit of the enzyme urease, and which is present in subsp. *damselae* strains but absent in subsp. *piscicida* strains. With that multiplex PCR approach, presence of *Ph. damselae* subsp. *piscicida* in a sample would be indicated by the amplification of the 267-bp 16S rRNA gene fragment, and the lack of amplification of a 448-bp fragment of the *ureC* gene. Presence of *Ph. damselae* subsp. *damselae* DNA in the sample would produce amplification of the two fragments. Unfortunately, this multiplex PCR protocol has been tested only with pure cultures. Further genetic research conducted with the two subspecies of *Ph. damselae* will help to find subspecies-specific gene targets that allow additional alternatives for the discrimination of the two subspecies.

Photobacterium damselae subsp. *damselae* (*Vibrio damsela*)

Photobacterium damselae subsp. *damselae* (formerly *Vibrio damsela*), is a halophilic bacterium that has been reported to cause wound infections and fatal disease in a variety of marine animals and humans (Buck *et al.* 1991, Clarridge and Zighelboim-Daum 1985, Morris *et al.* 1982). This species was reclassified as *Photobacterium damselae* subsp. *damselae*, sharing species epithet with the marine bacterium *Photobacterium damselae* subsp. *piscicida* (formerly *Pasteurella piscicida*), since these two taxa show more than a 70% homology in DNA-DNA hybridization studies (Gauthier *et al.* 1995), as well as a complete homology at the 16S rRNA sequence level (Gauthier *et al.* 1995, Osorio *et al.* 1999). About 15 DNA sequences from *Ph. damselae* subsp. *damselae* are currently deposited in the EMBL database. In addition, most of the sequences available for subspecies *piscicida* actually correspond to genes that also have their counterpart in subspecies *damselae*.

In 1981, Love *et al.* demonstrated that *Ph. damselae* subsp. *damselae* is the causative organism of skin ulcers on the damselfish *Chromis punctipinnis*, and since them this bacterium has also been recognized as a pathogen for other fish, such as turbot, seabream, seabass, yellowtail and sharks, reptiles, mollusks, and crustacean species (Fouz *et al.* 1992). This subspecies of *Ph. damselae* does not have the economical importance of subsp. *piscicida* in sea-farming. Maybe because of this, specific protocols reported for the detection of subsp. *damselae* in fish samples were actually a direct consequence of the design of molecular methods for the detection of the subspecies *piscicida*. Thus, 16S rRNA-gene based PCR detection of subsp. *damselae* has previously been described above for the subspecies *piscicida* (Osorio *et al.* 1999). Another approach, also described previously (Osorio *et al.* 2000a), consists of a multiplex PCR protocol for both 16S rRNA and *ureC* genes. An additional PCR application to *Ph. damselae* subsp. *damselae* is that reported by Osorio *et al.* (2000b), using as a target the *dly* gene, coding for the hemolysin "damselysin." Fouz *et al.* (1993) reported the correlation found between the degree of virulence and hemolytic activity in *Ph. damselae* isolated from fish. The *Ph. damselae* subsp. *damselae* hemolytic toxin gene (*dly*) has been cloned and expressed in *Escherichia coli* (Cutter and Kreger 1990). Using this partial gene as a probe, these authors found that not all the *Ph. damselae* subsp. *damselae* isolates contained the *dly* gene in the chromosome, but only those strains showing a highly- and intermediate hemolytic activity. Nucleotide sequence of *dly* gene coding for the toxin damselysin, was used by Osorio *et al.* (2000b) to design a primer set, (dly-5'/dly-3') (see Table 3), flanking a 567-bp gene fragment. In addition, this 567-bp PCR fragment was digoxigenin-labelled and used as a probe in dot-blot hybridization. With these primers and with the probe, it was shown that not all the *Ph. damselae* subsp. *damselae* strains harbour the *dly* gene, regardless of their hemolytic phenotype. The simultaneous use of primers for 16S rRNA *Ph. damselae* gene, as well as primers for subsp. *damselae ureC* and *dly* genes was carried out in a Multiplex PCR assay in our laboratory, demonstrating that this approach may serve to both distinguish subsp. *piscicida* from subsp. *damselae* strains and, among this latter subspecies, it serves to give additional information about the presence of a possible virulence gene marker such as the *dly* gene (see Figure 2).

Aeromonas salmonicida subsp. *salmonicida*

Aeromonas salmonicida subsp. *salmonicida* is one of the most important fish pathogens, causing economicly devastating losses in cultivated salmonids in fresh and marine waters (Toranzo *et al.* 1991), being the causative agent

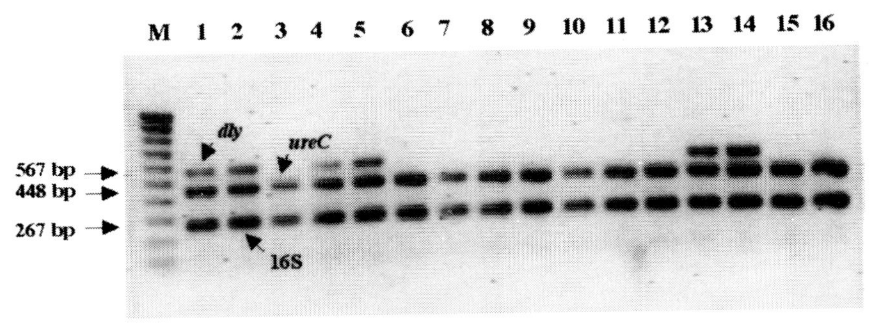

Figure 2. Results of Multiplex PCR using simultaneously primers for 16S rRNA, *ureC* and *dly* genes. Lanes: M: DNA marker (1 Kb). 1-16, *Ph. damselae* subsp. *damselae* strains: RG-91, RM-71, LD-07, CDC 2227, ATCC 33539, ATCC 35083, 158, 162, 192, PG 801, J3G 801, 309, RG-214, RG-153, 340, 238. Molecular sizes (in bp) are indicated on the left.

of typical furunculosis (Bernoth *et al.* 1997). It also affects a variety of non-salmonid fish, and shows a widespread distribution. It is as well one of the oldest fish pathogens described, the first report of this bacterium being that by Emmerich and Weibel in 1894. *Aeromonas salmonicida* subsp. *salmonicida* includes the so-called "typical" strains. The "non-typical" *A. salmonicida* strains are included within three subspecies, *masoucida, achromogenes* and *smithia*.

The absence of an efficient selective medium, and the slow growth characteristics of this bacterium which allow other species to overgrow *A. salmonicida*, make DNA-based detection methods a very worthy approach for epidemiological studies of furunculosis. In addition, the existence of a viable-but-non-culturable state (Enger 1997), as well as the possibility of vertical transmission of *A. salmonicida* (Austin and Adams 1996), also support the need of culture-independent, molecular detection protocols.

Our knowledge about *A. salmonicida* genome has increased substantially in recent years, and as many as 76 entries of DNA sequences from this species are currently available in the EMBL database. Both DNA probes and PCR primers have been generously selected by many authors, attempting to achieve a successful detection of this bacterium from a variety of samples. In the following paragraphs emphasis will be placed on currently available knowledge about identification and detection of *A. salmonicida* by DNA-based molecular tools.

Application of RAPD-based fingerprinting methodologies has shown, on one hand, that this species is very homogeneous (Miyata *et al.* 1995). However, there is evidence that a certain genetic heterogeneity can be evidenced by RAPD analysis in *A. salmonicida* strains isolated from marine fish (El Morabit 1999).

The first published report of PCR and DNA probe application to the detection of *A. salmonicida* was that by Hiney *et al.* (1992). A specific DNA probe for *A. salmonicida* was described by these authors, consisting of a fragment isolated from a genomic library of this bacterium by differential hybridization. This 455-bp DNA fragment was radiolabelled and used as a probe for slot blot hybridization with purified genomic DNA from different *A. salmonicida* strains, as well as from other bacterial genera and species. The probe was demonstrated to be species specific, and no cross-reactivities were found with any other bacteria assayed. Slot blot analysis with serial dilutions of *A. salmonicida* purified genomic DNA revealed that 10 ng of DNA could be detected with the designed probe after 8 h of autoradiography.

In the same work, the authors determined the nucleotide sequence of the probe (EMBL accession number X64214), and two 20-base regions were selected as targets for PCR primers (see Table 3), flanking a 423-bp DNA fragment. When serial dilutions of *A. salmonicida* DNA were used as a template for a PCR reaction with these primers, and the PCR products were slot blotted and probed with 1 μg of the radiolabelled probe described above, the sensitivity of the detection was increased about 10^6 times, i.e., a positive result could be obtained when starting from as little as 10 fg of DNA in the PCR reaction.

This procedure, though showing a high specificity and a very good sensitivity, has the disadvantage of using radiolabelled probes. Moreover, the authors used purified DNA from *A. salmonicida* pure cultures as the template for both the hybridization and the PCR reaction. In this sense, the suitability of this target DNA sequence should be proven in field experiments, and a protocol for its detection in natural samples would be desirable.

Also in 1992, Gustafson *et al.* reported the first case of the successful detection of *A. salmonicida* from fish and water samples with the PCR reaction. In this case, the target DNA sequence was a 421-bp internal fragment of the surface array protein gene (*vapA*). The nucleotide sequence of this gene was determined by Chu *et al.* (1991), and is available in the EMBL database under accession number M64655. Primers AP-1 and AP-2 were selected within this sequence (see Table 3). When as little as 100 fg of purified chromosomal DNA were used as the template, the 421-pb PCR product could be detectable after agarose electrophoresis. No amplification was observed with a variety of bacterial genera and species included as negative controls. When homogenized rainbow trout tissues and feces samples were seeded with serial dilutions of a bacterial suspension of *A. salmonicida*, the PCR protocol described by Gustafson *et al.* (1992) proved to be effective for the detection of as few as 10 CFU per mg of tissue homogenate. After filtering fish farms tankwater, incubating this filter in TSB broth, and analyzing aliquots of this broth by PCR, the authors were also able to detect the presence of *A. salmonicida*, with a detection level of as few as 5 CFU per 500 ml of filtered water. These primers described by Gustafson *et al.* (1992) have been tested in our laboratory with a collection of *A. salmonicida* strains isolated from marine fish in Galicia (Northwest Spain), as well as reference strains, producing the expected PCR product in all the isolates (see Figure 3).

A similar methodology was applied by O'Brien *et al.* (1994) to detect *A. salmonicida* from water, feces and effluent samples from a fish tank containing moribund fish, in a hatchery rearing Atlantic salmon (*Salmo salar*) smolts. These authors employed the *A. salmonicida* specific DNA probe isolated from a genomic library of the pathogen by Hiney *et al.* (1992). Moreover, they also put to the test the PCR primer pair previously cited Hiney *et al.* (1992). In order to guarantee that PCR inhibitors were eliminated from the fecal and effluent samples, the authors included a purification step including the use of Sephadex G-200 columns. The suitability of the DNA template for PCR amplification was tested by carrying out a preliminary PCR reaction with primers targeted to universally conserved

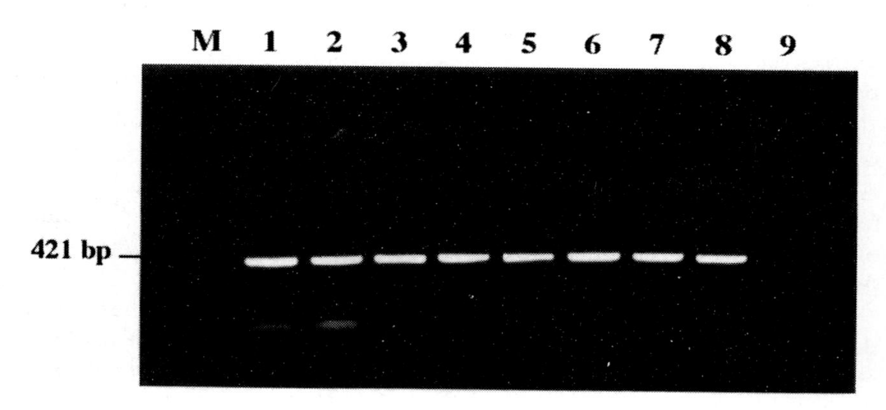

Figure 3. Results of PCR amplification of the 421-bp fragment of *vapA* gene in eigth *Aeromonas salmonicida* strains using primers AP1 and AP-2 (Gustafson *et al.* 1992; see table 3). M: Amplisize Molecular Ruler (50-2000 bp ladder; Bio-Rad). **1-8**: *A. salmonicida* strains: NCMB 2261, ACR 72-1, ASF1-1, EW22-1, RSP 43-1, L9-S1, EW 9-1, NF 19-1. **9**: negative control (no DNA). Molecular sizes (in bp) are indicated on the left.

regions within the 16S rRNA genes, where amplification of the desired fragment indicates the efficient removal of possible inhibitors. With these approaches, O'Brien *et al.* (1994) report the successful detection of *A. salmonicida* in feces and particulate matter, where traditional plate culture techniques had failed to identify the presence of this bacterium. The detection limit achieved with the PCR was about 200 *A. salmonicida* genome equivalents per g of sample, which can be considered a useful detection level. The nondestructive manner of sampling reported in the work of O'Brien *et al.* (1994) represents a good option for epidemiological studies of *A. salmonicida* in fish farms, since the presence of the pathogen in the environment of the fish farm can be evidenced without the need of killing fish.

A similar approach was later reported by Mooney *et al.* (1995). By this time, it became evident that the previously described specific DNA probe for *A. salmonicida* isolated from a genomic library (Hiney *et al.* 1992) was part of a DNA locus contained in a 6.4 Kb cryptic plasmid known to occur in 90% of *A. salmonicida* subsp. *salmonicida* isolates (Toranzo *et al.* 1991, Sørum *et al.* 1993). With this information available, Mooney *et al.* (1995) used this probe to determine the presence of *A. salmonicida* in wild Atlantic salmon. They examined blood samples from 61 fish and demonstrated the presence of the pathogen in 87% of the cases. An interesting contribution of their work is that they utilized a non-destructive sampling procedure. They extracted between 10 and 100 µl of blood from the dorsal vein of wild salmon, and DNA was extracted with lysis buffer containing SDS and proteinase K, further phenol-chloroform separation and final ethanol precipitation. A first PCR reaction was assayed using those primers previously described by Hiney *et al.* (1992), which amplify a 423-bp DNA segment. Two microlitres of these amplification products were then used as a DNA template for a second amplification reaction (nested-PCR), where the primers were newly designed so that they amplify a 270-bp internal DNA piece within that 423-bp PCR fragment amplified in the first PCR reaction. In addition, a short 20-mer oligodeoxynucleotide internal to this 270-bp nested-PCR fragment, was radioactively end-labelled and used to probe nylon membranes to which the PCR products from both the first and the nested-PCR reactions had been transferred. It was observed that some PCR-negative samples after the first round, when transferred and probed, showed positive hybridization. In addition, when the second PCR round was carried out, it was observed that 87% of 61 wild salmon individuals tested positive for the presence of *A. salmonicida*.

Miyata *et al.* (1996), employed the Randomly Amplified Polymorphic DNA (RAPD) technique to generate a pattern of amplification bands from *Aeromonas salmonicida* subsp. *salmonicida*, which were then cloned and their specificity assayed by DNA-DNA hybridization. This led Miyata *et al.* (1996) to the isolation of a 680-bp DNA fragment which was found to be specific for subsp. *salmonicida*. A PCR primer set flanking a 512-bp fragment within that 680-bp RAPD-generated band was used to detect subsp. *salmonicida* from pure cultures as well as from kidney samples of experimentally infected amago salmon (*Oncorhynchus rhodurus* subsp. *macrostomus*). The authors followed a DNA extraction protocol that included homogenization of kidney in TE buffer followed by 5 min boiling of 50 µl of homogenate, treatment with 50 µl phenol. DNA from the aqueous phase was then precipitated with 3M sodium acetate and absolute ethanol. After 30 cycles of PCR, positive amplification was corroborated for samples from eight fish that had been intraperitoneally injected with 5 ´ 10^4 CFU of *A. salmonicida*. When serial dilutions of pure chromosomal DNA were used for PCR, amplification of the expected 512-bp fragment was observed with as little as 100 fg of template DNA, which is in good agreement with the detection limit found by Gustafson *et al.* (1992) using the *vapA* gene as a PCR target.

More recently, Høie *et al.* (1997) used 16S gene and plasmid sequences as PCR targets for the detection of *A. salmonicida* subsp. *salmonicida* in kidney and gills of infected Atlantic salmon (*Salmo salar*). Two primer pairs flanking a 271-bp fragment of the 16S rRNA gene and a 710-bp fragment of a 5.0 Kb *A. salmonicida* plasmid respectively (see Table 3) were employed. When serial dilutions of an *A. salmonicida* cell suspension were mixed with kidney and gill samples, 20 and 200 CFU were detected in 10 µl PCR template by the 16S rDNA and the plasmid primers respectively. This corresponds to 10^3 and 10^4 CFU in 100 ml kidney tissue suspension.

Analysis of 25 Atlantic salmon by both PCR and agar cultivation showed 13/25 PCR-positive samples, whereas only 6/25 proved to be positive by agar cultivation. In addition, testing of kidney samples from feral brood fish showed 24/72 PCR-positive samples, whereas only 6/72 were positive for *A. salmonicida* by agar cultivation.

Flexibacter maritimus

Flexibacter maritimus is the causative agent of flexibacteriosis in marine fish. Several other names, as "salt water columnaris disease", "gliding bacterial disease of sea fish", "black patch necrosis" or "eroded mouth syndrome", also designate the disease caused by this bacterium. The organism was first isolated in Japan from cultured juvenile sea bream in Japan, and subsequently characterized by Wakabayashi *et al.* (1986) who designated the species as *Flexibacter maritimus,* thus belonging to the group of gliding bacteria. It has been responsible of many outbreaks associated with high fish mortalities and significant economic losses, and it is now considered a potentially limiting factor for the culture of many species of marine fish of commercial value (Devesa *et al.* 1989, Santos *et al.* 1999). *Flexibacter maritimus* causes ulcerative lessions in a variety of marine fish. In Europe, it has been isolated from Dover sole (*Solea solea*), sea bass (*Dicentrarchus labrax*), turbot (*Scophthalmus maximus*), Atlantic salmon (*Salmo salar*), and coho salmon (*Oncorhynchus kisutch*). In Japan, the disease has been reported from Japanese flounder (*Paralichthys olivaceous*), Red Sea bream (*Pagrus major*), and Black Sea bream (*Acanthopagrus schlegeli*). It has also been isolated in Southern California, USA, from white sea bass (*Atractoscion nobilis*), Pacific sardine (*Sardinops sagax*), and northern anchovy (*Engraulis mordax*).

This bacterium needs a specific medium for its growth, which must contain seawater, as well as a low concentration of nutrients (oligotrophic medium) (Pazos *et al.* 1996). One of the major problems in the study of this bacterium is the difficulty of distinguishing it from other phylogenetically related and phenotypically similar species, particularlly those of the genera *Flavobacterium* and *Cytophaga. Flavobacterium columnare* and *Flavobacterium psychrophilum* have been isolated from diseased fish, and also constitute primary fish pathogens. The 16S rRNA gene sequences of these three gliding bacteria pathogenic for fish are presently available in public databases. These sequences have been used to determine the phylogenetic classification of the three species, and also constitute the basis for the design of different DNA-based detection methods.

To detect *Flexibacter maritimus* by using the PCR, Toyama *et al.* (1996) chose the 16S rRNA gene as a target. They selected a pair of primers, flanking a 1088 bp fragment within the 16S gene, which could differentiate *F. maritimus* from the related species *Flavobacterium branchiophilum* and *Flavobacterium columnare,* as well as from several other fish pathogenic bacteria. With the protocol described by Toyama *et al.* (1996), it is possible to selectively differentiate *F. maritimus* pure cultures from those of other species, but unfortunately no application to field experiments, i.e, PCR detection in fish samples, is described by these authors.

Bader *et al.* (1998) selected a pair of *Flexibacter maritimus* species-specific PCR primers from unique sequence stretches within the 16S rRNA gene sequence (EMBL accession number M64629) of this species, delimiting a 400-bp DNA fragment. Unfortunately, the authors only tested these primers with pure cultures of collection strains and field isolates, but no information is available on how would they work when applied to environmental and/or fish samples, which presumably may contain DNA from a variety of genetically related bacteria.

The unique PCR protocol tested for the detection of *F. maritimus* in field conditions existing to date is that by Cepeda *et al.* (2002). These authors have successfully reported a specific and sensitive PCR-based protocol for the rapid detection of *F. maritimus* in pure and mixed bacterial cultures, as well as from a variety of fish tissues including kidney, spleen and brain. This protocol uses a nested-PCR approach where a first amplification reaction is carried out with a pair of universal primers that recognize conserved regions of the eubacterial 16S rRNA gene. PCR products generated during this first reaction are then diluted 100 or 1000-fold and used as DNA templates in a second PCR reaction, where the primers are those described previously by Toyama *et al.* (1996), which flank a 1088-bp fragment internal to 16S gene. DNA purification from fish samples was carried out using the Dynabeads DNA DIRECT™ Universal (Dynal, Oslo, Norway).

Renibacterium salmoninarum

The gram-positive bacterium *Renibacterium salmoninarum* is the causative agent of bacterial kidney disease (BKD), a prevalent disease of cultured and wild salmonids (Fryer and Sanders 1981). It is of great importance, causing clinical and subclinical infections in both farmed and wild salmonid populations in Europe, North and South America, and Japan (Evenden *et al.* 1993). This bacterium can be transmitted vertically (from parent to progeny) and horizontally (from fish to fish), and it is known that the fecal-oral route of horizontal transmission contributes significantly to the prevalence of BKD in salmon farms (Balfry *et al.* 1996).

Research on *R. salmoninarum* physiology and genetics is limited due to the technical difficulties regarding its culture in laboratory conditions. Its primary isolation from infected fish is not always successful and, in addition, this bacterium presents a very slow generation time of 24 h (Fryer and Sanders 1981), which makes traditional culture-dependent diagnosis procedures rather useless for fast detection of the pathogen. Current diagnosis strategies developed for *R. salmoninarum* include, on the one hand, antibody-based methodologies. On the other hand, DNA-based methods have also been reported. Our knowledge about the *R. salmoninarum* genome, though still scarce, is increasing in the last years. By April 2001, 102 entries of DNA sequences were available in the EMBL database, most of them corresponding to sequences of the intergenic transcribed spacer regions of the ribosomal operon, as well as ribosomal genes. Special interest has been placed in the gene coding for the 57-kDa major soluble antigen (p57) of *R. salmoninarum* (EMBL accession n° Z12174), which has been extensively used in DNA-based detection protocols for this bacterium.

DNA fingerprinting methodologies have been applied to this bacterial pathogen to gain some knowledge of its intraspecies diversity. Grayson *et al.* (1999) used the RAPD technique to assess the molecular variation between *R. salmoninarum* isolates from different geographical locations, and reported that a reproducible differentiation between isolates is possible. More recently, Grayson *et al.* (2000) carried out a RAPD analysis on 60 isolates with different places and dates of isolation, identifying as many as 21 banding patterns, thus corroborating that RAPD analysis constitutes a valid molecular tool for intraspecies typing of *R. salmoninarum.*

One of the first approaches to the application of molecular protocols for the detection of *R. salmoninarum*, consisted of the construction of a recombinant DNA library, from which a 149 base-pair fragment was isolated as species-specific for this bacterium, and used as a DNA probe for dot-blotting (Etchegaray *et al.* 1991). This probe did not cross-react with other species of gram-negative bacteria tested, but it showed cross-reactivity with *Corynebacterium striatum*.

In 1993, Mattsson *et al.* described the use of a DNA probe targeted to a short region of the 16S rRNA of *R. salmoninarum*. The radiolabelled probe was tested on isolated bacterial RNA applied onto nitrocellulose filters. It was shown that the probe reacted selectively with *R. salmoninarum*, whereas no cross-reactivities were observed with other bacterial pathogens of salmonid fish. However, some positive hybridization signal was observed with several *Arthrobacter* species. In addition, kidney samples from naturally infected fish were lysed and the extracts were dot-blotted onto nitrocellulose filters. Samples that had been proven to be positive for *R. salmoninarum* by an indirect fluorescence antibody test (IFAT) were also positive with this probe. The use of a DNA probe complementary to the rRNA transcripts has the advantage that, due to the large amount of ribosomes in growing cells, there are a high number of potential hybridization targets. Is in this species where we find an interesting example of how useful DNA-based diagnosis methodologies can be when it comes to prevent dissemination of bacterial fish pathogens: it is the case of detection of the presence of the bacterium in salmonid egg stocks (Brown *et al.* 1994). *R. salmoninarum* can be transmitted vertically, i.e. from parent to offspring via the egg (Evelyn *et al.* 1986), and the bacterium has been identified on both the surface and the inside of coho salmon eggs. Brown *et al.* (1994) designed a PCR protocol to allow detection of *R. salmoninarum* within individual salmonid eggs. A pair of primers, RS1 and RS2 (see Table 3), flanking a 501-bp DNA fragment of the gene encoding the 57-kDa p57 antigen were chosen. For DNA extraction from salmonid eggs, samples were homogenized in STE buffer (100mM NaCl, 100mM EDTA, 50 mM Tris-HCl, pH 8.0), and lysed with 10 µl of 20% sodium dodecyl sulfate (SDS). A phenol-chloroform extraction followed by an additional chloroform step was carried out and, after treatment with RNase, DNA was precipitated with ethanol and redissolved in TE buffer (50mM Tris-HCl pH 8.0, 10 mM EDTA). Salmonid eggs injected with numbers of *R. salmoninarum* cells ranging from 2 to 2000 cells per egg, all gave positive amplification of the expected fragment by PCR. Detection of as few as 2 *R. salmoninarum* cells per egg constitutes an important achievement for broodstock screening in salmonid fish farming. Other bacterial species have been reported to produce a 57-kDa protein that reacts with polyclonal antiserum against *R. salmoninarum*. However, in another report, Brown *et al.* (1995) corroborated that primers targeting the 501-bp fragment of *R. salmoninarum* p57 gene do not cross-react with those other bacterial species.

Application of PCR for the detection of *R. salmoninarum* in salmon tissues was reported by León *et al.* (1994). These authors used a *R. salmoninarum*-specific cloned DNA fragment as the basis for the design of a primer pair (see Table 3) flanking a 149-bp DNA piece. In addition, that cloned DNA fragment was labelled with photobiotin and used as a confirmative probe for the PCR products. Primers used in that study did not show cross-reactions with the other bacterial fish pathogens tested. When *R. salmoninarum* diluted DNA was used as a PCR template to assess the sensitivity of the PCR, it was calculated that about 22 *R. salmoninarum* cells can be detected per PCR tube. Samples of as little as 1 mg of salmon kidney tissue were treated with Chelex 100 resin at 95–100°C for 15 min for DNA extraction, and subjected to further PCR amplification. In this case, it was confirmed that amplification of the expected PCR band from kidney tissues of diseased coho salmon occurred, whereas no amplification took place with samples from rainbow trout with no history of BKD that were used as control.

Magnússon *et al.* (1994) found that a nested reverse-transcription PCR of a stretch of the 16S rRNA of *R. salmoninarum* was a sensitive approach to detect this bacterium from salmon ovarian fluid. A pair of primers targeted to conserved regions of the 16S rDNA gene were initially used for cDNA synthesis and preamplification. Then, *R. salmoninarum* species-specific 16S primers F176 and R492 (see Table 3) were used for the PCR. The specificity of the PCR products was tested by transferring them to nylon membranes and probing with a radiolabelled oligonucleotide specific for the *R.*

salmoninarum 16S rRNA gene. When this nested reverse-transcription PCR was carried out with samples of salmon ovarian fluid seeded with *R. salmoninarum*, the detection limit was close to 1 single cell per reaction tube. This assay was also performed with ovarian fluid samples of naturally infected fish, and it was found that the sensitivity of the PCR was higher than that obtained with an ELISA. Since standard methods of lysing bacteria work very unreliably on *R. salmoninarum* cells in tissue samples, Magnússon *et al.* (1994) described a lytic method that included lysis with achromopeptidase, as well as treatment with hexadecyltrimethylammonium bromide. Though it worked well with ovarian fluid samples, the same procedure applied to detection in kidney samples showed inhibition of the PCR reaction.

The gene coding for the p57 protein was also chosen by McIntosh *et al.* (1996) as a PCR target. In this case, a pair of primers (G6480 and G6481) (see Table 3) flanking a 376-bp region of this gene were selected. These authors developed a rapid DNA extraction protocol, and used lymphocytes as a DNA source to reduce the inhibitory effects of kidney tissue components on the PCR efficiency. For this purpose, lymphocytes were obtained from the second half of kidneys from artificially infected rainbow trout (*Oncorhynchus mykiss*), and lysed with distilled water and vigorous vortexing. The lysates were further treated with Instagene matrix (Bio-Rad). The detection limit was evaluated by inoculating lymphocyte lysates with *R. salmoninarum* suspensions. A value of $5x10^3$ cells per ml, equivalent to approximately 12 cells per PCR tube, was successfully detected with this approach. The use of lymphocytes rather than other tissues, such as kidney, seems to be a better choice for PCR detection of this pathogen. In fact, *R. salmoninarum* is an intracellular pathogen capable of surviving and multiplying within the cytoplasm of macrophages. McIntosh *et al.* (1996) reported that the use of lymphocytes results in a increase in the sensitivity of detection compared with the use of kidney homogenates. Miriam *et al.* (1997) performed a PCR assay for detection of the bacterium in both ovarian fluid and kidney tissue of Atlantic salmon (*Salmo salar* L.), and compared the results with those obtained by culture methods. As other authors did, different primer pairs were chosen within the p57 gene (see Table 3). A primer pair FL7-RL11 flanking a 349-bp fragment, a pair FL7-RL5 amplifying a 372-bp fragment, and a pair FL10-RL11 flanking a 149-bp fragment of the p57 gene. This 149-bp fragment was labelled with psoralen biotin and used as a probe. When primer combinations FL7-RL11 and FL7-RL5 were tested on *R. salmoninarum* DNA extracted from bacterial pure cultures using the Instagene matrix (Bio-Rad), the expected amplification fragments of 349- and 372-bp respectively, were obtained from between 4 and 40 cells per PCR tube. These PCR fragments were subsequently used as templates for the internal primers FL10 and RL11, producing the expected 149-bp DNA band, thus confirming that the correct *R. salmoninarum* sequences were amplified.

Kidney tissue homogenates and ovarian fluid samples were deposited on glass fiber filters, dried, and then DNA was extracted from the filters by using the Instagene matrix (Bio-Rad). DNA was also extracted in the same way from presumed negative kidney tissue homogenates and ovarian fluid samples seeded with serial dilutions of a *R. salmoninarum* cell suspension. As could be expected, sensitivity of PCR carried out with these extracts was lower compared with that obtained for pure cultures. In addition, the authors report that the decreased sensitivity was dependent on the concentration of extracted kidney tissue homogenate used as a template. To increase the sensitivity, the internal 149-bp probe described above was used to detect the PCR products amplified from samples with primers FL10 and RL11. With this assay, sensitivity was improved to a level of between one and four extracted cells. After testing many tissue samples, the authors report that this PCR approach was more sensitive than the culture method used to identify subclinically infected fish, since many PCR-positive samples had been identified as negative by the culture methods.

Recent reports on *R. salmoninarum* DNA-based detection are also based on the p57 gene. Chase and Pascho (1998) developed a nested PCR method to amplify a 383-bp fragment of this gene in the first PCR round and a 320-bp fragment in the nested PCR, and applied this protocol to the detection of the bacterium in the salmonid kidney. However, sequences of the primers are not described in their publication. When using only the first pair of primers in a conventional single PCR reaction, 1000 *R. salmoninarum* cells per reaction in kidney tissue were detected. However, the use of a second PCR step allowed detection of as few of 10 cells per reaction in kidney tissue. These authors also report that the use of the QIAamp tissue kit for DNA purification (Qiagen Inc., USA) produces good results.

Similarly, a nested reverse transcriptase PCR assay was described by Cook and Lynch (1999) to amplify a 349-bp fragment of the gene coding for the p57 antigen. These authors selected a pair of primers LP3 and UPI (see Table 3) for the reverse transcription and first PCR round. For the second PCR round, the primers used were those previously described by Miriam *et al.* (1997), flanking a 349-bp p57 gene fragment. RNA was extracted from cultured bacteria, salmonid kidney tissue and ovarian fluid seeded with the pathogen, as well as from kidney tissue of experimentally challenged and commercially raised fish, and then that RNA was amplified by RT PCR and further by PCR. When this protocol was carried out with RNA extracts of between 1 and 10 *R. salmoninarum* cells in seeded samples, it was possible to detect amplification of the specific 349-bp fragment in acrylamide gel electrophoresis followed by silver staining.

This reverse transcriptase PCR assay provides an important advance in *R. salmoninarum* detection protocols, in a similar way as that reported previously by Magnússon *et al.* (1994). Since mRNA is now the target to be detected, rather than DNA, it is more plausible that a positive PCR result actually means the presence of viable *R. salmoninarum* cells.

Piscirickettsia salmonis

In the late 1980s an infectious disease of unknown origin started to cause high mortalities in salmon aquaculture facilities in southern Chile. A microbial agent was isolated (Fryer *et al.* 1990) and further characterized as a novel intracellular organism (Fryer *et al.* 1992). Nowadays *Piscirickettsia salmonis* is responsible for high mortalities (60-90%) of different salmonid fish species cultured in seawater in Chile. From 1992 onwards, it has also been detected in Canada, Norway, Ireland and Scotland, causing mortalities in Atlantic salmon (Almendras and Fuentealba 1997).

P. salmonis is an intracellular parasitic microorganism, requiring for its isolation the use of cell lines similar to those utilized in the culture of fish viruses. Epizootics caused by *P. salmonis* generally occur a few weeks after transferring salmon to seawater, which suggests that horizontal transfer is one of the main routes of infection. However, vertical transmission of *P. salmonis* has also been demonstrated in certain cases.

Knowledge of *P. salmonis* at the genome level is rather scarce. About 22 entries of DNA sequences are currently available in the EMBL database, most of them corresponding to the ribosomal genes and their respective intergenic transcribed spacers.

Mauel *et al.* (1996) developed a nested PCR protocol for the detection of *P. salmonis* from tissues of infected fish. For the first amplification reaction, universal 16S rDNA primers were used, flanking a 1540 bp fragment. For the second amplification reaction, *P. salmonis*-specific primers were utilized to amplify a 476-bp fragment of the 16S rDNA gene. DNA extraction from salmonid kidney and spleen was performed by digestion at 65°C in a lysis buffer (50mM KCl, 10mM Tris pH 7.8, 2.5 mM Cl_2Mg, 0.1% gelatin, 0.45% NP40, 0.45% Tween 20, and 1mg ml^{-1} proteinase K), followed by boiling for 10 min, quenching on ice and further centrifugation to remove cell debris. For the PCR reactions, 5 µl of the supernatant were used. For the second amplification, 3 µl from the first reaction were used as a template. It was demonstrated that this nested PCR procedure was specific for *P. salmonis* isolates, and did not yield amplification products with a collection of 12 other species of fish pathogenic bacteria that were tested. This nested PCR made possible the detection of as few as 1 *P. salmonis* tissue culture infectious dose 50 ($TCID_{50}$). This sensitive approach thus represents a valuable tool for identification of this infectious agent, even below the detection limits of microscopic examination.

Another molecular approach applied to the detection of *P. salmonis* is that described by Marshall *et al.* (1998). These authors applied a PCR-based approach to detect the presence of amplifiable *P. salmonis* DNA in fish that do not display signs of disease that consisted of the screening of minute samples of fish serum (less than 5 μl). For this purpose, PCR primers were selected within sequences of the internal transcribed spacer region (ITS) and the flanking 23S rRNA gene. Three primers were chosen and used in two different combinations: primer pair RTS1/RTS2 (flanking a 91-bp fragment), and primer pair RTS1/RTS4 (flanking a 283-bp fragment) (see Table 3). Less than 0.5 mg of fish tissue or 5 to 20 μl of serum were mixed with 100 μl of Instagene matrix (Bio-Rad) and boiled for 15 min. After vortexing and centrifugation, supernatants were used as templates for PCR reactions. With this procedure, positive PCR detection was observed when a series of field-collected tissue and serum samples from animals displaying visible evidence of disease were examined. Positive amplification was also obtained from tissue of asymptomatic trout. The assay of fish serum samples by PCR has the advantage of being a nonlethal screening procedure. Thus, the work reported by Marshall *et al.* (1998) constitutes a valuable tool for the efficient detection of *P. salmonis* in fish.

Vibrio cholerae

Vibrio cholerae is the causative agent of cholera, a severe diarrhoeal disease that occurs most frequently in epidemic form (Wachsmuth *et al.* 1994). The bacterium colonizes the mucosal surface of the human small intestine and secretes cholera toxin. This toxin stimulates secretion of water and electrolytes by the cells of the intestinal epithelium, thus leading to the severe watery diarrhea that is characteristic of cholera.

Ingestion of the organism with contaminated seafood or water is a crucial step in the disease process. Raw or incompletely cooked shellfish as well as other seafoods are implicated in outbreaks of cholera. This means that a molecular approach for a fast and accurate detection of the presence of this important pathogen in seafood and/or water samples is desirable. At the same time, molecular procedures are needed for the diagnosis of *V. cholerae* in clinical samples.

The species *V. cholerae* includes both pathogenic and nonpathogenic strains, and they vary in their virulence gene content. Genes for virulence factors, capsular polysaccharides and new antigens are constantly being transferred between strains and biotypes of *V. cholerae* by means of phage and pathogenicity islands. Strains belonging to serogroup O1 biotype El Tor, and serogroup O139 have been described as the causative agents of diarrhea (Albert *et al.* 1993, Ramamurthy *et al.* 1993).

V. cholerae infection can be diagnosed by conventional bacteriologic techniques, such as culturing stool on a selective medium, followed by biochemical testing of colonies and its further confirmation by a slide agglutination test with specific antiserum (Ansaruzzaman *et al.* 1995). In addition, commercial immunodiagnostic tests are available (Qadri *et al.* 1995).

The genome project of *V. cholerae* was completed recently (Heidelberg *et al.* 2000), which constitutes a milestone in marine microbiology. However, DNA sequences of many *V. cholerae* genes were already known before that, which made possible their use as targets for DNA-based detection protocols. Many molecular tools for the detection of *V. cholerae* from a variety of samples have been published worldwide. Thus, it is not the intention of the authors to summarize all of them, but to point out those that can be readily found in the literature, thereby facilitating the retrieval of further information.

In 1992, Fields *et al.* developed a PCR protocol for the detection of toxigenic *V. cholerae* O1 strains from the Latin American cholera epidemic. For this purpose, a pair of PCR primers, CTX2 and CTX3 (see Table 3) flanking a 564-bp fragment of the cholera toxin A subunit gene (*ctxA*), were selected. These primers were tested with a collection of *V. cholerae* O1 strains isolated from patients, food and water. It was found that this PCR approach gave positive results with all the strains that had also been proven to be toxigenic by the routine diagnostic enzyme-linked immunosorbent assay.

Later, Koch *et al.* (1993) used PCR to detect the presence of the cholera toxin operon in a variety of food samples, including oysters, crabmeat and shrimp. Samples were spiked with *V. cholerae*, homogenized followed by a 6- to 8-h enrichment step in alkaline peptone water. Aliquots of the crude homogenates were then boiled for 5 min, frozen and thawed, and directly used for the PCR reaction. The selected primers (P1 and P3, see Table 3) flanked a 777-bp DNA fragment of the *ctxAB V. cholerae* operon. According to Fields *et al.* (1992) reducing the proportion of oyster homogenate to 1% (wt/vol) in alkaline peptone water helps to avoid PCR inhibition. Oysters artificially contaminated with 10 *V. cholerae* CFU/g produced the specific PCR band after an 8-h enrichment step. The specificity of the amplified band was tested by hybridization with a nonisotopically labeled probe corresponding to an internal portion of the 777-bp fragment of the *ctxAB* operon.

Two radiolabeled DNA probes specific for the *V. cholerae* O139 serogroup were developed by Nair *et al.* (1995), selecting a chromosomal region specific for the lipopolysaccharide O side chain synthesis of *V. cholerae* O139. These two probes did not hybridize with any non-O139 *V. cholerae* strain nor with a collection of other *Vibrio* species tested.

Later, Falklind *et al.* (1996) sequenced a *V. cholerae* DNA fragment isolated by arbitrary PCR. This genome fragment was shown to contain ORFs with homology to glycosyltransferases from other bacterial species, and was found to be specific for *V. cholerae*. Two PCR primer pairs (O139-1/O139-2, and O139-3/O139-4; see Table 3) were selected within the obtained sequence, to be used in a nested-PCR assay to identify *V. cholerae* in bacterial cultures. It was shown that the expected PCR fragment was amplified uniquely in *V. cholerae* O139 strains whereas the other cholera strains or other gram-negative bacteria tested were negative.

Primers O139-1 and O139-2 designed by Falklind *et al.* (1996) were later used by Albert *et al.* (1997) to screen diarrheal stool specimens for the presence of *V. cholerae* O139. With this PCR approach it was shown that all the culture-positive stool specimens analyzed were positive by the PCR, while the culture-negative samples were also negative by the PCR.

Recently, Chun *et al.* (1999) analyzed the sequence of the intergenic spacer regions of *V. cholerae* and *V. mimicus*, and proposed a primer pair (prVC-F/prVCM-R, see Table 3) to be used in the species-specific PCR-based identification of *V. cholerae*. However, very recently, Vieira *et al.* (2001), after analyzing a large number of *V. mimicus* and *V. cholerae* strains, reported that the PCR-based identification method proposed by Chun *et al.* (1999) can fail to discriminate between the two species.

Vibrio parahaemolyticus

Vibrio parahaemolyticus is considered to be the causative agent of as many as 50 to 70% of all reported cases of diarrhea associated with the consumption of raw or partially-cooked seafood in the summer months. Like other species of vibrios, *V. parahaemolyticus* is a naturally occurring member of the marine microbiota, and thus typical indicators of fecal pollution are not always adequate to indicate its presence. These situations emphasize the need for rapid and sensitive methods to detect the presence of pathogenic *V. parahaemolyticus* strains in ready-to-eat seafood products.

The EMBL database has more than 110 entries of DNA sequences from *V. parahaemolyticus*, including a variety of structural, regulatory and virulence factor-related genes. Researchers have paid special attention to the different hemolysin genes that are present in this species. Thus, PCR screening for the presence of hemolysin genes can help to identify pathogenic strains.

DNA fingerprinting techniques have been applied in order to gain some knowledge on the intraspecific genetic diversity of *V. parahaemolyticus*. For instance, Marshall *et al.* (1999) report a comparison of different molecular methods for typing of *V. parahaemolyticus*. This study included analysis of the RFLP patterns of the ribosomal genes (ribotyping) and of the *fla* locus, as well as the PFGE and enterobacterial repetitive intergenic consensus sequence (ERIC) PCR.

Lee *et al.* (1995) sequenced a *V. parahaemolyticus* DNA fragment (pR72H) and selected a pair of primers VP33/VP32 (See Table 3) for PCR-based detection of this bacterium. When crude bacterial lysates from *V. parahaemolyticus* pure cultures were used for the PCR reaction, a sensitivity limit of 10 cells was achieved. However, when chromosomal DNA was purified and used as a template, the detection limit was lowered to as little as 2.6 fg of DNA, which is equivalent to ca. 1 *V. parahaemolyticus* cell. The specificity of these primers was tested with 124 *V. parahaemolyticus* strains and 50 strains of other vibrios and related genera. When oyster samples were contaminated with as few as 9 CFU/g, and subjected to an enrichment step in tryptic soy broth containing 3% NaCl for 3 h at 35°C, positive PCR detection of *V. parahaemolyticus* was confirmed.

The *gyrB* gene of *V. parahaemolyticus*, coding for the b-subunit of the enzyme DNA girase was cloned and sequenced by Venkateswaran *et al.* (1998) and used as the basis for a PCR protocol for the detection of this bacterium in shrimp. Primers (see Table 3) flanking a 285-bp DNA fragment were tested with a collection of 117 *V. parahaemolyticus* strains and with 90 strains of other *Vibrio* species and 60 strains of non-*Vibrio* species, showing complete specificity for *V. parahaemolyticus*. As little as 4 pg of purified DNA, and as few as 5 CFU were detected per 100 ml PCR mixtures. Also, when shrimp homogenates were inoculated with as few as 1.5 CFU of *V. parahaemolyticus* per g, the bacterium was succesfully detected with PCR after an enrichment step of 18 h.

The *V. parahaemolyticus tl* gene was used by Brasher *et al.* (1998) for the detection of this species in shellfish, together with other four bacterial pathogens by a multiplex-PCR approach. As reported by other authors, boiling of oyster homogenates in the presence of 18% (wt/vol) Chelex 100 (Bio-Rad) produced DNA suitable for further PCR amplification.

The use of DNA probes for the detection of the thermolabile hemolysin gene of *V. parahaemolyticus* was recently reported by McCarthy *et al.* (1999). They evaluated both alkaline phosphatase- and digoxigenin-labelled probes. Specificity of the probe was corroborated by testing a collection of *V. parahaemolyticus* strains as well as other *Vibrio* and non-vibrio species. The two differently-labelled probes were shown to be equally useful for the detection of this bacterium.

Kim *et al.* (1999) used the nucleotide sequence of *toxR* gene as a target for PCR detection, as well as for a DNA colony-hybridization test. *toxR* is a regulatory gene that is present in many *Vibrio* species, and whose degree of sequence similarity between species is much lower than that of the 16S rRNA gene (Osorio and Klose, 2000). The *toxR* probe described by Kim *et al.* (1999) was shown to be suitable for identification of the 373 *V. parahaemolyticus* strains tested. However, closely related *Vibrio* species, such as *V. alginolyticus*, gave weakly positive results.

The PCR protocol was based on a primer set flanking a 399-bp *toxR* gene fragment (see Table 3). Although certain non-specific amplification bands were produced with some *V. vulnificus* strains, this *toxR*-based PCR protocol can be considered as a specific and rapid method for identification of *V. parahaemolyticus* strains.

A multiplex PCR approach reported by Bej *et al.* (1999) was successfully used to detect *V. parahaemolyticus* by simultaneous amplification in a single PCR reaction of partial *tl*, *tdh* and *trh* genes, which code for thermolabile hemolysin, thermostable direct hemolysin and thermostable direct hemolysin-related hemolysin respectively. The 111 *V. parahaemolyticus* strains tested, showed positive amplification of the *tl* gene. Sixty of the isolates showed positive amplification of the *tdh* gene fragment, and 43 for the *trh* gene fragment. This procedure thus gives additional information about clinically significant *V. parahaemolyticus* isolates, since it enables detection of total and hemolysin-producing strains. Application of this multiplex PCR protocol to oyster tissue homogenates pre-enriched in alkaline peptone water showed detection of at least 10^2 cells per 10 g of homogenate using the *tdh* gene as a target, and at least 10 cells when targets were *tl* and *trh* genes. More recently, the *tdh* gene of this species was chosen as a target for a nonisotopic DNA hybridization protocol (McCarthy *et al.* 2000).

Future Perspectives

Molecular techniques of diagnosis are becoming a very useful tool for the detection and identification of microorganisms. The use of these techniques is widespread in human microbiology laboratories, both in clinical and research applications. However, its use in fish microbiology still needs additional research to solve the many problems encountered. First of all, many bacterial pathogens that affect sea-farming are still poorly studied in terms of their DNA sequence. For many marine bacteria, only the 16S rRNA gene sequence is available, if at all. Sometimes, when a given gene target is selected to design PCR primers or DNA probes, how similar that gene is between related species or strains remains unknown. In addition, genetic research is needed to guarantee that a selected gene target is present in all the strains to be detected.

One of the main limiting steps when applying the PCR or RT-PCR to fish and environmental samples is obtaining good quality DNA and RNA, free from inhibitors. Thankfully, during recent years, we have assisted in designing an increasing variety of commercial DNA and RNA purification kits. It is expected that new improvements in this field will help to overcome the most common problems encountered in sample nucleic acid extraction.

Any laboratory that works with bacteria associated with sea-farming, or aquaculture in general, and which intends to give answers and solutions to a demanding fish-farm industry, should be prepared to screen a single sample for the presence of several potential pathogens simultaneously. This means that with a small volume of fish tissue, water or sediment, one laboratory can detect in a matter of hours, whether those samples are free from a list of important bacterial pathogens. This can be possible if, for each bacterial pathogen and each type of sample, DNA-extraction protocols, selected DNA targets and specific PCR primers are extensively validated.

Another improvement that needs to be brought into general use in the molecular diagnostic microbiology of fish is the ability to carry out non-lethal diagnostic analyses. This means that nucleic acid targets can be obtained without killing the fish, which is of great interest for the screening of valuable broodstock individuals. Though some approaches have been reported for the analysis of minute blood or serum samples of fish (Marshall *et al.* 1998), this methodology has not yet been shown to be valid for detection of all the important bacterial species as well as for all the cultured fish species.

There is another important idea that should be considered in the interpretation of the results obtained in PCR-based diagnosis. PCR protocols designed to detect DNA targets do not actually demonstrate the presence of viable target bacterial cells. Thus, if RNA is the target to be detected, rather than DNA, it is more plausible that a positive PCR result actually means the presence of viable cells. Detection of bacterial RNA by RT-PCR, rather than DNA, may give a more accurate idea of the presence of potentially active cells.

Acknowledgements

The authors wish to thank the Ministerio de Ciencia y Tecnología for grants MAR99-0478, and AGL2000-0492, and to EEC for the FEDER project 1FD97-0156. Carlos R. Osorio thanks the University of Santiago for a postdoctoral research fellowship.

References

Albert, M.J., A.K. Siddique, M.S. Islam, A.S.G. Faruque, M. Ansaruzzaman, S.M. Faruque, and R.B. Sack. 1993. Large outbreak of clinical cholera due to *Vibrio cholerae* non-O1 in Bangladesh. *Lancet* 341: 704.

Albert, M.J., D. Islam, S. Nahar, F. Qadri, S. Falklind, and A. Weintraub. 1997. Rapid detection of *Vibrio cholerae* O139 Bengal from stool specimens by PCR. *J. Clin. Microbiol.* 35: 1633-1635.

Alm, E.W., D.B. Oerther, N. Larsen, D.A. Stahl, and L. Raskin. 1996. The oligonucleotide probe database. *Appl. Environ. Microbiol.* 27: 977-985.

Almendras, F.E. and I.C. Fuentealba. 1997. Salmonid rickettsial septicemia caused by *Piscirickettsia salmonis*: a review. *Dis. Aquat. Org.* 29: 137-144.

Alsina, M., J. Martínez-Picado, J. Jofre, and A.R. Blanch. 1994. A medium for presumptive identification of *Vibrio anguillarum*. *Appl. Environ. Microbiol.* 60: 1681-1683.

Ansaruzzaman, M., M. Rahman, A.K.M.G. Kibriya, N.A. Bhuiyan, M.S. Islam, and M.J. Albert. 1995. Isolation of sucrose late-fermenting and nonfermenting variants of *Vibrio cholerae* O139 Bengal: implications for diagnosis of cholera. *J. Clin. Microbiol.* 33: 1339-1340.

Ansorge, W., B. Sproat, J. Stegemann, and C. Schwager. 1986. A non-radioactive automated method for DNA sequence determination. *J. Biochem. Biophys. Methods* 13: 315-323.

Aoki, T., I. Hirono, T. De Castro, and T. Kitao. 1989. Rapid identification of *Vibrio anguillarum* by colony hybridization. *J. Appl. Ichthyol.* 5: 67-73.

Aoki, T., I. Hirono, and A. Hayashi. 1995. The fish-pathogenic bacterium *Pasteurella piscicida* detected by the polymerase chain reaction. Pages 347-353 in *Diseases in Asian Aquaculture II*, M. Shariff, J.R. Arthur and R.P Subasinghe, eds., Fish Health Section, Asian Fisheries Society, Manila.

Aoki, T., D. Ikeda, T. Katagiri, and I. Hirono. 1997. Rapid detection of the fish-pathogenic bacterium *Pasteurella piscicida* by polymerase chain reaction targeting nucleotide sequences of the species-specific plasmid pZP1. *Fish Pathology* 32: 143-151.

Aono, E., H. Sugita, J. Kawasaki, H. Sakakibara, T. Takahashi, K. Endo, and Y. Deguchi. 1997. Evaluation of the polymerase chain reaction method for identification of *Vibrio vulnificus* isolated from marine environments. *J. Food Protect.* 60: 81-83.

Argenton, F., S. De Mas, C. Malocco, L. Dalla Valle, G. Giorgetti, and L. Colombo. 1996. Use of random DNA amplification to generate specific molecular probes for hybridization tests and PCR-based diagnosis of *Yersinia ruckeri*. *Dis. Aquat. Org.* 24: 121-127.

Arias, C.R., E. Garay, and R. Aznar. 1995. Nested PCR method for rapid and sensitive detection of *Vibrio vulnificus* in fish, sediments, and water. *Appl. Environ. Microbiol.* 61: 3476-3478.

Arias, C.R., L. Verdonck, J. Swings, E. Garay, and R. Aznar. 1997. Intraspecific differentiation of *Vibrio vulnificus* biotypes by amplified fragment length polymorphism and ribotyping. *Appl. Environ. Microbiol.* 63: 2600-2606.

Austin, B., and C. Adams. 1996. Fish pathogens. Pages 197-243 in *The Genus Aeromonas*, B. Austin, M. Altwegg, P.J. Gosling and S. Joseph, eds., John Wiley and Sons Ltd., West Sussex, England.

Bader, J.A. and E.B. Shotts Jr. 1998. Identification of *Flavobacterium* and *Flexibacter* species by species-specific polymerase chain reaction primers to the 16S ribosomal RNA gene. *Aquatic Animal Health* 10: 311-319.

Balfry, S.K., L.J. Albright, and T.P.T. Evelyn. 1996. Horizontal transfer of *Renibacterium salmoninarum* among farmed salmonids via the fecal-oral route. *Dis. Aquat. Org.* 25: 63-69.

Bej, A.K. and M.H. Mahbubani. 1992. Application of the polymerase chain reaction in environmental microbiology. *PCR Methods Appl.* 1: 151-155.

Bej, A.K., D.P. Patterson, C.W. Brasher, M.C.L. Vickery, D.D. Jones, and C.A. Kaysner. 1999. Detection of total and hemolysin-producing *Vibrio parahaemolyticus* in shellfish using multiplex PCR amplification of *tl*, *tdh* and *trh*. *J. Microbiol. Meth.* 36: 215-225.

Bernoth, E.-M., A. E. Ellis, P. J. Midtlyng, G. Olivier, and P. Smith, eds., 1997. *Furunculosis*, Academic Press., London, UK.

Biosca, E.G., C. Amaro, C. Esteve, E. Alcaide, and E. Garay. 1991. First record of *Vibrio vulnificus* biotype 2 from diseased European eels (*Anguilla anguilla*). *J. Fish Dis.* 14: 103-109.

Biosca, E.G., J.D. Oliver, and C. Amaro. 1996. Phenotypic characterization of *Vibrio vulnificus* biotype 2, a lipopolysaccharide-based homogeneous O serogroup within *Vibrio vulnificus*. *Appl. Environ. Microbiol.* 62: 918-927.

Blake, P., R. Weaver, and D. Hollis. 1980. Diseases of human (other than cholera) caused by vibrios. *Annu. Rev. Microbiol.* 34: 341-367.

Brasher, C.W., A. Depaola, D.D. Jones, and A.K. Bej. 1998. Detection of microbial pathogens in shellfish with multiplex PCR. *Current Microbiol.* 37: 101-107.

Brauns, L.A., M.C. Hudson, and J.D. Oliver. 1991.Use of the polymerase chain reaction in detection of culturable and nonculturable *Vibrio vulnificus* cells. *Appl. Environ. Microbiol.* 57: 2651-2655.

Brauns, L.A. and J.D. Oliver. 1995. Polymerase chain reaction of whole cell lysates for the detection of *Vibrio vulnificus*. *Food. Biotechnol.* 8: 1-6.

Brown, L.L., G.K. Iwama, T.P.T. Evelyn, W.S. Nelson, and R.P. Levine. 1994. Use of the polymerase chain reaction (PCR) to detect DNA from *Renibacterium salmoninarum* within individual salmonid eggs. *Dis. Aquat. Org.* 18: 165-171.

Brown, L.L., T.P.T. Evelyn, G.K. Iwama, W.S. Nelson, and R.P. Levine. 1995. Bacterial species other than *Renibacterium salmoninarum* cross-react with antisera against *R. salmoninarum* but are negative for the p57 gene of *R. salmoninarum* as detected by the polymerase chain reaction (PCR). *Dis. Aquat. Org.* 21: 227-231.

Buck, J.D., N.A. Overstrom, G.W. Patton, H.F. Anderson, and J.F. Gorzelany. 1991. Bacteria associated with stranded cetaceans from the northeast USA and southwest Florida Gulf coasts. *Dis. Aquat. Org.* 10: 147-152.

Cepeda, C. and Y. Santos. 2000. Rapid and low-level toxic PCR-based method for routine identification of *Flavobacterium psychrophilum. Internatl. Microbiol.* 3: 235-238.

Cepeda, C., S.G. Márquez, and Y. Santos. 2002. Detection of *Flexibacter maritimus* from fish tissue by rapid nested PCR amplification. *Appl. Environ. Microbiol.* (submitted).

Cerdá-Cuellar, M., J. Jofre, and A.R. Blanch. 2000. A selective medium and a specific probe for detection of *Vibrio vulnificus. J. Appl. Microbiol.* 66: 855-859.

Chase, D.M. and R.J. Pascho. 1998. Development of a nested polymerase chain reaction for amplification of a sequence of the p57 gene of *Renibacterium salmoninarum* that provides a highly sensitive method for detection of the bacterium in salmonid kidney. *Dis. Aquat. Org.* 34: 223-229.

Chu, S., S. Cavaignac, J. Feutrier, B.M. Phipps, M. Kostrzynska, W.W. Kay, and T.J. Trust. 1991. Structure of the tetragonal surface virulence array protein and gene of *Aeromonas salmonicida. J. Biol. Chem.* 266: 15258-15265.

Chun, J., A. Huq, and R.R. Colwell. 1999. Analysis of 16S-23S rRNA intergenic spacer regions of *Vibrio cholerae* and *Vibrio mimicus. Appl. Environ. Microbiol.* 65: 2202-2208.

Clarridge, J.E. and S. Zighelboim-Daum. 1985. Isolation and characterization of two hemolytic phenotypes of *Vibrio damsela* associated with a fatal wound infection. *J. Clin. Microbiol.* 21: 302-306.

Coleman,S.S., D.M. Melanson, E.G. Biosca, and J.D. Oliver. 1996. Detection of *Vibrio vulnificus* biotypes 1 and 2 in eels and oysters by PCR amplification. *Appl. Environ. Microbiol.* 62: 1378-1382.

Coleman, S.S. and J.D. Oliver. 1996. Optimization of conditions for the polymerase chain reaction amplification of DNA from culturable and nonculturable cells of *Vibrio vulnificus. FEMS Microbiol. Ecol.* 19: 127-132.

Cook, M. and W.H. Lynch. 1999. A sensitive nested reverse transcriptase PCR assay to detect viable cells of the fish pathogen *Renibacterium salmoninarum* in atlantic salmon (*Salmo salar* L.). *Appl. Environ. Microbiol.* 65: 3042-3047.

Cutter, D.L. and A.S. Kreger. 1990. Cloning and expression of the damselysin gene from *Vibrio damsela. Infect. Immun.* 58: 266-268.

Devesa, S., J.L. Barja, and A.E. Toranzo. 1989. Ulcerative skin and fin lesions in reared turbot, *Scophthalmus maximus* (L.). *J. Fish Dis.* 12: 323-333.

Dorsch, M., D. Lane, and E. Stackebrandt. 1992. Towards a phylogeny of the genus *Vibrio* based on 16S rRNA sequences. *Int. J. Syst. Bacteriol.* 42: 58-63.

Egidius, E., R. Wiik, K. Andersen, K.A. Hoff, and B. Hjeltnes. 1986. *Vibrio salmonicida* sp. nov., a new fish pathogen. *Int. J. Syst. Bacteriol.* 36: 518-520.

El Morabit, A. 1999. Caracterización de cepas de *Aeromonas salmonicida* ssp. *salmonicida* aisladas de peces marinos. Master Thesis, University of Santiago, Spain.

Emmerich, R. and E. Weibel. 1894. Uber eine durch Bakterien erzeugte Seuche unter den Forellen. *Arch. Hyg. Bakteriol* 21: 1-21 1894.

Enger, Ø. 1997. Survival and inactivation of *Aeromonas salmonicida* outside the host—A most superficial way of life. Pages 159-177 in *Furunculosis,* E.-M. Bernoth, A.E. Ellis, P.J. Midtlyng, G. Olivier, and P. Smith, eds., Academic Press., London, UK.

Erlich, H.A. 1989. *PCR technology, principles and applications for DNA amplification,* Stockton Press, New York.

Etchegaray, J.P., M.A. Martínez, M. Krauskopf, and G. León. 1991. Molecular cloning of *Renibacterium salmoninarum* DNA fragments. *FEMS Microbiol. Lett.* 79: 61-64.

Evenden, A.J., T.H. Grayson, M.L. Gilpin, and C.B. Munn. 1993. *Renibacterium salmoninarum* and bacterial kidney disease—the unfinished jigsaw. *Annu. Rev. Fish Dis.* 3: 87-104.

Evelyn, T.P.T., L. Prosperi-Porta, and J.E. Ketcheson. 1986.Experimental intra-ovum infection of salmonid eggs with *Renibacterium salmoninarum* and vertical transmission of the pathogen with such eggs despite their treatment with erythromycin. *Dis. Aquat. Org.* 1: 197-202.

Falklind, S., M. Stark, M.J. Albert, M. Uhlen, J. Lundeberg, and A. Weintraub. 1996. Cloning and sequencing of a region of *Vibrio cholerae* O139 Bengal and its use in PCR-based detection. *J. Clin. Microbiol.* 34: 2904-2908.

Fields, P.I., T. Popovic, K. Wachsmuth, and Ø. Olsvik. 1992. Use of polymerase chain reaction for detection of toxigenic *Vibrio cholerae* O1 strains from the Latin American cholera epidemic. *J. Clin. Microbiol.* 30: 2118-2121.

Fluit, A.C., R. Torensma, M.J. Visser, C.J. Aarsman, M.J. Poppelier, B.H. Keller, P. Klapwijk, and J. Verhoef. 1993. Detection of *Listeria monocytogenes* in cheese with the magnetic immuno-polymerase chain reaction assay. *Appl. Environ. Microbiol.* 59: 1289-1293.

Fouz, B., J.L. Larsen, B. Nielsen, J.L. Barja, and A.E. Toranzo. 1992. Characterization of *Vibrio damsela* strains isolated from turbot *Scophthalmus maximus* in Spain. *Dis. Aquat. Org.* 12: 155-166.

Fouz, B., J.L. Barja, C. Amaro, C. Rivas, and A.E. Toranzo. 1993. Toxicity of the extracellular products of *Vibrio damsela* isolated from diseased fish. *Curr. Microbiol.* 27: 341-347.

Fratamico, P.M., F.J. Schultz, and R.L. Buchanan. 1992. Rapid isolation of *Escherichia coli* O157 : H7 from enrichment cultures of foods using an immunomagnetic separation method. *Food Microbiol.* 9: 105-113.

Fryer, J.L., and J.E. Sanders. 1981. Bacterial kidney disease of salmonid fish. *Ann. Rev. Microbiol.* 35: 273-298.

Fryer, J.L., C.N. Lannan, L.H. Garcés, J.J. Larenas, and P.A. Smith. 1990. Isolation of a rickettsiales-like organism from diseased coho salmon (*Oncorhynchus kisutch*) in Chile. *Fish Pathol.* 25: 107-114.

Fryer, J.L., C.N. Lannan, S.J. Giovannoni, and N.D. Wood. 1992. *Piscirickettsia salmonis* gen. nov., sp. nov., the causative agent of an epizootic disease in salmonid fishes. *Int. J. Syst. Bacteriol.* 42: 120-126.

Gauthier, G., B. Lafay, R. Ruimy, V. Breittmayer, J.L. Nicolas, M. Gauthier, and R. Christen. 1995. Small-Subunit rRNA Sequences and Whole DNA. Relatedness Concur for the Reassignment of *Pasteurella piscicida* (Snieszko *et al.*) Janssen and Surgalla to the Genus *Photobacterium* as *Photobacterium damsela* subsp. *piscicida* comb. nov. *Int. J. Syst. Bacteriol.* 45: 139-144.

Giovannetti, L. and S. Ventura. 1995. Application of total DNA restriction pattern analysis to identification and differentiation of bacterial strains. Pages 181–200 in *Diagnostic Bacteriology Protocols*, J. Howard and D.M. Whitcombe, eds., Humana Press Inc., Totowa, New Jersey.

González, S.F., C.R. Osorio, C. Cepeda, and Y. Santos. 2000. Diagnóstico de la vibriosis por PCR. III Aquatic Environmental Microbiology Meeting, Spanish Society for Microbiology, Santiago de Compostela, May 2000.

Grayson, T.H., L.F. Cooper, F.A. Atienzar, M.R. Knowles, and M.L. Gilpin. 1999. Molecular differentiation of *Renibacterium salmoninarum* isolates from worldwide locations. *Appl. Environ. Microbiol.* 65: 961-968.

Grayson, T.H., F.A. Atienzar, S.M. Alexander, L.F. Cooper, and M.L. Gilpin. 2000. Molecular diversity of *Renibacterium salmoninarum* isolates determined by Randomly Amplified Polymorphic DNA analysis. *Appl. Environ. Microbiol.* 66: 435-438.

Grimont, F. and P.A.D. Grimont. 1986. Ribosomal ribonucleic acid gene restriction patterns as potential taxonomic tools. *Ann. Inst. Pasteur/Microbiol.* 137B: 165-175.

Grimont, F. and P.A.D. Grimont. 1995. Determination of rRNA gene restriction patterns. Pages 181-200 in *Diagnostic Bacteriology Protocols*, J. Howard and D.M. Whitcombe, eds., Humana Press Inc., Totowa, New Jersey.

Gustafson, C.E., C.J. Thomas, and T.J. Trust. 1992. Detection of *Aeromonas salmonicida* from fish by using polymerase chain reaction amplification of the virulence surface array protein gene. *Appl. Environ. Microbiol.* 58: 3816-3825.

Heidelberg, J.F., J.A. Eisen, W.C. Nelson, R.A. Clayton, M.L. Gwinn, R.J. Dodson, D.H. Haft, E.K. Hickey, J.D. Peterson, L. Umayam, S.R. Gill, K.E. Nelson, T.D. Read, H. Tettelin, D. Richardson, M.D. Ermolaeva, J. Vamathevan, S. Bass, H. Qin, I. Dragoi, P. Sellers, L. McDonald, T. Utterback, R.D. Fleishmann, W.C. Nierman, O. White, S.L. Salzberg, H.O. Smith, R.R. Colwell, J.J. Mekalanos, J.C. Venter, and C.M. Fraser. 2000. DNA sequence of both chromosomes of the cholera pathogen *Vibrio cholerae*. *Nature* 406: 477-483.

Hill, W.E., S.P. Keasler, M.W. Trucksess, P. Feng, C.A. Kaysner, and K.A. Lampel. 1991. Polymerase chain reaction identification of *Vibrio vulnificus* in artificially contaminated oysters. *Appl. Environ. Microbiol.* 57: 707-711.

Hill, W.E. 1996. The polymerase chain reaction: applications for the detection of foodborne pathogens. *Critical Reviews in Food Science and Nutrition* 36: 123-173.

Hillier, A.J. and B.E. Davidson. 1995. Pulsed field gel electrophoresis. Pages 149-164 in *Diagnostic Bacteriology Protocols*, J. Howard and D.M. Whitcombe, eds., Humana Press Inc., Totowa, New Jersey.

Hiney, M., M.T. Dawson, D.M. Heery, P.R. Smith, F. Gannon, and R. Powell. 1992. DNA probe for *Aeromonas salmonicida*. *Appl. Environ. Microbiol.* 58: 1039-1042.

Høie, S., M. Heum, and O.F. Thoresen. 1997. Evaluation of a polymerase chain reaction-based assay for the detection of *Aeromonas salmonicida* subsp. *salmonicida* in Atlantic salmon *Salmo salar*. *Dis. Aquat. Org.* 30: 27-35.

Høi. L., A. Dalsgaard, J.L. Larsen, J.M. Warner, and J.D. Oliver. 1997. Comparison of ribotyping and randomly amplified polymorphic DNA PCR for characterization of *Vibrio vulnificus*. *Appl. Environ. Microbiol.* 63: 1674-1678.

Innis, M.A., D.H. Gelfand, J.J. Sninsky, and T.J. White. 1990. *PCR protocols, a guide to methods and applications*, Academic Press, San Diego.

Jablonski, E., E.W. Moomaw, R.H. Tullis, and J.L. Ruth. 1986. Preparation of oligonucleotide-alkaline phosphatase conjugates and their use as hybridization probes. *Nucleic Acids Res.* 14: 6115-6128.

Jaykus, L.-A., R. De Leon, and M.D. Sobsey. 1996. A virion concentration method for detection of human enteric viruses in oysters by PCR and oligoprobe hybridization. *Appl. Environ. Microbiol.* 62: 2074-2080.

Kim, E-H. and T. Aoki. 1993. The structure of the chloramphenicol resistance gene on a transferable R plasmid from the fish pathogen, *Pasteurella piscicida*. *Microbiol. Immunol.* 37: 705-712.

Kim, E-H. and T. Aoki. 1994. The transposon-like structure of IS*26*-tetracycline, and kanamycin resistance determinant derived from transferable R plasmid of fish pathogen, *Pasteurella piscicida*. *Microbiol. Immunol.* 38: 31-38.

Kim, E-H., and T. Aoki. 1996a. Sulfonamide resistance gene in a transferable R plasmid of *Pasteurella piscicida*. *Microbiol. Immunol.* 40: 397-399.

Kim, E-H. and T. Aoki. 1996b. Sequence analysis of the florfenicol resistance gene encoded in the transferable R-plasmid of a fish pathogen, *Pasteurella piscicida*. *Microbiol. Immunol.* 40: 665-669.

Kim, Y.B., J. Okuda, C. Matsumoto, N. Takahashi, S. Hashimoto, and M. Nishibuchi. 1999. Identification of *Vibrio parahaemolyticus* strains at the species level by PCR targeted to the *toxR* gene. *J. Clin. Microbiol.* 37: 1173-1177.

Kitao, T., T. Aoki, M. Fukudome, K. Kawano, Y. Wada, and Y. Mizuno. 1983. Serotyping of *Vibrio anguillarum* isolated fom diseased freshwater fish in Japan. *J. Fish Dis.* 6: 175-181.

Koch, W.H., W.L. Payne, B.A. Wentz, and T.A. Cebula. 1993. Rapid polymerase chain reaction method for detection of *Vibrio cholerae* in foods. *Appl. Environ. Microbiol.* 59: 556-560.

Lane, D.J. 1991. 16S/23S sequencing. Pages 115-175 in *Nucleic Acid Techniques in Bacterial Systematics*, E. Stackebrandt and M. Goodfellow, eds., New York: John Wiley and Sons, Chichester, United Kingdom.

Lee, C.Y., S.F. Pan, and C.H. Chen. 1995. Sequence of a cloned pR72H fragment and its use for detection of *Vibrio parahaemolyticus* in shellfish with the PCR. *Appl. Environ. Microbiol.* 61: 1311-1317.

Lee, S.E., S.Y. Kim, S.J. Kim, H.S. Kim, J.H. Shin, S.H. Choi, S.S Chung, and J.H. Rhee. 1998. Direct identification of *Vibrio vulnificus* in clinical specimens by nested PCR. *J. Clin. Microbiol.* 36: 2887-2892.

Lee, J.Y., Y.B. Bang, J.H. Rhee, and S.H. Choi. 1999. Two-stage nested PCR effectiveness for direct detection of *Vibrio vulnificus* in natural samples. *J. Food Science* 64: 158-162.

León, G., N. Maulén, J. Figueroa, J. Villanueva, C. Rodríguez, M.I. Vera, and M. Krauskopf. 1994. A PCR-based assay for the identification of the fish pathogen *Renibacterium salmoninarum. FEMS Microbiol. Lett.* 115: 131-136.

Linder, K. and J.D. Oliver. 1989. Membrane fatty acid and virulence changes in the viable but nonculturable state of *Vibrio vulnificus*. *Appl. Environ. Microbiol.* 55: 2837-2842.

Lippke, J.A., M.N. Strzempko, F.F. Raia, S.L. Simon, and C.K. French. 1987. Isolation of intact high-molecular-weight DNA by using guanidine isothiocyanate. *Appl. Environ. Microbiol.* 53: 2588-2589.

Love, M., D.T. Fisher, J.E. Hose, J.J. Farmer, F.W. Hickman, and G.R. Fanning. 1981. *Vibrio damsela*, as a marine bacterium, causes skin ulcers on the damselfish *Chromis punctipinnis*. *Science* 214: 1140-1141.

Magariños B., J.L. Romalde, I. Bandín, B. Fouz, and A.E. Toranzo. 1992. Phenotypic, antigenic and molecular characterization of *Pasteurella piscicida* strains isolated from fish. *Appl. Environ. Microbiol.* 58: 3316-3322.

Magariños, B., J.L. Romalde, J.L. Barja, and A.E. Toranzo. 1994. Evidence of a dormant but infective state of the fish pathogen *Pasteurella piscicida* in seawater and sediment. *Appl. Environ. Microbiol.* 60: 180-186.

Magariños, B., A.E. Toranzo, and J.L. Romalde. 1996. Phenotipic and pathobiological characteristics of *Pasteurella piscicida*. *Ann. Rev. Fish. Dis.* 6: 41-64.

Magariños, B., C.R. Osorio, A.E. Toranzo, and J.L. Romalde. 1997. Applicability of ribotyping for intraspecific classification and epidemiological studies of *Photobacterium damselae* subsp. *piscicida*. *Syst. Appl. Microbiol.* 20: 634-639.

Magariños, B., A.E. Toranzo, J.L. Barja, and J.L. Romalde. 2000. Existence of two geographically-linked clonal lineages in the bacterial fish pathogen *Photobacterium damselae* subsp. *piscicida* evidenced by random amplified polymorphic DNA analysis. *Epidemiol. Infect.* 125: 213-219.

Magnússon, H.B., O.H. Frigjónsson, O.S. Andrésson, E. Benediktsdóttir, S. Gudmundsdóttir and V. Andrésdóttir. 1994. *Renibacterium salmoninarum*, the causative agent of bacterial kidney disease in salmonid fish, detected by nested reverse transcription-PCR of 16S rRNA sequences. *Appl. Environ. Microbiol.* 60: 4580-4583.

Marchesi, J.R., T. Sato, A.J. Weightman, T.A. Martin, J.C. Fry, S.J. Hiom, and W.G. Wade. 1998. Design and evaluation of useful bacterium-specific PCR primers that amplify genes coding for bacterial 16S rRNA. *Appl. Environ. Microbiol.* 64: 795-799.

Marshall, S., S. Heath, V. Enriquez, and C. Orrego. Minimally invasive detection of *Piscirickettsia salmonis* in cultivated salmonids via the PCR. *Appl. Environ. Microbiol.* 64: 3066-3069.

Marshall, S., C.G. Clark, G. Wang, M. Mulvey, M.T. Kelly, and W.M. Johnson. 1999. Comparison of molecular methods for typing *Vibrio parahaemolyticus*. *J. Clin. Microbiol.* 37: 2473-2478.

Martinez-Picado, J., A.R. Blanch, and J. Jofre. 1994. Rapid detection and identification of *Vibrio anguillarum* by using a specific oligonucleotide probe complementary to 16S rRNA. *Appl. Environ. Microbiol.* 60: 732-737.

Martinez-Picado, J., M. Alsina, A.R. Blanch, M. Cerdá, and J. Jofre. 1996. Species-specific detection of *Vibrio anguillarum* in marine aquaculture environments by selective culture and DNA hybridization. *Appl. Environ. Microbiol.* 62: 443-449.

Mattsson, J.G., H. Gersdorf, E. Jansson, T. Hongslo, U.B. Gobel, and K.-E. Johansson. 1993. Rapid identification of *Renibacterium salmoninarum* using an oligonucleotide probe complementary to 16S rRNA. *Molec. Cell. Probes* 7: 25-33.

Mauel, M.J., S.J. Giovannoni, and J.L. Fryer. 1996. Development of polymerase chain reaction assays for detection, identification, and differentiation of *Piscirickettsia salmonis*. *Dis. Aquat. Org.* 26: 189-195.

McCarthy, S.A., A. Depaola, D.W. Cook, C.A. Kaysner, and W.E. Hill. 1999. Evaluation of alkaline phosphatase- and digoxigenin-labelled probes for detection of the thermolabile hemolysin (*tlh*) gene of *Vibrio parahaemolyticus*. *Lett. Appl. Microbiol.* 28: 66-70.

McCarthy, S.A., A. Depaola, C.A. Kaysner, W.E. Hill, and D.W. Cook. 2000. Evaluation of nonisotopic DNA hybridization methods for detection of the *tdh* gene of *Vibrio parahaemolyticus*. *J. Food Protection* 63: 1660-1664.

McIntosh, D., P.G. Meaden, and B. Austin. 1996. A simplified PCR-based method for the detection of *Renibacterium salmoninarum* utilizing preparations of rainbow trout (*Oncorhynchus mykiss*, Walbaum) lymphocytes. *Appl. Environ. Microbiol.* 62: 3929-3932.

Mileham, A.J. 1995. Identification of microorganisms using random primed PCR. Pages 257-267 in *Diagnostic Bacteriology Protocols*, J. Howard and D.M. Whitcombe, eds., Humana Press Inc., Totowa, New Jersey.

Miller D.N., J.E. Bryant, E.L. Madsen, and W.C. Ghiorse. 1999. Evaluation and optimization of DNA extraction and purification procedures for soil and sediment samples. *Appl. Environ. Microbiol.* 65: 4715-4724.

Miriam, A., S.G. Griffiths, J.E. Lovely, and W.H. Lynch. 1997. PCR and probe-PCR assays to monitor broodstock atlantic salmon (*Salmo salar* L.) ovarian fluid and kidney tissue for presence of DNA of the fish pathogen *Renibacterium salmoninarum*. *J. Clin. Microbiol.* 35: 1322-1326.

Miyata, M., T. Aoki, V. Inglis, T. Yoshida, and M. Endo. 1995. RAPD analysis of *Aeromonas salmonicida* and *Aeromonas hydrophila*. *J. Appl. Bacteriol.* 79: 181-185.

Miyata, M., V. Inglis, and T. Aoki. 1996. Rapid identification of *Aeromonas salmonicida* subspecies *salmonicida* by the polymerase chain reaction. *Aquaculture* 141: 13-24.

Mooney, J., E. Powell, C. Clabby, and R. Powell. 1995. Detection of *Aeromonas salmonicida* in wild Atlantic salmon using a specific DNA probe test. *Dis. Aquat. Org.* 21: 131-135.

Morris, J.G Jr., R. Wilson, D.G. Hollis, R.E. Weaver, H.G. Miller, C.O. Tacket, F.W. Hickman, and P.A. Blake. 1982. Illness caused by *Vibrio damsela* and *Vibrio hollisae*. *Lancet* 1: 1294-1297.

Morris, J.G Jr., A.C. Wright, D.M. Roberts, P.K. Wood, L.M. Simpson, and J.D. Oliver. 1987. Identification of environmental *Vibrio vulnificus* isolates with a DNA probe for the cytotoxin-hemolysin gene. *Appl. Environ. Microbiol.* 53: 193-195.

Muroga, K., Y. Jo, and M. Nishibuchi. 1976. Pathogenic *Vibrio* isolated from cultured eels. Characteristics and taxonomic status. *Fish Pathol.* 11: 141-145.

Nair, G. B., P.K. Bag, T. Shimada, T. Ramamurthy, T. Takeda, S. Yamamoto, H. Kurazono, and Y. Takeda. 1995. Evaluation of DNA probes for specific detection of *Vibrio cholerae* O139 Bengal. *J. Clin. Microbiol.* 33: 2186-2187.

Naka, H., I. Hirono, and T. Aoki. 1999. Genome analysis in *Pasteurella piscicida*. Fourth symposium on diseases in Asian aquaculture: Aquatic animal health for sustainability, Cebu City, Philippines, 22-26 November 1999.

O´Brien, D., J. Mooney, D. Ryan, E. Powell, M. Hiney, P.R. Smith, and R. Powell. 1994. Detection of *Aeromonas salmonicida*, causal agent of furunculosis in salmonid fish, from the tank effluent of hatchery-reared atlantic salmon smolts. *Appl. Environ. Microbiol.* 60: 3874-3877.

Oliver, J. D., L. Nilsson, and S. Kjelleberg. 1991. Formation of nonculturable *Vibrio vulnificus* cells and its relationship to the starvation state. *Appl. Environ. Microbiol.* 57: 2640-2644.

Olsvik, O., T. Popovic, E. Skjerve, K. Cudjoe, E. Hornes, J. Ugelstad, and M. Uhlén. 1994. Magnetic separation techniques in diagnostic microbiology. *Clin. Microbiol. Rev.* 7: 43-54.

Osorio, C.R., M.D. Collins, A.E. Toranzo, J.L. Barja, and J.L. Romalde. 1999. 16S rRNA gene sequence analysis of *Photobacterium damselae* and nested PCR method for the rapid detection of the causative agent of fish pasteurellosis. *Appl. Environ. Microbiol.* 65: 2942-2946.

Osorio, C.R. 2000. Caracterización genética de *Photobacterium damselae*: estudio del operón ribosómico, microevolución y diagnóstico molecular. Ph.D. Thesis. University of Santiago de Compostela, Spain.

Osorio, C.R., A.E. Toranzo, J.L. Romalde, and J.L. Barja. 2000a. Multiplex PCR assay for *ureC* and 16S rRNA genes clearly discriminates between both subspecies of *Photobacterium damselae*. *Dis. Aquat. Org.* 40: 177-183.

Osorio, C.R., J.L. Romalde, J.L. Barja, and A.E. Toranzo. 2000b. Presence of phospholipase-D (*dly*) gene coding for damselysin production is not a pre-requisite for pathogenicity in *Photobacterium damselae* subsp. *damselae*. *Microb. Pathogen.* 28: 119-126.

Osorio, C.R. and K.E. Klose. 2000. A region of the transmembrane regulatory protein ToxR that thethers the transcriptional activation domain to the cytoplasmic membrane displays wide divergence among *Vibrio* species. *J. Bacteriol.* 182: 526-528.

Pazos, F., Y. Santos, A.R. Macías, S. Núñez, and A.E. Toranzo. 1996. Evaluation of media for the successful culture of *Flexibacter maritimus*. *J. Fish Dis.* 19: 193-197.

Pedersen, K., L. Grisez, R. Van Houdt, T. Tiainen, F. Ollivier, and J.L. Larsen. 1994. Extended serotyping scheme for *Vibrio anguillarum* with the definition of seven provisional O-serogroups. *Curr. Microbiol.* 38: 183-189.

Persing, D.H (ed). 1996. *PCR protocols for emerging infectious diseases*, ASM Press, Washington.

Porteous, L.A., J.L. Armstrong, R.J. Seidler, and L.S. Watrud. 1994. An effective method to extract DNA from environmental samples for polymerase chain reaction amplification and DNA fingerprinting analysis. *Current Microbiol.* 29: 301-307.

Powell, J.L., and M.W. Loutit. 1994a. Development of a DNA probe using differential hybridization to detect the fish pathogen *Vibrio anguillarum*. *Microb. Ecol.* 28: 365-373.

Powell, J.L. and M.W. Loutit. 1994b. The detection of fish pathogen *Vibrio anguillarum* in water and fish using a species-specific DNA probe combined with membrane filtration. *Microb. Ecol.* 28: 375-383.

Qadri, F., J.A.K. Hasan, J. Hossain, A. Chowdhury, Y.A. Begum, T. Azim, L. Loomis, R.B. Sack, and M.J. Albert. 1995. Evaluation of the monoclonal antibody-based kit Bengal SMART for rapid detection of *Vibrio cholerae* O139 synonym Bengal in stool samples. *J. Clin. Microbiol.* 33: 732-734.

Qiu, X., L. Wu, H. Huang, P.E. McDonel, A.V. Palumbo, J.M. Tiedje, and J. Zhou. 2001. Evaluation of PCR-generated chimeras, mutations, and heteroduplexes with 16S rRNA gene-based cloning. *Appl. Environ. Microbiol.* 67: 880-887.

Ramamurthy, T., S. Garg, R. Sharma, S.K. Bhattacharya, G.B. Nair, T. Shimada, T. Takeda, T. Karasawa, H. Kurazono, A. Pal, and Y. Takeda. 1993. Emergence of a novel strain of *Vibrio cholerae* with epidemic potential in southern and eastern India. *Lancet* 341: 703-704.

Rehnstam, A.-S., A. Norqvist, H. Wolf-Watz, and A. Hagström. 1989. Identification of *Vibrio anguillarum* in fish by using partial 16S rRNA sequences and a specific 16S rRNA oligonucleotide probe. *Appl. Environ. Microbiol.* 55: 1907-1910.

Romalde, J.L., and A.E. Toranzo. 2002. Molecular approaches for the study and diagnosis of salmonid streptococcosis. Pages 211-233 in *Molecular Diagnosis of Salmonid Diseases*. C. Cunnignham, ed., Kluwer Academic Publishers B.V., The Netherlands.

Rossen, L., P. Norskov, K. Holmstrom, and O.F. Rasmussen. 1992. Inhibition of PCR by components of food samples, microbial diagnostic assays and DNA-extraction solutions. *Int. J. Food Microbiol.* 17: 37-45.

Santos, Y., F. Pazos, and J.L. Barja. 1999. *Flexibacter maritimus*, causal agent of flexibacteriosus in marine fish. *ICES* identification leaflets for diseases and parasites of fish and shellfish. ICES ed., Denmark, n55, pp. 1-6.

Safarík, I., M. Safaríková, and S.J. Forsythe. 1995. The application of magnetic separations in applied microbiology. *J. Appl. Bacteriol.* 78: 575-585.

Simon, M.C., D.I. Gray, and N. Cook. 1996. DNA extraction and PCR methods for the detection of *Listeria monocytogenes* in cold-smoked salmon. *Appl. Environ. Microbiol.* 62: 822-824.

Smith, L.M., J.Z. Sanders, R.J. Kaiser, P. Hughes, C. Dodd, C.R. Connel, C. Heiner, S.B.H. Kent, and L.E. Hood. 1986. Fluorescence detection in automated DNA sequence analysis. *Nature* 321: 674-679.

Somerville, C.C., I.T. Knight, W.L. Straube, and R.R. Colwell. 1989. Simple, rapid method for direct isolation of nucleic acids from aquatic environments. *Appl. Environ. Microbiol.* 55: 548-554.

Sørensen, U.B.S., and J.L. Larsen. 1986. Serotyping of *Vibrio anguillarum*. *Appl. Environ. Microbiol.* 51: 593-597.

Sørum, H., A.B. Hvaal, M. Heum, F.L. Daae, and R. Wiik. 1990. Plasmid profiling of *Vibrio salmonicida* for epidemiological studies of cold-water vibriosis in Atlantic salmon (*Salmo salar*) and cod (*Gadus morhua*). *Appl. Environ. Microbiol.* 56: 1033-1037.

Sørum, H., J.H. Kvello, and T. Håstein. 1993. Occurrence and stability of plasmids in *Aeromonas salmonicida* ss *salmonicida* isolated from salmonids with furunculosis. *Dis. Aquat. Org.* 16: 199-206.

Speksnijder, A.G.C.L., G.A. Kowalchuk, S. De Jong, E. Kline, J.R. Stephen, and H.J. Laanbroek. 2001. Microvariation artifacts introduced by PCR and cloning of closely related 16S rRNA gene sequences. *Appl. Environ. Microbiol.* 67: 469-472.

Steffan, R.J., J. Goksøyr, A.K. Bej, and R.M. Atlas. 1988. Recovery of DNA from soils and sediments. *Appl. Environ. Microbiol.* 54: 2908-2915.

Steffan, R.J., and R.M. Atlas. 1988. DNA amplification to enhance detection of genetically engineered bacteria in environmental samples. *Appl. Environ. Microbiol.* 54: 2185-2191.

Steffan, R.J., and R.M. Atlas. 1991. Polymerase chain reaction: applications in environmental microbiology. *Annu. Rev. Microbiol.* 45: 137-161.

Suau, A., R. Bonnet, M. Sutren, J.-J. Godon, G.R. Gibson, M.D. Collins, and J. Doré. 1999. Direct analysis of genes encoding 16S rRNA from complex communities reveals many novel molecular species within the human gut. *Appl. Environ. Microbiol.* 65: 4799-4807.

Tebbe, C.C. and W. Vahjen. 1993. Interference of humic acids and DNA extracted directly from soil in detection and transformation of recombinant DNA from bacteria and a yeast. *Appl. Environ. Microbiol.* 59: 2657-2665.

Thyssen, A., S.Van Eygen, L. Hauben, J. Goris, J. Swings, and F. Ollevier. 2000. Application of AFLP for taxonomic and epidemiological studies of *Photobacterium damselae* subsp. *piscicida*. *Int. J. Syst. Evol. Microbiol.* 50: 1013-1019.

Tison, D.L., M. Nishibuchi, J.D. Greenwood, and R.M. Seidler. 1982. *Vibrio vulnificus* biogroup II: new biogroup pathogenic for eels. *Appl. Environ. Microbiol.* 44: 640-646.

Tomoyasu, T. 1992. Development of the immunomagnetic enrichment method selective for *Vibrio parahaemolyticus* serotype K and its application to food poisoning study. *Appl. Environ. Microbiol.* 58: 2679-2682.

Toranzo, A.E., and J.L. Barja. 1990. A review of the taxonomy and seroepizootiology of *Vibrio anguillarum*, with special reference to aquaculture in the northwest of Spain. *Dis. Aquat. Org.* 9: 73-82.

Toranzo, A.E., Y. Santos, S. Nuñez, and J.L. Barja. 1991. Biochemical and serological characteristics, drug resistance and plasmid profiles of Spanish isolates of *Aeromonas salmonicida*. *Fish Pathol.* 26: 55-60.

Toranzo, A.E., Y. Santos, and J.L. Barja. 1997. Immunization with bacterial antigens: *Vibrio* infections. *Fish Vaccinology*, Dev. Biol. Standard. Ed. Karger, Germany, Vol. 90, pp. 93-105.

Toyama, T., K.K. Tsukamoto, and H. Wakabayashi. 1996. Identification of *Flexibacter maritimus*, *Flavobacterium branchiophilum* and *Cytophaga columnaris* by PCR targeted 16S ribosomal DNA. *Fish Pathol.* 31: 25-31.

Trost, P.A., W.E. Hill, C.A. Kaysner, and M.M. Wekell. 1993. Detection of three pathogenic *Vibrio* species by using the polymerase chain reaction. *FDA Laboratory Information Bull.*, #3733.

Trust, T. 1986. Pathogenesis of the infectious diseases of fish. *Annual Rev. Microbiol.* 40: 479-502.

Tsai, Y.-L. and B.H. Olson. 1992. Detection of low numbers of bacterial cells in soils and sediments by polymerase chain reaction. *Appl. Environ. Microbiol.* 58: 754-757.

Venkateswaran, K., N. Dohmoto, and S. Harayama. 1998. Cloning and nucleotide sequence of the *gyrB* gene of *Vibrio parahaemolyticus* and its application in detection of this pathogen in shrimp. *Appl. Environ. Microbiol.* 64: 681-687.

Vermunt, A.E., A.A. Franken, and R.R Beumer. 1992. Isolation of salmonellas by immunomagnetic separation. *J. Appl. Bacteriol.* 72: 112-118.

Vickery, M.C.L., N. Harold, and A.K. Bej. 2000. Cluster analysis of AP-PCR generated DNA fingerprints of *Vibrio vulnificus* isolates from patients fatally infected after consumption of raw oysters. *Lett. Appl. Microbiol.* 30: 258-262.

Vieira, V.V., L.F.M. Teixeira, A.C.P. Vicente, H. Momen, and C.A. Salles. 2001. Differentiation of environmental and clinical isolates of *Vibrio mimicus* from *Vibrio cholerae* by multilocus enzyme electrophoresis. *Appl. Environ. Microbiol.* 67: 2360-2364.

Volossiouk, T., E.J. Robb, and R.N. Nazar. 1995. Direct DNA extraction for PCR-mediated assays of soil organisms. *Appl. Environ. Microbiol.* 61: 3972-3976.

Wachsmuth, I.K., P.A. Blake, and O. Olsvik. (eds.). 1994. *Vibrio cholerae* and cholera. Molecular to Global Perspectives. ASM Press, Washington, USA.

Wakabayashi, H., M. Hikida, and K. Masumura. 1986. *Flexibacter maritimus* sp. nov., a pathogen of marine fishes. *Int. J. Syst. Bacteriol.* 36: 396-398.

Walsh, P.S., D.A. Metzger, and R. Higuchi. 1991. Chelex 100 as a medium for simple extraction of DNA for PCR-based typing from forensic material. *Biotechniques* 10: 506-513.

Warner, J.M. and J.D. Oliver. 1998. Randomly amplified polymorphic DNA analysis of starved and viable but nonculturable *Vibrio vulnificus* cells. *Appl. Environ. Microbiol.* 64: 3025-3028.

Warner, J.M., and J.D. Oliver. 1999. Randomly amplified polymorphic DNA analysis of clinical and environmental isolates of *Vibrio vulnificus* and other *Vibrio* species. *Appl. Environ. Microbiol.* 65: 1141-1144.

Welsh, J., and M. McClelland. 1990. Fingerprinting genomes using PCR with arbitrary primers. *Nucleic Acids Res.* 18: 7213-7218.

Wernars, K., C.J. Heuvelman, T. Chakraborty, and S.H.W. Notermans. 1991. Use of the polymerase chain reaction for direct detection of *Listeria monocytogenes* in soft cheese. *J. Appl. Bacteriol.* 70: 121-126.

Widjojoatmodjo, M.N., A.C. Fluit, R. Torensma, B.H. Keller, and J. Verhoef. 1991. Evaluation of the magnetic immuno PCR assay for rapid detection of *Salmonella*. *Eur. J. Clin. Microbiol. Infect. Dis.* 10: 935-938.

Wiik, R., K. Andersen, F.L. Daae, and K.A. Hoff. 1989. Virulence studies based on plasmid profiles of the fish pathogen *Vibrio salmonicida*. *Appl. Environ. Microbiol.* 55: 819-825.

Williams, J.G.K., A.R. Kubelik, K.J. Livak, J.A. Rafalski, and S.V. Tingey. 1990. DNA polymorphisms amplified by arbitrary primers are useful as genetic markers. *Nucleic Acids Res.* 18: 6531-6535.

Wilson, I.G., J.E. Cooper, and A. Gilmour. 1991.Detection of enterotoxigenic *Staphylococcus aureus* in dried skimmed milk: use of the polymerase chain reaction for amplification and detection of staphylococcal enterotoxin genes *entB* and *entCl* and the thermonuclease gene *nuc*. *Appl. Environ. Microbiol.* 57: 1793-1798.

Wilson, I.G. 1997. Inhibition and facilitation of nucleic acid amplification. *Appl. Environ. Microbiol.* 63: 3741-3751.

Woese, C.R. 1987. Bacterial evolution. *Microbiol. Rev.* 51: 221-271.

Wolff, C., D. Hornschemeyer, D. Wolff, and K. Kleesiek. 1995. Single-tube nested PCR with room temperature stable reagents. *PCR Methods Applic.* 4: 376-379.

Wright, A.C., G.A. Miceli, W.L. Landry, J.B. Christy, W.D. Watkins, and J.G. Morris, Jr. 1993. Rapid identification of *Vibrio vulnificus* on nonselective media with an alkaline phosphatase-labeled oligonucleotide probe. *Appl. Environ. Microbiol.* 59: 541-546.

Yamamoto, K., A.C. Wright, J.B. Kaper, and J.G. Morris, Jr. 1990. The cytolisin gene of *Vibrio vulnificus*: sequence and relationship to *Vibrio cholerae* El Tor hemolysin gene. *Infect. Immun.* 58: 2706-2709.

Yourno, J. 1992. A method for nested PCR with single closed reaction tubes. *PCR Methods Applic.* 2: 60-65.

Zhao, J., and T. Aoki. 1989. A specific DNA hybridization probe for detection of *Pasteurella piscicida*. *Dis. Aquat. Org.* 7: 203-210.

Index

Acanthopagrus schlegeli 288
Acenerogobius caninus 56
Acinetobacter 187, 205
Actaeodes tomentosus 125
Aeromonas 63, 187
Aeromonas hydrophila 148
Aeromonas salmonicida 144-145, 148, 150-151, 255-256, 259, 283-288
Alexandrium 61, 107, 123
Alexandrium acatenella 57
Alexandrium andersoni 123
Alexandrium catenella 57-58, 61, 123
Alexandrium lusitanicum 114
Alexandrium minutum 57-58, 60-61, 123
Alexandrium polyedra 57, 61
Alexandrium tamarensis 57-58, 61, 123
Alteromonas 187
Alteromonas putrefaciens 205
Amago salmon 287
4-Aminopyridine 71-86
Amnesic shellfish poisoning (ASP) 89-98
Anabaena circinalis 123
Anchovy 288
Anemonia 6
Anguilla japonica 273
Anise 176
Anthopleura 6
Anthopleurins 6
Antimicrobial peptides 141-152
Aphanizomenon flos-aquae 57, 61, 64, 123
Aquaculture 141, 148, 151, 157, 253-300
Aristichthys noblis 161, 173
Arthroacter spp. 290
Aspergillus spp. 217
Asteropecten latespinosis 56
Asteropecten polyacanthus 56, 63
Asteropecten scoparius 56
Atelopus chiriquiensis 56
Atergatopsis germaini 56, 57, 59, 125
Atergatus dilitatus 125
Atergatus floridus 56, 57, 59, 125
Atergatus integerrimus 125
Atlantic salmon 166, 288
Atractoscion nobilis 288
Aulacomya ater 58
Austrobvenus stutchburyi 15
Azaspiracid 31-49

Babylonia formosae 56
Babylonia japonica 56
Bacillus cereus 190
Bacillus sp. 62, 63, 187
Biointoxication 109
Biosensor 131
Birgus latro 59
Bivalves 89
Botulism 204-205, 208-209, 215, 225
Brevetoxin-ELISA 103
Brevetoxins 1-24, 90, 102, 128
Busycon contrarium 15
Butter clam 71, 95

Calappa calappa 125
Campylobacter jejuni 190
Cancer magister 125, 126
Capelin 159, 166, 169, 171, 175
Carcinoscorpius rotundicauda 56, 59
Carcinus maenas 147
Cardiorespiratory infirmity 74
Carpilius maculatus 125
Charonia sauliae 56, 63
Chattonella antiqua 1
Chattonella marina 1
Chikuwa 241
Chinook salmon 272
Chionoecetes bairdi 125
Chionoecetus sp. 126
Chlamys farreri 58
Cholera toxin 41-49
Chromis punctipinnis 282
Ciguatera fish poisioning (CFP) 89, 100
Ciguatoxins 90, 100-114, 128
Clione cancellata 15
Clostridium 187
Clostridium botulinum 190, 204-205, 208-209, 215, 225
Clupus harengus 159
Coho salmon 145, 146, 149, 288
Colostethus inguinalis 56
Colourimetric enzyme immunoassays 91-103

Contoxins 5-6, 8
Conus textile 8
Cornestes 219
Corynebacterium striatum 290
Crassostrea gigas 58, 60
Crassostrea virginica 15
Crayfish 159, 174
Cryoprotectant 241
Croaker 229
Crustacea 121-134
Cyanobacteria 122
Cynops ensicauda 56
Cynops pyrrhogaster 56
Cyprinus carpio 146
Cytophaga 288
Cytotoxicity assays 112

Daldorfia horrida 125
Damselfish 282
dc-ELISA 130
Demania alcala 57, 59, 125
Demania reynaudi 56, 57, 59, 125
Demania toxica 59, 125
Dendrobates pumilio 7
Dermestes 219
Dermestes maculatus 191
Diarrhetic shellfish poisoning (DSP) 89, 92
Diatoms 89
Dicentrarchus labrax 272, 288
Dinoflagellate 1, 5, 10, 53-54, 89, 107, 122-124
Dinophysistoxins 90, 92
DNA-based diagnostics 253-300
Domoic acid 90, 98
Dover sole 288
Dromidiopsis sp. 125
DSP-Check 93
Dungeness crab 126

Edwardsiella ictaluri 148-149
Electrophysiological assays 115
ELISA 91-103
Emerita analoga 125
Engraulis mordax 288
Enterobacteriaceae 187
Enjymatic hydrolysis 159
Eriphia scabricula 59, 125
Eriphia sebana 59, 125
Escherichia 63
Escherichia coli 44, 190, 269, 272, 283
Ethmostigmus rubripes 130
Etisus rhynchophorous 125
Etisus splendidus 125
Euzanthus exsculptus 125

Fibrocapsa japonica 1
Fish disease control 141-152
Fish protein concentrates 159
Fish protein hydrolysates 157-177
Fish spoilage 183-232
Fish vaccine 176
Flavobacterium 205
Flavobacterium branchiophilum 288
Flavobacterium columnare 288
Flavobacterium psychrophilum 288
Flexibacter maritimus 255-256, 259, 288-289
Flexibacteriosis 288
Fluorimetric assays 114-115
Furunculosis 255
Fusarium spp. 217

Gadus morhua 159
Gambierdiscus toxicus 10
Gonyaulax 71
Gonyaulax catenella 122
Gonyautoxins 57, 107, 114-115
Grapsus albolineatus 59, 125
Gymnodinium 107
Gymnodinium breve 1
Gymnodinium brevis 2
Gymnodinium catenatum 57, 61, 98, 123
Gymnodinium sp. 114

Hake 159
Hemifusua ternatanus 56
Herring 159, 170, 173
Heterosigma akashiwo 1
Homarus americanus 125
Horseshoe crab 144, 146

Ichthyotoxic activity 12
Ictalurus punctatus 145, 149
Immune stimulant 176
Immunoassays 89, 91, 104
Isurus oxyrinchus 159

Jania sp. 123, 126

Kamaboko 241
Karenia brevis 2, 16

Lactobacillus spp. 187, 205, 208
Lordoglyphus spp. 219
Lateral flow immunochromatography (LFI) 97, 99
Lepidospsetta bilineata 202
Leptodius sanguineus 59, 125
Leuconostoc 228
Limulus polyphemus 146

Lineus fuscoviridis 56
Listeria 214, 216, 269
Listeria monocytogenes 190
Lithodes aequispina 126
Lobster 159, 171, 173-174, 208
Lophozozymus octodentatus 125
Lophozozymus pictor 56, 57, 59, 124, 125
Lyngbya wollei 123

Mad Cow Disease 177
Maitotoxin 90
Mallotus villosus 159, 166
Marine biotoxins 89
Membrane immunobead assay (MIA) 101
Mercenaria sp. 15
Meretrix lamarckii 58
Meretrix lusoria 60
Merluccius productus 159
Metopograpsus frontalis 125
Microalgae 122
Micrococcus 187
Misgurin 145
Misgurnus anguillicaudatus 145
Modori 246
Molluscs 89
Moraxella sp. 62, 187
Morone americanus 278
Morone chrysops 149
Morone saxatilis 149, 278
Mouse bioassays 90
Multiplex PCR 262
Musca domestica 132
Mussels 31-32
Mya arenaria 58
Myticins 144, 147
Mytilins 144, 147
Mytilus edulis 31, 32, 58
Mytilus galloprovincialis 147

Nassarius condoidalis 56
Natica lineata 56
Natica tumidus 56
Natica vitellus 56
Necrobia spp. 219
Nemipterus spp. 229
Neosaxitoxin 57
Neoxanthias impressus 59, 125
Neurotoxic shellfish poisoning (NSP) 1, 89, 102
Niotha clathrata 56
Nonoplex echo 56
Octopus maculosa 56, 60
Okadaic acid 90, 92
Okadaic acid ELISA kit 93
Oncorhynchus kisutch 145, 288
Oncorhynchus mykiss 146, 292
Oncorhynchus rhodurus 287
Oncorhynchus tshawytscha 272
Orechromis mossambicus 174
Oyster 15

Palytoxin 57
Panulirus spp. 159
Paralithodes sp. 126
Paralytic shellfish poisoning (PSP) 5, 31, 53, 89, 95, 107, 121-134
Paralytic shellfish toxins 53-64, 107-116, 121-134
Parasagitta elegans 56
Parasilurus asotus 145
Pardaxins 144, 150
Pasteurella piscicida 278-282
Pasteurellosis 255
Pecten albicans 58
Pediococcus 228
Penaeidins 144, 147
Penaeus indicus 224
Penaeus vannamei 147
Penicillium spp. 217
Peptides, antimicrobial 141-152
Percnon planissimum 125
Perna canaliculus 15
Photobacterium 63
Photobacterium damselae 255-256, 258, 264, 278-283
Photobacterium phosphoreum 187
Phymodius ungulatus 125
Pilodius areolatus 125
Pilumnus pulcher 125
Pilumnus vespertilio 59, 125
Pimpinella anisum 176
Pink perch 229
Piscirickettsia salmonis 255-256, 260, 294-295
Placopecten magellanicus 58
Planocera multitentaculata 56
Planocera reticulata 56
Plastein 174
Platypodia granulosa 59, 125
Platypodia pseudogranulosa 125
Plesiomonas 63
Pleurocidin 144, 147, 149-151
Pleuronectes americanus 144
Polinices didyna 56
Polyphemusins 144, 147, 151
Pomacanthus semicirculatus 56
Pomfret 224
Portunus pelagicus 125

Power spectrographic analysis 83
Proteus 187
Protogonyaulax 123
Pseudomonas 187, 205, 207
Pseudomonas sp. 62, 63
Pseudopotamilla occelata 56
Ptychodiscus brevis 2
Puffer fish 5, 53, 54, 60, 63
Purple clam 60
Pyrodinium 107
Pyrodinium bahamense 57, 58, 61, 123
Pyrodinium phoneus 57, 61

Radicidation 215
Rainbow trout 146, 199, 292
Rapana rapiformis 56
Rapana venosa 56
RAPD 264, 290
Raphidophycean flagellates 1
Rastrelliger kanagurta 204
Receptor binding assay 111
Red calcareous alga 123
Red tides 1
Renibacterium salmoninarum 255-256, 260, 264, 289-294
Rock sole 202
Ruditapes phililinarum 58

Salmo salar 159, 287, 288
Salmonella spp. 190, 205, 214-216, 269
Sardine 159, 168
Sardina longiceps 198
Sardina pilchardus 159, 186, 198
Sardinops sagax 288
Saxidomus 71
Saxidomus giganteus 58, 95
Saxiphilin 112, 129-130
Saxitoxins 5, 8, 57, 71-86, 90, 95-97, 107, 109-115, 121-134
Saxitoxin test 96
Scarus gibbus 56
Scarus ovifrous 56
Sceloporus poinsetti 130
Schizophrys aspera 125
Sciaenid spp. 229
Scophthalmus maximus 288
Scrippsiella trochoidea 114
Sea bass 278, 288
Sea bream 278, 281, 288
Seriola quinqueradiata 278, 280
Shellfish 89,121
Shewanella putrefaciens 187
Shrimp 199, 208, 212-214, 224
Solea solea 288
Soletellina diphos 58, 60
Solid-phase immunobead assay (S-PIA) 101
Sparus aurata 278, 281
Squalus acanthias 159
Staphylococcus aureus 190
Staphylococcus sp. 62, 205
Streptococcus 205, 228
Streptococcus iniaie 149
Striped bass 149, 278
Stromateus cinereus 224
Surimi technology 241-250

Tachyplesins 144, 147, 151
Tachypleus tridentatus 146
Takifugu rubripes 63
Tanner crab 126
Taricha torosa 56
Tetrodotoxin (TTX) 5, 8, 53-64, 122
Thalamita sp. 59, 125
Thalamita stimpsoni 125
Thalamita wakensis 125
Tigriopus californicus 125
Tilapia 199
Tridacna crocea 58
Triturus granulosa 56
Tubalanus punctatus 56
Turbo marmorata 58
Turbot 288
Tutufa lissostona 56

Umborium suturale 63
Urophycis chuss 159

Vibrio 63, 187, 214-216
Vibrio alginolyticus 62, 63, 272, 298
Vibrio anguillarum 144, 145, 148-151, 255-256, 258, 270-273
Vibrio cholerae 41, 255-257, 261, 263, 269, 297-299
Vibrio damsela 62, 282-283
Vibrio mimicus 297
Vibrio ordalii 273
Vibrio parahaemolyticus 41, 63, 190, 255, 257, 261, 263, 269, 297-299
Vibrio vulnificus 41-49, 255-256, 263-264, 273-278
Voltage dependent sodium channel 122

Water-holding capacity 168-169
White perch 278

Yellowtail 278, 280
Yersinia enterocolitica 190
Yersinia ruckeri 148

Yoneichthys nebulosus 56

Zeuxis scalaris 56
Zeuxis siquijorensis 56
Zeuxis sufftatus 56
Zosimus aeneus 56, 57, 59, 125, 126